Organic Free Radicals

Organic Free Radicals

William A. Pryor, EDITOR

Boyd Professor of Chemistry
Louisiana State University

A symposium sponsored by

the Division of Organic

Chemistry at the 174th

Meeting of the American

Chemical Society, Chicago,

Illinois, August 29–

September 2, 1977.

ACS SYMPOSIUM SERIES **69**

AMERICAN CHEMICAL SOCIETY

WASHINGTON, D. C.   1978

Library of Congress CIP Data

Main entry under title:
Organic free radicals.

(ACS symposium series; 69 ISSN 0097–6156)

Includes bibliographies and index.

1. Radicals (Chemistry)—Congresses. 2. Chemistry,
Physical organic—Congresses.
I. Pryor, William A. II. American Chemical Society.
Division of Organic Chemistry. III. Series.

QD471.068   547'.1'224   78-1672
ISBN 0-8412-0421-7

118169

ACS Symposium Series

Robert F. Gould, *Editor*

FOREWORD

The ACS SYMPOSIUM SERIES was founded in 1974 to provide a medium for publishing symposia quickly in book form. The format of the SERIES parallels that of the continuing ADVANCES IN CHEMISTRY SERIES except that in order to save time the papers are not typeset but are reproduced as they are submitted by the authors in camera-ready form. As a further means of saving time, the papers are not edited or reviewed except by the symposium chairman, who becomes editor of the book. Papers published in the ACS SYMPOSIUM SERIES are original contributions not published elsewhere in whole or major part and include reports of research as well as reviews since symposia may embrace both types of presentation.

CONTENTS

vii

PREFACE

The chapters in this volume are grouped into six sections. In the first, Walling provides a fascinating overview of the first forty years of free radical chemistry. His talk on this subject at the Chicago meeting ended with a heartfelt standing ovation.

The next nine chapters, grouped under the subheading of "Initiator Chemistry," are concerned with various aspects of research on topics of current interest dealing with the energetics and detailed mechanisms of radical production. The first chapter in this section summarizes Paul Bartlett's most recent results on photooxidation and the identity of the elusive epoxidation reagent. Other chapters treat MAH reactions, peroxide and perester reactions, cyclic endoperoxides, diazenes, cage effects, and radiation.

The third section deals with structures and rearrangement reactions of free radicals. This field has seen great advances in recent years, especially because of the use of electron spin resonance techniques.

In the section entitled, "Propagation Reactions of Free Radicals," topics ranging from molecular orbital calculations to detailed studies of eléctron transfer and radical displacement reactions are discussed. The reactions of sulfur, phosphorous, and nitrogen radicals are reviewed also.

Four presentations of the kinetics and mechanisms of radical ion reactions are presented in the section on radical ion chemistry. Again, electron spin resonance and other physical techniques play a large role.

The final three chapters cover termination reactions. The structural dependence of the rate of oxy-radical terminations, the technique of spin trapping, and pyridinyl radicals are discussed.

The chapters in this book present a review of contemporary research on radical reactions by U.S. and Canadian chemists. These chapters are meant to provide a broad-brush treatment of the research of these contributors, and an easily read introduction to their current interests. I and the other authors also hope that these chapters attest to the pleasure we have all had from our associations with Paul Bartlett and Cheves Walling.

Louisiana State University   WILLIAM A. PRYOR
Baton Rouge, Louisiana
January 17, 1978

INTRODUCTION

In late August, 1977, in connection with its 174th National Meeting in Chicago, the American Chemical Society sponsored a three-day symposium entitled, "Organic Free Radicals." At this symposium, 27 well-known radical chemists presented their current research in a program dedicated to the contributions that Paul D. Bartlett and Cheves Walling have made to free radical chemistry. Many more speakers could have participated, but the program was limited by the time made available by the ACS. Furthermore, some speakers were unable to attend because of conflicts with other international meetings held at about the same time.

The papers in this volume are based on talks given at the symposium —with one exception. Owing to an unfortunate accident, Paul Bartlett was laid up with a leg in a cast, and could not attend. However, he was present in spirit, and he agreed to include in this volume the talk that he would have given had Nature, on a dark night in Munich the month before, been slightly kinder.

The breadth and excitement of the current research in the free radical field is, I believe, represented excellently by these papers. And they all, both in explicit acknowledgments and in subtle and personal ways that are less obvious, demonstrate the enormous debt that today's radical chemists owe to these two pioneers.

Radical chemistry has had a tempestuous and interesting history. During the late 1800's, most chemists felt that free radicals could not exist. We were led from this wilderness by Moses—the remarkable Moses Gomberg, who demonstrated that the triphenylmethyl radical can be produced in rather concentrated solutions, and who, with clever and insightful intuition, worked out some of the chemistry of this still-interesting radical.

In subsequent years, a succession of brilliant physical chemists interested in the fundamental laws of chemical kinetics began to interpret their results in terms of radical reactions. In 1918, J. A. Christiansen, K. F. Herzfeld, and M. Polanyi independently suggested a radical chain process for the H_2-Br_2 reaction. In 1925, H. S. Taylor postulated the occurrence of the ethyl radical to rationalize a gas-phase photolysis. In 1931, Norrish suggested that radicals occur in the photolysis of carbonyl compounds. And then in 1939, in a very influential paper, F. Paneth showed that small alkyl radicals could be produced in a flowing gas stream and that their approximate lifetimes could be measured. In 1935, F. O. and K. K. Rice published their book, "The Aliphatic Free Radicals," and in 1946 E. W. R. Steacie published "Atoms and Free Radical Reactions."

Organic chemists apparently had little interest in radical chemistry in these early years. However, the enormous utility of the vinyl polymers and the critical need for synthetic rubber brought about by the second World War dramatically changed that.

The first suggestion that vinyl polymerization occurs by a radical chain mechanism was made by H. S. Taylor in 1927, and independently by H. Staudinger in 1932. In 1937, P. J. Flory developed the model further and introduced the concept of chain transfer. These developments may have appeared quite academic, but during the World War the Japanese first threatened, and then captured the areas in the Far East that supplied America's natural rubber. At first we began a stockpiling program, but our supply of natural rubber in 1941 was only two-thirds of one year's consumption. Germany already had developed a synthetic rubber based on the radical-initiated copolymerization of styrene and butadiene, and, because of a scientific exchange program between German and American industries, our government knew of this work. Thus, in the early 1940's, a huge and vitally important effort began to use and improve the process and product from a vinyl polymerization.

Paul Bartlett and Cheves Walling became involved in this effort to understand and use free radical chemistry in different ways. In 1941, Bartlett, then at Harvard as a young faculty member, began a long and fruitful program of research sponsored by the Pittsburgh Plate Glass Company. As Mike McBride relates in his dedication to Bartlett, even the earliest contributions from this program had the magic Bartlett touch—they were important, novel, and insightful. For example, I have used the Nozaki–Bartlett 1946 publications in my classes for years to illustrate the kinetics of initiator decompositions; these papers are as fresh and current today as they were the day they were published, and they form the cornerstone for all later studies of initiators. As is the Bartlett trademark, they are models of lucidity. The flow of fascinating research from the Bartlett group became a fixture and touchstone in the free radical field.

Cheves Walling, too, was in the forefront of the newly born study of vinyl polymerization process. As Earl Huyser relates in his dedication, Cheves obtained a PhD under Morris Kharash in 1939—a most propitious moment for a young man to arrive on the scene with a special talent for unraveling the mysteries of the kinetics and mechanisms of radical reactions. In his early years at the U. S. Rubber Company Laboratories, Cheves Walling contributed to the newly appreciated understanding of polymerization and copolymerization processes. Walling has become well known for his ability to see into the most complex and difficult system, and to reduce it to understandable equations. But it was Walling's book, "Free Radicals in Solution," published in 1957, that surely stands as his most remarkable contribution. Quite literally, this has served as the introduction to the field for a generation of radical chemists, and it remains a frequent source and cited reference in the field despite the fact that it is now over two decades old.

If I may be excused the pleasure of some personal remarks, both of these men played vital roles in my own entreé into the radical field. After receiving a PhD at Berkeley (and never having even heard the word radical during my university days), I started a happy and profitable period with the California Research Corporation. After a short induction period in an exploratory polymer group, I was fortunate to be chosen to direct the one-man basic research group and I began a study of sulfur as an oxidant and the mechanism of the oxidation of methylaromatics to benzoic acids by sulfur. That process proved to be a radical one—at least according to my evidence—and I purchased and studied Walling's newly published book. In addition, Cheves himself was an intermittent consultant to the company, and for the first time I observed his astounding ability to analyze tough kinetic problems in critical detail.

My relationship with Paul Bartlett began with my first visit to Harvard and the tour through his laboratories in about 1961, when I was a young faculty member. During that tour of Bartlett's labs years ago, and in subsequent years whenever any critical academic problem came up, like many other young men I sought to find the answer to the critical question, "How does PD handle that?". Paul Bartlett has stocked the universities and laboratories of America with persons that have looked to him as a teacher, a mentor, and a father figure for a period of nearly 40 years.

Both Paul Bartlett and Cheves Walling have many friends and admirers, and I feel privileged to be counted in that number. And with these many admirers, I dedicate this book to them both.

December 9, 1977   WILLIAM A. PRYOR

To Paul Bartlett

DEDICATION TO PAUL BARTLETT

It is inadequate to characterize Paul Bartlett as a free radical chemist. His interests and contributions have ranged over the entire field of organic reaction mechanisms. Only about one-fourth of his 225 research publications deal directly with free radicals. Another fourth deal with the related topics of singlet oxygen and cycloadditions. Still, these contributions have assured his place as a founder of the modern school of organic free radical chemistry. Of 27 participants in this symposium, seven were his graduate students, two were his postdoctoral associates, and another four were students of his students or post-docs. Obviously his impact on the practitioners has been as strong as on the field itself.

Bartlett did not begin his career with radical chemistry; he did not even participate directly in the emergence of the field during the 1930's. After graduating from Amherst in 1928 and earning his doctorate under Conant at Harvard in 1931 he rusticated in New York and Minnesota before returning to Cambridge as instructor in 1934. During the rest of the decade he became a leader in the study of ionic reactions and rearrangements. In 1938 this work brought him the first of many honors, the ACS Award in Pure Chemistry.

At the beginning of the war the Pittsburgh Plate Glass Company agreed to support Bartlett's laboratory in what became a long and fruitful program of research on vinyl polymerization. He entered the free radical field by proving with Saul Cohen that both phenyl and benzoyloxy radicals from halobenzoyl peroxide initiators were incorporated into polystyrene. Soon after this he and Nozaki began investigating the role of solvent in induced decomposition of dibenzoyl peroxide, but most of his dozen publications during the forties concerned the kinetics of polymerization itself and the role of inhibitors and retarders. Some of the most striking of these described his work with Swain, and later with Kwart and Broadbent, using light intermittency to measure absolute rate constants for steps in the polymerization of vinyl acetate. The rotating sector is a good example of Bartlett's ability and instinct to use a new, incisive approach that cuts to the heart of a problem. Another is his work with Tate using the kinetic isotope effect on the polymerization of dideuteroallyl acetate to establish the mode of chain transfer and termination.

Such clean experiments are Bartlett's hallmark, but with Hammond and Kwart he fearlessly entered the thicket of inhibitors and retarders and through the fifties followed its path to separate investigations of sulfur chemistry and the complex chemistry of iodine with styrene.

His work with Leffler on the decomposition of phenylacetyl peroxide appeared in 1950 and inaugurated new lines of investigation, focusing on the mechanism of initiator homolyses and on the properties of the free radicals they generate. The latter led to important stereochemical insights through studies of triptycyl, norbornyl, and decalyl radicals, but the former had still wider impact. He developed tert-butyl peresters which were a more reliable tool than diacyl peroxides and could be used to generate radicals over a broad temperature range. With Hiatt, Rüchardt, and their successors he demonstrated the place of simultaneous bond rupture and of substituent effects in radical formation. One of the most widely recognized contributions from this research was Bartlett's rationale for the correlation between the entropy and enthalpy of activation in terms of restricted rotation in the transition state for concerted perester homolysis.

Toward the end of the fifties Bartlett began a systematic study of the cage effect, popularizing galvinoxyl as an efficient scavenger, and naming it after Galvin Coppinger. In subsequent work with Nelson, McBride, Engel, and Porter he clarified the mechanism for photosensitized decomposition of azoalkanes and used these compounds to demonstrate the importance in radical-pair reactions of electron spin correlation and matrix rigidity.

The work with Porter on cyclic azo compounds coupled Bartlett's free radical studies into his ambitious effort to delineate the border between concerted and biradical cycloaddition mechanisms. Few demonstrations in chemistry have matched the clarity of his 1963 stereochemical proof with Montgomery that the 2 + 2 cycloaddition of dichlorodifluoroethylene to 2,4-hexadiene involves a biradical intermediate. Through the following decade Bartlett led the biradical parade. His long series of investigations on cycloaddition became particularly important in demonstrating nature's resourcefulness in coping with the supposedly inviolable constraints of orbital symmetry.

For the last ten years Bartlett has concentrated on the world's most important diradical, oxygen. This effort has its roots in his earlier studies of radical-chain autoxidation—Traylor's experiments with oxygen-36, Günther's with kinetics, and Guaraldi's with EPR—that together demonstrated the importance of tetroxide and trioxide in the chains involving cumyl- and tert-butyl peroxy radicals. Until recently it appeared that photosensitized oxidations could be explained efficiently by nonradical mechanisms involving singlet oxygen. But now Bartlett is uncovering important free radical pathways in ketone-sensitized photooxidation. As so often before, the definitive publications in a complex field will be his. The only perceptible change over a period of forty years is that the word that used to come from the Erving Professor at Harvard now comes from the Welch Professor at Texas Christian.

Bartlett's contributions have spanned free radical chemistry as they have spanned mechanistic chemistry as a whole. Perhaps the truest testimony to his importance in the field can be borne by those of us who missed him at this Chicago symposium. Of course we were sorry that a broken ankle had kept our friend from the occasion that was to honor him and Cheves Walling, and we felt cheated at not hearing the latest news on his photooxygenation studies. But most of all we missed seeing him in the front row, intently following all of our talks, and, through his questions, helping us see new possibilities in our own work.

Yale University   J. MICHAEL McBRIDE
New Haven, Connecticut

To Cheves Walling

DEDICATION TO CHEVES WALLING

For well over a quarter of a century, the name Cheves Walling has been synonymous with free radical chemistry. His contributions to this field of chemistry during its early years of development alone would have earned him this distinction. While it is an impressive experience to have a pioneer in any area of endeavor present at a symposium recognizing his contributions, it is even more impressive, as it is with Cheves Walling, when the recipient of such recognition is regarded as one of the more active and influential contributors to that field. Those of us who attend symposia and conferences in the general realm of mechanistic organic chemistry are accustomed to finding Cheves occupying a seat in the front row of the audience and to the fact that he can be counted on to contribute to (and often generate) the discussion that follows the presentation of a paper. This tradition was maintained in the best sense at this Chicago Free Radical Symposium.

From the beginning of his professional career, Cheves Walling has been involved intimately with free radical chemistry. After obtaining the AB degree from Harvard in 1937, he joined the research group of Professor M. S. Kharasch at the University of Chicago. Although his tenure as a graduate student was remarkably short (PhD in 1939), it provided the opportunity for him to participate in a significant period of the development of modern free radical chemistry. With Kharasch and Frank R. Mayo, he coauthored some of the early publications that demonstrated that the peroxide effect was a general phenomenon in the addition reactions of hydrogen bromide. In 1940, Walling and Mayo published a survey of free radical addition reactions in *Chemical Reviews*—that did much to establish the free radical chain mechanism of these additions as a general reaction of carbon–carbon unsaturated linkages.

Following his formal education, Cheves Walling spent over a decade as an industrial research chemist. Most notable were the years (1943–1949) when he was a research chemist at the General Research Laboratories of the U.S. Rubber Company. It was during this period that contributions were made by Walling and his co-workers (principal among them Frank Mayo, his colleague from his graduate school days in Chicago) that resulted in the elucidation of some of the fundamental kinetic principles of vinyl polymerization reactions. Among the important contributions made by the U.S. Rubber group was the research in the areas of chain transfer and the copolymerization reactions of vinyl monomers. The results of these investigations helped to unravel what appeared at the time to be a complex problem. The work by this group undoubtedly played a significant role in the fantastic growth of the polymer industry in succeeding years. A particularly exciting aspect of the copolymerization work by Walling and his collaborators during this period was the recognition of the resonance, steric, and polar factors that determine the reactivities of free radicals in chain propagating reactions thereby bringing free radical chemistry into the arena of mechanistic physical organic chemistry. The interpretive survey of copolymerization reactions by Walling and Mayo that appeared in *Chemical Reviews* in 1950 served as a basic reference manual for polymer scientists and free radical chemists for many years. The years he spent as an industrial research chemist had their influence on Cheves throughout his professional life in that not only has he been a successful industrial consultant but he always has been able to maintain a meaningful rapport with industrial chemists and their particular challenges.

In 1952, Cheves Walling joined the faculty of Columbia University as Professor of Chemistry, a position he held (serving as Department Chairman from 1963–1966) until 1970 when he became Distinguished Professor of Chemistry at the University of Utah. The research contributions to free radical chemistry by him and his students since the early fifties have been consistently innovative and incisive. An impressive aspect of his research is the breadth of areas in which publications by Walling and his co-workers have appeared. Basic and authoritative articles can be found concerning peroxide chemistry, free radical chemistry of organophosphorus compounds and sulfur compounds, the role of metal ions in free radical reactions, free radical cyclization reactions, halogenation reactions, and the use of high pressure techniques in investigating free radical reactions. His abiding interest in the chemistry of alkyl hypochlorites for over two decades has resulted not only in development of a useful synthetic reaction and information about alkoxyl radicals, but these studies also have provided insight into such subtle aspects of free radical chemistry as the nature of solvent and polar effects on radical reactions and the stereochemistry of allylic radicals. His book "Free Radicals in Solution," which appeared in 1957, must be regarded as one of the more influential publications in modern organic chemistry. In addition to being a comprehensive and interpretive survey of free radical chemistry up to the time of its publication, it both delineated the important areas of future investigations and established the criteria of meaningful research in this field. His scientific accomplishments have earned him membership in the National Academy of Sciences and the American Academy of Arts and Sciences. In 1971, he was the recipient of the James Flack Norris Award in Physical Organic Chemistry and was chosen Alumni Medalist of 1975 by the University of Chicago Alumni Association.

In addition to his scientific achievements, Cheves Walling has given generously of his time and talents in a variety of services to scientific professional organizations. The list of committees in which he has served or is serving presently is extensive and includes the Advisory Board of the Petroleum Research Fund, the Board of Directors of the Gordon Research Conferences, as well as various committee assignments in the American Chemical Society and the National Research Council. Noteworthy has been his work as Chairman of the Committee on Professional Training of the American Chemical Society during the period 1966–1973, a committee on which he serves presently as a consultant. Indicative of his continuing active concern and involvement with his profession is the fact that in 1975 he became editor of the *Journal of the American Chemical Society*, one of the world's most prestigious scientific periodicals. His high scientific and scholarly standards have been and are reflected in these efforts for his profession.

Before I became acquainted with him over two decades ago, Cheves Walling was described to me as a "chemist's chemist." During the time that I was privileged to work with him, it was quite apparent to me why he had earned this particular description. The many important contributions he has made to the chemical profession in succeeding years and the fact that his present activities indicate that there are many more to come give added significance to this unique title for a unique individual in the scientific community.

University of Kansas   EARL S. HUYSER
Lawrence, Kansas
December 13, 1977

xxiii

History

Forty Years of Free Radicals

CHEVES WALLING

Department of Chemistry, University of Utah, Salt Lake City, UT 84112

In a four-day symposium covering most of the developments of free radical chemistry, one retrospective paper seems appropriate to give some picture of the history of our field and how it got where it is. My title, "Forty Years of Free Radicals," was chosen for two reasons. First, although free radical chemistry really dates back to Moses Gomberg's discovery of triphenylmethyl, and the idea of small free radicals as intermediates in high temperature gas phase reactions had had considerable development in the early '30's, 1937 was a crucial year in the recognition that free radicals might be important in ordinary liquid-phase organic chemistry. It was in that year that Hey and Waters suggested that the arylation of aromatics by benzoyl peroxide was a radical process (1) and also that Kharasch proposed that the abnormal addition of hydrogen bromide, which he and Mayo had recognized in 1933 was a radical chain process and thus discovered that he was a free radical chemist (2).

Second, it was the year that I arrived at the University of Chicago, and, on the recommendation of Max Tishler, looked up Morris Kharasch as a possible research sponsor and was thus trapped by this new field. There I worked on some aspects of the abnormal addition problem, showing in part that it was not restricted to terminal olefins, and, with inate caution declined an invitation to investigate the chemistry of acetyl peroxide. Some idea of the rather rudimentary state of our concepts at the time is shown by the fact that at the Boston ACS meeting in September, 1939 I gave the first explanation, based on the energetics of chain propagation steps, why radical chain reactions had been observed with HBr, but not with other halogen acids. That ACS meeting is probably best remembered as the occasion when a large number of attendees came down with food poisoning from eating contaminated seafood at a now-defunct hotel in Swamscott. Fortunately I missed that celebration, but I spoke the next morning to a rather small audience. The interpretation was subsequently published in a review with Frank Mayo (3).

1937 was not long before World War II, and the war led to a

major turning-point in free radical chemistry. On the one hand,
orthodox academic research was drastically curtailed; on the
other, the Japanese occupation of S.E. Asia and the East Indies
cut off our supply of natural rubber. The development of a
synthetic rubber industry, almost from scratch, became a high-
priority need, and polymer chemistry suddenly became a hot
subject. A massive research program was launched in academic,
industrial, and Government laboratories, all under the coordination of the Rubber Reserve Corp. In a way, the situation was a
small scale preview of our present crisis over energy sources.
    As it turned out, the short term solution, the immediate
production of large quantities of synthetic rubber, depended
primarily on engineering and modest improvements in German technology, knowledge of which was available large because of prior
information exchange between the Standard Oil Co. of New Jersey
(now Exxon) and the I.G. They got information about butyl
rubber, while we learned about the manufacture of styrene-
butadiene copolymer (GR-S), considerably the better part of the
bargain. The research program, to be honest, contributed only
modestly to war-time rubber production, but, as we shall see, had
far-reaching later consequences.
    I got involved in this myself in early 1943 when I left
Dupont and went to work for the U.S. Rubber Co. (now Uniroyal) on
the top floor of an old silk mill in Passaic, New Jersey. U.S.
Rubber had decided to launch a fundamental study of polymer
chemistry to support their active applied work, and I joined the
group headed by Frank Mayo, whom I'd known at Chicago. This was
a marvellously exciting time. All sorts of new things were being
learned and there were many groups in the field exchanging information quite freely. In particular Herman Mark was at Brooklyn
Polytech where he was teaching Americans the extensive European
lore on polymer chemistry and conducting regular Saturday
morning seminars on the latest developments. Facilities at
Brooklyn Poly were a bit primitive. I remember one investigation
involving a high-pressure bomb which would hardly have met
present standards of safety or equal opportunity. The bomb was
in an alcove beween two buildings, hidden from the sidewalk by
a thin piece of galvanized iron, and the controls could only be
read from a toilet in the men's room. Nevertheless, the seminars
were a great success, and attracted participants from all over
the northeast.
    At the point when I arrived, the Mayo group had already
recognized and analyzed two fundamental problems, the kinetics
of chain transfer and of copolymerization, and had obtained the
first data showing their analysis worked. Unless polymer
chemistry was to be a purely empirical art, understanding of both
was critical. Chain transfer largely determines polymer molecular
weight via the competition between chain growth and reaction of
a growing chain with solvent or other transfer agent:

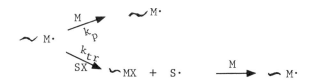

Similarly, copolymer compositions are determined by two competitions:

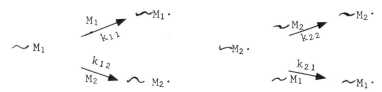

described by a copolymerization equation (4). In fairness, I should note that similar analyses were carried out by other workers - Alfrey, Goldfinger, and Tobolsky, among others. The success of the Mayo group lay in that we not only understood the theory, but provided the bulk of the experimental data. Frank set high experimental standards and one member of the group in particular, Fred Lewis, was a remarkably skillful devisor of new techniques. Equally important, we had the backup of a first class industrial analytical laboratory.

Beyond the practical import of these studies for polymer chemistry, the resulting data provided the first extensive body of information on the relation between structure and substrate reactivity in radical reactions and gave us the general picture we still use today: relative reactivities are determined by three factors - overall energetics (the usual picture in terms of resonance stabilization of reactants and products); steric hindrance (which for example explains the lowered reactivity of non-terminal olefins and much of the regioselectivity of radical additions); and the "polar effect" (increased reactivity between radicals with electron withdrawing groups and substrates with electron supplying groups, or vice versa). This last was quite unexpected and has proved to be quite general, although its detailed interpretation is still a subject of debate.

By about 1947 many of the major themes of free radical chemistry had emerged from this war-initiated polymer research: data on structure and reactivity and establishment of the utility of competitive kinetics for determining relative rates of chain propagation steps in chain reactions by product analyses as I've just described; the first good rate constants for elementary steps in chain processes - here the first really reliable results were on vinyl acetate polymerization from Paul Bartlett's laboratory (5); degradative chain transfer, first demonstrated by Bartlett and Altschul and showing how the length of kinetic

chains may be determined by the competition between radical
addition to double bonds and attack on allylic hydrogen (6); the
chemistry of inhibitors and of initiators - azo compounds and
particularly peroxides with all their intricate chemistry;
redox systems as radical sources. These are some of the examples
I can mention, as is the recognition of the peculiar properties
and advantages of emulsion polymerization. This last was the
subject of intensive investigation by many distinguished scientists including Harkins and Debye, and the final picture was
produced by Smith and Ewart (7). Ros Ewart worked in the lab
next to mine and we shared a driving pool. I recall the day
that he explained his solution to me on the way home from work;
I was so fascinated that I drove through a red light and paid a
$35 fine.

In the post-war period this new knowledge and enthusiasm
for radical reactions shifted from polymer to small molecule
chemistry. A logical bridge was the process of "telomerization,"
essentially polymerization carried out in the presence of such
reactive chain transfer agents that the resulting product
contained only a few monomer units per molecule. In retrospect,
the idea was rather obvious, since the radical addition of
mercaptans and HBr to olefins to give essentially monomeric
products was already known, but its extension to reactions
forming C-C bonds was stumbled on by several groups rather by
chance. Workers at Dupont investigating the effect of solvents
on the high pressure polymerization of ethylene discovered
and named the process, but since they were working at high
pressures produced relatively high molecular grease-like products
which proved of limited interest.

At U.S. Rubber, we lacked the high pressure equipment, but
Julian Little and his group quite independently undertook the
study of ethylene polymerization in solvents at a pressure of a
few atmospheres. To their considerable surprise, in CCl_4
ethylene was rapidly consumed, but no polymer precipitated.
Rather, distillation yielded a series of low molecular weight
products $Cl(C_2H_4)_nCCl_3$, (n=1-4). Again, apparently independently,
Kharasch and his students, perhaps lacking any pressure equipment
at all, tried the same reaction with 1-octene and CCl_4,
obtaining similar products.

In the next few years the field saw intensive research,
and it was shown that a large number of molecules with weak C-H
or C-1 bonds could be added to olefins to give a remarkable
variety of products. As a synthetic method the reaction has been
a bit disappointing since only a few of the products proved to be
of practical use, but it has had interesting ramifications,
including radical cyclizations, which my students investigated
at Columbia (8), e.g.

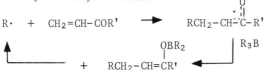

and which have been studied in detail by Julia (9). Another
example is the interesting alkylation of α,β-unsaturated carbonyl
compounds by trialkyl borons:

$$R\cdot \;+\; CH_2=CH-COR' \longrightarrow RCH_2-CH\dot{-}C-R'$$

discovered by H. C. Brown, although I should note that the reaction was described some time before its radical chain nature
was appreciated (10).

In 1947 I had the good fortune to attend the Faraday Society
Discussion on "The Labile Molecule," which, despite its title,
was largely concerned with free radical chemistry. This was an
enormously exciting occasion, since, in those days chemists
didn't make annual trips abroad, and, for the first time the
American participants were exposed first hand to the British out-
look and the considerable amount of research results which had
accumulated during and right after the war. One of the most
important ideas which·I got from the meeting and visiting British
laboratories was the potential significance of free radical
autoxidation reactions, which had been brilliantly opened up by
Farmer and his colleagues at the laboratories of the British
Rubber Producers Association. Again, this was a subject originally investigated for practical reasons (BRPA was concerned with the
rapid aging of natural rubber in air) but which had far-reaching
results. When we got back to the U.S., we tried to interest our
employer in the field with only modest success, but it certainly
has had enormous ramifications since.

Since the intermediates in autoxidation are almost always
peroxides, the growth of the field in turn focused attention on
both their radical and polar reactions with some spectacular
successes, as in the phenol synthesis from cumene via an intermediate peroxide rearrangement. As Dr. Porter's paper at this
symposium on peroxides related to prostaglandins shows, Nature
has known for a long time that the field was interesting, although
it still receives neglible attention in organic textbooks.

Another area which received prominent attention at the 1947
Faraday Society Discussion was that of the role of free radicals
in redox reactions. The unpredictable effects of trace metals on

organic reactions had long been known, but for the first time the
subject began to develop a rational interpretation. Again, the
next few years were a period of rapid development, with Kharasch
being an early leader in the U.S. Typical of his studies was
the copper catalyzed reaction of perester with olefins (11).

$$ R-C \overset{O}{\underset{OO-tBu}{}} \quad + \quad \bigcirc \quad \overset{Cu}{\longrightarrow} \quad \bigcirc^{OCOR} \quad + \quad t-BuOH $$

At first there were conflicting ideas about mechanism, and I recall working out a scheme which I considered plausible in the
middle '50's and describing it in a seminar at the Shell Development Co. This caused considerable confusion, since Jay Kochi and
his colleagues there had worked out the same approach in considerable detail, but were restrained from publishing it because
of industrial secrecy. Kochi of course has gone on to make
fundamental contributions to the study of redox systems, identi-
fying a variety of paths for transition metal-radical reactions
(12).
        Another area of radical chemistry which blossomed in the
50's with very practical results to synthetic chemists was free
radical halogenation. In 1950 Mayo and I proposed that polar
effects played an important role in halogen selectivity (13). A
bit later Russell demonstrated remarkable solvent effects in
chlorination which should still repay further study (14), and
Poutsma showed that halogen-olefin systems initiate their own
chains via molecule assisted homolysis (15). Most useful, I
think, has been the work on what I call halogen carriers and
intramolecular halogenation. The classic case of a potential
halogen carrier is N-bromosuccinimide (NBS). By 1950 most
chemists believed that the chain carrier in NBS bromination was
the succinimide radical (the Bloomfield mechanism). In the late
'60's the Goldfinger mechanism (bromine atom chains) took its
place, chiefly on the basis of relative reactivity results with
benzyl- and allylic substrates. More recently, the succinimide
chain has resurfaced, and recent results are described at this
symposium by Phil Skell. From my own point of view, this time
sequence was fortunate. In 1952 the Bloomfield mechanism was in,
and it seemed to me that chlorination via a t-butoxy radical
chain using t-butyl hypochlorite might be a useful analog.
My expectations were realized: t-butyl hypochlorite proved to be
a most useful allylic halogenating agent, and a convenient
starting material for investigating alkoxy radical chemistry (16).
With that done, a bit later I was able to participate in dis-
crediting the Bloomfield mechanism for NBS (17), and Jim
McGuiness in my group was able to show that chlorine atom chains
could be observed with t-butyl hypochlorite as well (18).
        We were led into intermolecular halogenation through Padwa's
demonstration of very clean regiospecific intermolecular halogena-

tion in long-chain hypochlorites to yield δ-chloroalcohol (<u>19</u>).
Many other workers have taken independent advantage of this tendency of radicals to undergo intramolecular hydrogen transfer
through 6-membered cyclic transition states to yield highly
specific products as in the Barton reaction (<u>20</u>). Most recently,
the concept has been extended by Breslow to highly specific
substitutions at much more remote sites in the steroid series.(<u>21</u>)
      Time doesn't permit my following all the other threads in
the development of free radical chemistry, and I will just mention
two. In the middle '50's photochemistry was pursued principally
by two groups: a considerable number of physical chemists like
W. A. Noyes, Jr, studying small molecules in the gas phase, and
a few organic chemists living in hot countries with little
equipment but a lot of sunshine (Schönberg and Mustapha in Egypt
for example) who put things on the roof to see what would happen.
About 1960 physical organic chemists discovered photochemistry,
I think for two reasons. Gas chromatography had come in, so
it was possible to analyze expeditiously complex mixtures of
rather small molecules, and George Hammond and his students
showed quite clearly that the photoreduction of benzophenone
was a triplet process which could be studied conveniently by
competitive kinetics (<u>22</u>). Triplet states are essentially biradicals and much of photochemistry turned out to be free radical
chemistry. I've only had a mild hand in this area; Morton Gibian
and I showed the close similarity in reactions of benzophenone
triplets and t-butoxy radicals (<u>23</u>), but two of my former students,
Al Padwa and Pete Wagner, have made substantial contributions to
the field.
      In the 1950's electron spin resonance (esr) equipment became
available, and, for the first time, it became possible to <u>see</u>
free radicals at very low concentrations. George Fraenkel at
Columbia had built one of the first highly sensitive instruments,
and in 1954 he and I were able to detect free radicals in poly-
merizing methyl methacrylate and show they had the expected
structure (<u>24</u>). Unfortunately, for a long time, most of the
people who knew how to operate esr spectrometers were interested
primarily in interpreting esr fine-structure, and couldn't care
less about reactions. Fortunately this is now changed, and
starting with R. O. C. Norman's work on flow systems, many groups
are applying the technique to the identification of radical intermediates and the measurement of elementary rate constants. Keith
Ingold has told us about some recent examples here.
      Where are we now? The years 1943-63 were probably the golden
age of free radicals in terms of exponential growth of the field
and introduction of new concepts. However, there is still
activity. Aside from this symposium at this meeting, I count
15-20 papers in the organic section involving radical reactions,
and a considerably larger number in the meeting as a whole. Free
radicals permeate our world - they make smog in Los Angeles, and
destroy ozone in the stratosphere. On a sunny day there are

roughly 10^6 hydroxyl radicals/cc. in the air we breathe, and they've been detected in interstellar space (more encouraging to future space travellers, there is also lots of ethanol). If I try to look ahead, I'd say that some of our important unsolved problems lie, first, in the borderline between radical and polar reactions - when can the two have common rate determining transition states and when do reactions go via single electron transfer, rather than via two electron displacements? Ashby, for example, has shown that both paths occur in Grignard reactions.

Second, and closely related, there is much to be learned about the role of radicals in oxidation-reduction processes, and organometallic chemistry and in the chamistry of radical ions. Here, as an example, my colleagues and I have been able to greatly strengthen the concept that radical cations are intermediates in a variety of radical reaction with aromatic molecules.

Third, what is the role of radical reactions in biochemical processes? Do enzymes dare use them? Plainly they sometimes do, but how often?

Radical chemistry continues to be a rewarding area, and, if the past is any guide to the future, radical intermediates will continue to turn up in unexpected places. In the meantime I can only regret that textbooks are unwilling to give them equal time with carbonium ions and molecular orbitals.

LITERATURE CITED

1. Hey, D. H. and Waters, W. A., Chem. Revs., (1937), 31, 169.
2. Kharasch, M. S.; Engelmann, H; and Mayo, F. R., J. Org. Chem. (1937), 2, 288.
3. Mayo, F. R. and Walling, C., Chem. Revs., (1940), 27, 351.
4. Mayo, F. R., and Lewis, F. M., J. Am. Chem. Soc., (1944), 66, 1594.
5. Bartlett, P. D. and Swain, C. G., J. Am. Chem. Soc., (1945), 67, 2273.
6. Bartlett, P. D., and Altshul, R., J. Am. Chem. Soc., (1945), 67, 812, 816.
7. Smith, W. V. and Ewart, R. W. J. Chem. Phys., (1948), 16, 592.
8. Walling, C.; Cooley, J.; Ponaras, A.; and Racah, E., J. Am. Chem. Soc., (1966), 88, 5361.
9. Julia, M., Accnts. Chem. Res., (1971), 4, 386.
10. Brown, H. C.; Rogic, M. M.; Rathke, M. W.; and Kabalka, G. W., J. Am. Chem. Soc., (1967), 89, 5709.
11. Kharasch, M. S.; Pauson, P.; and Nudenberg, W., J. Org. Chem. (1953), 18, 322.
12. Kochi, J. K. in "Free Radicals," J. K. Kochi, Ed., John Wiley and Sons, New York, (1973), Chapter 11.
13. Mayo, F. R. and Walling, C., Chem. Revs, (1950), 46, 192.
14. Russell, G. A., J. Am. Chem. Soc., (1958), 80, 4987.
15. Poutsma, M. L., J. Am. Chem. Soc., (1965), 87, 2161.
16. Walling, C., Pure & Appl. Chem., (1967), 15, 69.

17. Walling, C., and Rieger, A. L., J. Am. Chem. Soc., (1963),
     85, 3129.
18. Walling, C., and McGuinness, J. A., J. Am. Chem. Soc., (1969),
     91, 2053.
19. Walling, C., and Padwa, A., J. Am. Chem. Soc., (1963), 85,
     1597.
20. Barton, D. H. R., and Beaton, J. M., J. Am. Chem. Soc.,
     (1960), 82, 4083.
21. Breslow, R.; Corcoran, R.; Dale, J. A.; Liu, S. N.; and
     Kalicky, P., J. Am. Chem. Soc., (1974), 96, 1973.
22. Hammond, G. S. and Moore, W. M., J. Am. Chem. Soc., (1959),
     81, 6334.
23. Walling, C., and Gibian, M. J., J. Am. Chem. Soc., (1965),
     87, 3361.
24. Fraenkel, G. K.; Hirshon, J. M.; and Walling, C., J. Am.
     Chem. Soc., (1954), 76, 3606.

RECEIVED December 23, 1977.

Initiator Chemistry and the Initiation Step

Free Radical Aspects of Photooxidation

PAUL D. BARTLETT

Texas Christian University, Fort Worth, TX 76129

Photo-oxidation is defined, for our present
purpose, as oxidation by molecular oxygen under the
influence of light. The process may result in any
product more oxidized than the starting material;
and it may originate in the absorption of light by
the oxygen, by the substrate, or by a third molecule
(a "sensitizer") whose function may be to transfer
excitation energy or to involve the reactants in
short-lived reactive complexes or covalent intermediates. Many investigators of photo-oxidation may
have entertained the hope of discovering a common
mechanistic thread underlying all the kinds of
photo-oxidation; but the current mood is more one
of appreciation of the versatility of Nature, who
does not reserve her subtleties only for complex
biological systems, but lavishes them on even some
of the smallest molecules that we have.

In principle, any compound that can be thermally
oxidized can be oxidized by some application, direct
or indirect, of light. But the things that are
unique about photo-oxidation are seen mainly with
unsaturated starting materials, and it is with them
that the hope of sorting out mechanisms reaches a
maximum. Included in such photo-oxidations are
reactions leading to three general results:

1. The O-O bond remains intact in the product,
which may be an allylic hydroperoxide (1) (1), a
dioxetane (2) (2), or an "endoperoxide" (3) (3),
depending on the nature of the starting material.

$$2$$

$$3$$

2. The product contains only one of the atoms of the oxygen molecule, as in the case of an epoxide (4) (4), or a sulfoxide (5) (5, 6)

$$4$$

$$5$$

3. Cleavage products or rearrangement products are found, often attributable to an initial hydroperoxide (7) or peroxide (8) undergoing further reaction under the influence of light or thermally (9):

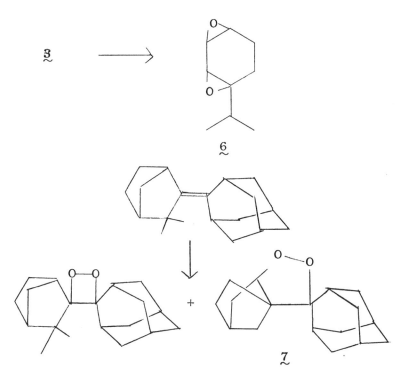

The Development of Mechanistic Criteria

Not only is there a ramification of products from any particular photo-oxidation method, but there are some clear cases where one and the same product may be formed by two or more mechanisms. The study of this field has therefore demanded the development of some criteria for at least assigning observed oxidations to mechanistic categories.

General criteria for singlet oxygen:

1. The formation of singlet oxygen requires the input of energy, and a photo-system generating singlet oxygen ceases to operate when the light is turned off. It thus differs from a chain reaction operating under thermal initiation.

2. Quenchers, typified by β-carotene and by 1,4-diaza(2.2.2)bicyclooctane (DABCO), powerfully inhibit reactions in which singlet oxygen is the reactant, while normally leaving free-radical chain reactions and ground-state thermal processes unaffected.

3. On the other hand, reactions depending on singlet oxygen do not respond specifically to chain inhibitors, such as hydroquinones or stable free radicals which strongly interfere with thermal chain reactions.

4. Certain compounds, such as tetramethylethylene or diphenylisobenzofuran, which are very reactive to singlet oxygen, may serve as indicators of its presence when introduced into a photo-oxidation where the participation of singlet oxygen is suspected. The absence of the normal singlet oxygen reaction products of these agents is a generally reliable criterion of the absence of the excited species.

5. The ability of singlet oxygen to attack a double bond is a sensitive function of electron availability in the pi system being attacked. Some photo-oxidations actively involve electron-poor double bonds which are unreactive toward singlet oxygen; this can be a straightforward mechanistic criterion.

Special Criteria for Certain Kinds of Oxidations.

For a reaction leading to hydroperoxides, a useful criterion of mechanism is the product distribution when 1,2-dimethylcyclohexene is oxidized by that method. Autoxidation under initiation by free radicals leads to introduction of the hydroperoxy group at the 3-position to the extent of 54% (10), this mode of reaction being undetectably small when either thermally- or photo-generated singlet oxygen is the reactant.

Criteria that have been used in the case of dioxetane formation include stereospecificity of the reaction, comparison of reactivities with those toward known singlet oxygen sources, and the presence or absence of termination products from any possible chain-carrying species.

Application of such criteria as these has shown that among photo-oxidations retaining the O-O bond, hydroperoxides can be formed either from singlet oxygen or from radical chain reactions, while there is a growing number of cases of dioxetanes or endoperoxides being apparently formed in non-singlet oxygen reactions. The oxidation of 2-phenylnorbornene in the dark gave products which could have involved dioxetane formation, but could also be accounted for in part by a Mayo chain. (11)

The Mechanistic Problem in Photo-epoxidation.

Epoxides can be produced stereospecifically
from olefins by compounds so constituted that they
can concertedly deliver an O atom and leave a stable
molecule behind. Peroxyacids are prime examples of
this capability. (12)
Substituent effects on this kind of concerted,
stereospecific epoxidation indicate that the peroxy-
acid functions as an electrophile; electron-donating
groups activate the olefin while electron-withdrawing
groups strongly deactivate it. When the electron
flow at the transition state is reversed by using
electron-poor olefins and peroxy anions for epoxidation, the reaction is stepwise and stereo-equilibrating and both cis- and trans-olefins (e.g.,
benzalacetophenone) yield trans-epoxide. (13, 14)
What might appear the simplest model of a
concerted epoxidation, the addition of a free gas-
phase oxygen atom to an olefin, proves to be in fact
a stepwise process as a result of the triplet character of ground-state oxygen atoms. (15)

Without the indicated "collisional deactivation"
(c.d.), the epoxide and carbonyl isomers are largely
replaced in the product by complex mixtures of cleavage products. The product figures of Cvetanovic
extrapolated to infinite pressure, allow an estimate
(16) that the energy-rich cis-biradical rotates to
the trans-conformation 2.9 times as fast as it undergoes ring closure to the epoxide. Although this
corresponds to a retention index of 2.1, which is
quite normal for condensed-phase cycloadditions via
biradicals, the gas phase systems cannot be perfect
models of reactions in solution because of the high
vibrational excitation which in this case leads to
migration of hydrogen and methyl in amounts comparable
to the amount of ring closure.
Thus the distinction between concerted and

stepwise epoxidations is not a matter of neutral vs.
ionic reactants, nor even of spin-paired vs. triplet
systems, but rather has to do with the availability
of a transition state allowing simultaneous bonding
of the incoming oxygen atom to both carbon atoms of
the double bond. The criteria for such a concerted
mechanism are not understood on a level allowing
prediction.

A clearly stepwise mechanism for epoxide formation was demonstrated in detail by Mayo and co-
workers (18, 19). It was shown that in a chain
copolymerization of styrene and molecular oxygen,
any polymeric radical terminating in a styrene unit
has the possibility of cleaving off an epoxide
molecule from the end of the chain:

$$R\ (OOCH_2CHC_6H_5)_n\ \ OOCH_2\overset{\bullet}{C}HC_6H_5\ \longrightarrow$$

$$RO^{\bullet}\ +\ n\ CH_2O\ +\ n\ C_6H_5CHO\ +\ \overset{O}{\overset{\diagup\diagdown}{CH_2-CHC_6H_5}}$$

In the case of styrene, because of rapid β-fission
of the remaining chain, epoxide is a minor product,
reaching its maximum yield in comparison to aldehydic fission products at a low oxygen pressure assur-
ing a significant lifetime to the styrene-terminated
radical. An example was recently found in the case
of photo-oxidation of 9-methoxymethylenefluorene (8)
where the cleavage of such a radical to epoxide is
so favored over addition of oxygen that comparable
amounts of epoxide and fluorenone are attained at
and above room temperature (20) (Scheme 1):

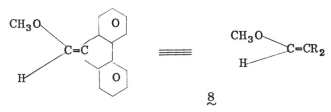

8

In a chain reaction involving peroxy radicals,
coupling of these radicals is a very rapid process
and can be the principal mechanism of chain termination. In the case of methoxymethylene fluorene,
however, peroxy radical coupling in normal course
does not lead to termination, but to fluorenone
and a chlorine atom to continue the chain (Scheme
2).

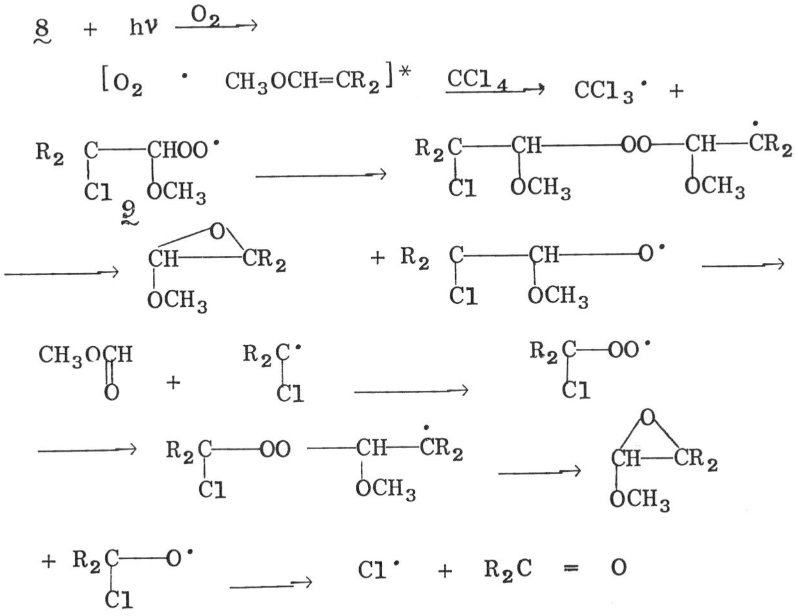

$\underset{\sim}{8}$ + hν $\xrightarrow{O_2}$

[O_2 · $CH_3OCH=CR_2$]* $\xrightarrow{CCl_4}$ CCl_3· +

$\underset{\underset{9}{\overset{|}{\underset{Cl}{\overset{\displaystyle\,}{}}}\ \underset{OCH_3}{\overset{\displaystyle\,}{}}}{R_2C\!-\!\!-\!CHOO}$· $\longrightarrow$ $R_2\underset{Cl}{\overset{\,}{C}}\!-\!\!-\!\underset{OCH_3}{\overset{\,}{CH}}\!-\!\!-\!OO\!-\!\!-\!\underset{OCH_3}{\overset{\,}{CH}}\!-\!CR_2$

$\longrightarrow$ $\underset{OCH_3}{\overset{\overset{\displaystyle O}{\diagdown\,\diagup}}{CH}}\!-\!\!-\!CR_2$ + $R_2\underset{Cl}{\overset{\,}{C}}\!-\!\!-\!\underset{OCH_3}{\overset{\,}{CH}}\!-\!\!-\!O$· $\longrightarrow$

$CH_3O\underset{\overset{\|}{O}}{CH}$ + $R_2\underset{Cl}{\overset{\,}{C}}$· $\longrightarrow$ $R_2\underset{Cl}{\overset{\,}{C}}\!-\!\!-\!OO$·

$\longrightarrow$ $R_2\underset{Cl}{\overset{\,}{C}}\!-\!\!-\!OO\!-\!\!-\!\underset{OCH_3}{\overset{\,}{CH}}\!-\!CR_2$ $\longrightarrow$ $\underset{OCH_3}{\overset{\overset{\displaystyle O}{\diagdown\,\diagup}}{CH}}\!-\!\!-\!CR_2$

+ $R_2\underset{Cl}{\overset{\,}{C}}\!-\!\!-\!O$· $\longrightarrow$ Cl· + R_2C = O

<u>Scheme 1.</u>

$2\,ClCR_2\!-\!\!-\!\underset{OCH_3}{\overset{\,}{CHOO}}$· $\longrightarrow$ $ClCR_2\ \underset{OCH_3}{\overset{\,}{CHOOOO}}\underset{OCH_3}{\overset{\,}{CH}}\!-\!CR_2Cl$

$\longrightarrow$ $2\ ClCR_2\ \underset{OCH_3}{\overset{\,}{CHO}}$·   + O_2

$ClCR_2\ \underset{OCH_3}{\overset{\,}{CHO}}$· $\longrightarrow$ $CH_3O\underset{\overset{\|}{O}}{CH}$ + $ClCR_2$·

$\longrightarrow$ $ClCR_2OO$· $\xrightarrow{\text{coupling, cleavage}}$

O_2 + $2\,ClCR_2O$· $\longrightarrow$ $R_2C{=}O$ + Cl·

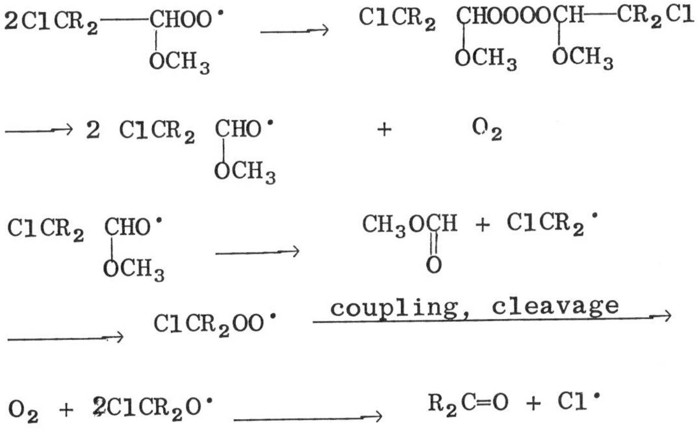

<u>Scheme 2.</u>

Coupling thus diverts intermediate radicals
from forming epoxide to forming fluorenone. Since
the activation energy for radical coupling is normally lower than for chain propagation, an explanation is at hand for the rather sharp reduction in
the epoxide/fluorenone ratio as the temperature is
lowered.
The compelling evidence in favor of a free
radical mechanism for this reaction consisted in
the fact that the results were identical when the
reaction was initiated in the dark by generating
the radical 9 from a hydroperoxide precursor with
lead tetraacetate, and in the determination of a
quantum yield of about 12 for the photo-initiated
reaction.
Thus free radicals may readily be involved in
a photo-epoxidation, and when they are they can be
detected by establishing the chain character of the
reaction and its response to initiators. If the
example of 9-methoxymethylene-fluorene is of general
significance, we may suspect that reactions leading
cleanly to epoxides as single products will not be
found to proceed through free radicals.

The Simultaneous Formation of Epoxides and Dioxetanes

The near-inseparability of conditions for the
formation of dioxetanes and of epoxides was first
observed when 7,7'-binorbornylidene was oxidized
with photosensitizers in a series of solvents (21).
The ratio of epoxide to dioxetane increased
28-fold on going from methylene chloride to benzene
with tetraphenylporphin as sensitizer, such solvents
as pinacolone, dioxan, and acetone occupying intermediate positions. Systematic investigation showed
that epoxide more often than not accompanies the
dioxetane from sensitized oxygenation of norbornene
(22) and of biadamantylidene (M. J. Shapiro, quoted
in (23); (26)). For each solvent-sensitizer combination, the ratio of epoxide to dioxetane increased
in the order: biadamantylidene < norbornene < binor-
bornylidene. This is also the decreasing order of
reactivity of these olefins toward singlet oxygen.
In the course of these and other investigations (24),
much negative evidence has been accumulated concerning
the mechanism. Since the hypothesis of an intermediate perepoxide was already an attractive one in
offering a symmetry-allowed path for stereospecific
dioxetane formation, it gained in attractiveness with

the apparent possibility of explaining the accompanying formation of epoxides.

However, although we do not yet understand the
full meaning of the observations, it is clear that
the almost ubiquitous epoxide products in these experiments do not come from singlet oxygen. For
example, in a series of controlled experiments by
Dr. M. J. Shapiro, biadamantylidene is converted
solely into dioxetane by the singlet oxygen generated
from the decomposition of triphenylphosphite ozonide
in methylene chloride at 0° in the dark, or in the
photosensitized oxidation with visible light by
crystal violet in the same solvent. On the other
hand, three sensitizer systems in carbon tetrachlor-
ide -- rose bengal on polymer beads, iodine, and
bromine -- have converted biadamantylidene entirely
into its epoxide. In one photosensitizer system in
benzene, the presence or absence of an ultraviolet
cutoff filter has made a difference of a factor of
17 in the epoxide/dioxetane ratio.

Such observations eliminate suggested mechanisms of deoxygenation of the perepoxide, whether by
solvent (27) or by singlet oxygen (23, 28, 29).
The powerful effect of tetracyanoethylene as an
additive (23) and of diphenyl sulfide as a solvent
(30) in shifting the product over to epoxide would
be compatible with deoxygenation of a perepoxide,
but in the case of the diphenyl sulfide a different
order of events seems more probable (5). Especially
striking is the observation of Jefford and Boschung
(26) that the amount of epoxide accompanying the
dioxetane varies greatly with the concentration of
the sensitizer: sometimes, as in the case of rose
bengal, it increases with increasing sensitizer
concentration, whereas with tetraphenylporphin it
declines. These facts, plus the wavelength dependence of the epoxide/dioxetane ratio, require that
the epoxide results from a photochemical sequence
independent of that producing singlet oxygen. The
conclusion is clear -- there is a photo-induced
sequence, of which many sensitizers are capable,
which leads to epoxide by a mechanism or mechanisms other than the direct transfer of energy to an
oxygen molecule.

It should be mentioned parenthetically that the
several roles which the perepoxide has been shown
not to perform have not served to eliminate this
intermediate from a possible central position on
the reaction path from singlet oxygen to dioxetane.
In deciding between the perepoxide as a way of

removing the orbital symmetry barrier and the chief
other possibility -- the 2s+2a cycloaddition --
McCapra's observation of polar-type rearrangements
to dioxolanes instead of dioxetanes (9) can be
interpreted as favoring the three-membered ring
intermediate with its obvious polar capabilities.
However, the rearrangement depends on the charge-
bearing disposition of the camphenyl skeleton in
the substrate, which might induce a polar rearrangement in any transition state en route to dioxetane.
      Frimer, Bartlett, Boschung and Jewett (32)
have recently observed by means of deuterium and
tritium isotope effects that the transition states
in the attack of singlet oxygen on 4-methyl-2,3
dihydro-4H-pyrans deviate from the "least motion"
path for the direct formation of either dioxetane
or allylic hydroperoxide, toward the type of transition state to be expected if perepoxide formation
were the rate determining step. Any final judgment
on the perepoxide as an intermediate in dioxetane
formation must await evidence as specific as that
which has eliminated it in the   competitive photo-
epoxidations.
      The commonly used photosensitizers for gener-
ating singlet oxygen undergo π,π^* activation; this
property is associated with their strong predisposi-
tion to transfer energy to other molecules rather
than entering into some of the more specific chemistry associated with n,π^* excited states. Yet
all these sensitizers have hetero-atoms with unshared electron pairs and hence, with light of the
appropriate energy, they should be capable of n,π^*
excitation.

Epoxidation with α-Diketones

      We have observed that a change in sensitizer
for olefin photo-oxidation from a π,π^* absorber to
benzophenone, benzil, and finally biacetyl (4)
shifts the reaction toward epoxidation at an accelerating rate, epoxidation with biacetyl being
general, fast, and of high yield. The epoxidation
does not respond to singlet oxygen quenchers, and
is effective on many olefins which are quite inert
to singlet oxygen. Although this reaction had little
in common with the chain epoxidation of methoxymethy-
lenefluorene, it was tempting at first to attribute
the epoxidation to the more localized electron distribution in an n,π^* excited state and to postulate
active intermediates having covalent interaction of

oxygen with the caronyl group.
An obvious possibility of the kind would be

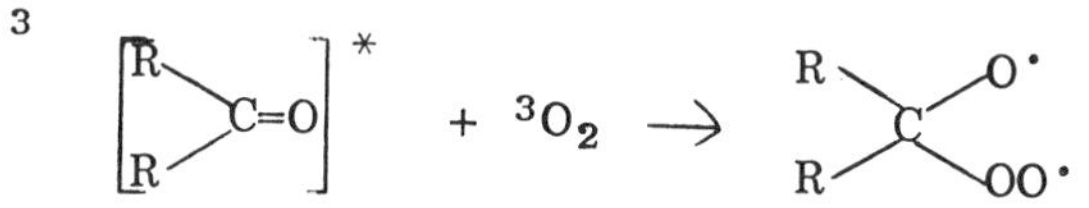

3

$$\begin{bmatrix} R \\ R \end{bmatrix} C{=}O \end{bmatrix}^* + {}^3O_2 \rightarrow$$

10   11

11 formed from 10 by attack of triplet oxygen. Although 11, as a singlet species, might cyclize to
12 which could be the same kind of one-O-donor as a
chelated peracid, yet 11 might also behave as a
biradical in reacting with an olefin with the end
result of donating a single oxygen atom to form the
observed epoxide.

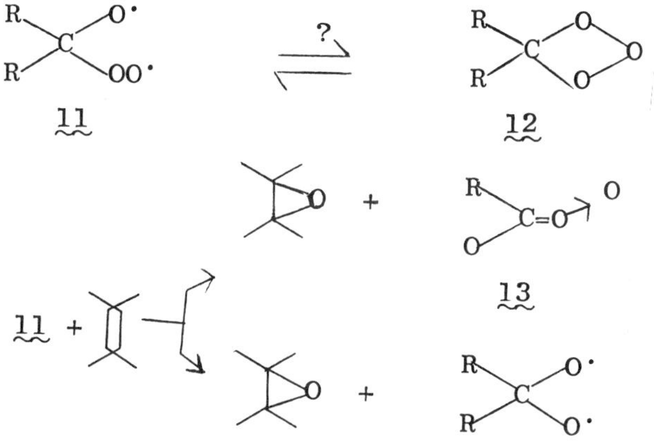

11   ?   12

13

11 +

14

The remaining species, 13 and 14, are alternate
forms of the caronyl oxide of the ketone. If the
Criegee structure 13 were formed in the process,
it would mean that the epoxide oxygen came not
from the oxygen molecule but from the keto oxygen
of the sensitizer. By either of these paths any
sensitizer keto group which becomes regenerated
must be either completely exchanged (in the case of
13) or half exchanged (in the case of 14) with the
oxygen from O_2.

   Dr. Johannes Becherer has tested this possibility of biradical involvement in the photo-epoxidation by carrying out the reaction with completely
18_O - labeled molecular oxygen, and unlabeled benzil
and biacetyl. Small samples in benezene containing

0.33 millimoles of $^{36}O_2$ and 0.20 millimoles each of
benzil or biacetyl and norbornene were irradiated
through pyrex with a 450-W. Hanovia mercury vapor
lamp. Under identical conditions, benzil on 15-
minute irradiation underwent about 5% isotopic O-
exchange in the absence of norbornene and 10% in
its presence. In the same length of time the norbornene was converted 70% into norbornene oxide
whose oxygen was 93% ^{18}O and 7% ^{16}O. A result like
this could not have been produced by any obvious
variant of the mechanism depicted above, but rather
requires a reaction path by which the oxygen of O_2
is conveyed into the epoxide without ever having
been covalently attached to the carbonyl group of
the sensitizer.
      The results with biacetyl are equally emphatic.
In the absence of olefin biacetyl is rapidly destroyed by light and oxygen: on irradiation for 15
minutes the yellow color of the alpha-diketone had
completely faded. The small amount of biacetyl
remaining showed 42% incorporation of isotopic
oxygen, still short of the calculated 62.5% for
equilibration incidental to the photo-destruction.
In the presence of equimolar norbornene after the
same irradiation, the much larger amount of recovered
diketone was only 8% exchanged, while there was a
90% conversion to norbornene oxide whose oxygen was
93% ^{18}O.

Ion Radicals in Photo-oxidation

      These experiments having eliminated partici-
pation of both the biradical 11 and the hypothetical
peracid analog 12, there remains a large area where
there are persistent indications of involvement of
odd-electron species in oxidation processes. Single
electron transfers occur in many of the same situations in which free radical initiation and photo-
sensitization occur. There have been cases of this
kind where superoxide radical ion, $O_2^{\underline{-}}$, has been
observed, and some of its interactions with singlet
oxygen have been studied.
      Jefford and Boschung (26) have interpreted
their extensive experiments on the photooxidation
of biadamantylidene in terms of an intervention
of superoxide anion, generated in a secondary reaction between singlet oxygen and various ground-
state sensitizers, especially rose bengal. In a
series of steps, the resulting RB^{+}·takes an electron

from Ad=Ad and the resulting cation radical Ad=Ad$^{+\cdot}$
reacts with triplet ground state oxygen to give the
cation radical of a perepoxide. This species in
turn is converted by a second superoxide anion into
the often-discussed pair, epoxide +ozone. The
advantage of this sequence over the direct process
via perepoxide and singlet oxygen is that it
provides a picture of the special role played by
each sensitizer in influencing the relative amounts
of dioxetane (still formed via singlet oxygen and
olefin) and epoxide. The authors show that DABCO, as
a singlet oxygen quencher, stops both kinds of
photo-oxidation, consistent with the view that singlet oxygen is a precursor of the epoxidizing reagent
itself.

Specific tools used in these experiments included di-t-butyl-p-cresol as a radical inhibitor
to interrupt selectively the radical chain, and the
use of phenylglyoxylic acid to react specifically
with superoxide anion (33).

Other probes that have been used for the occurrence of superoxide anion radical (36) in the
presence of singlet oxygen include sulfite anion
(34), selectively oxidized by superoxide, and superoxide dismutase (34, 35), which efficiently converts superoxide ion into oxygen and hydrogen
peroxide.

In seeking a framework for the epoxidizing action of the alpha-diketones as photosensitizers, it
is suggestive that these compounds are themselves
very effective electron-acceptors, giving rise to
the well characterized semidiones (37).

In terms of the Jefford-Boschung framework, we
might then say that a sensitizer whose excited state
is high enough in electron affinity (such as an
α-diketone) can simplify the photo-oxidation sequence
by taking an electron directly from the olefinic
substrate rather than transferring any excitation
energy to oxygen, and so can initiate an epoxidation
sequence without intervention of singlet oxygen.
Epoxidation could be completed in the absence of both
superoxide ion and singlet oxygen if the perepoxide
cation radical, 15, could react with triplet oxygen
to a species (16) which is converted by collision

$$R_2C \diagdown \quad \diagup O^{+}\text{-}O\cdot$$
$$R_2C \diagup$$

$$\left[R_2C \diagdown \quad \diagup^{+} O\text{-}OOO\cdot \atop R_2C \diagup \right]$$

15   16

with the semidione into epoxide and ozone. Thus the
inclusion of ion radicals among the reactive intermediates for photo-oxidation greatly widens the
mechanistic possibilities among which a decision must
be made.

Another sensitizer which might be expected to
generate a radical anion on excitation in the presence of an electron donor is 9,10-dicyanoanthracene
(31). Although Foote and co-workers found this to
be a sensitizer for oxygenation without any evidence of singlet oxygen being involved, yet its
action in producing benzophenone from tetraphenylethylene suggested that the oxidation product was a
dioxetane; no epoxidation was observed.

Cation radicals are strongly implicated in the
series of catalysts discovered by Barton and co-
workers (38) which, some photochemically and some
thermally, produce peroxides from ergosteryl acetate
and other dienes in methylene chloride at -78°.
These catalysts, including Lewis acids, carbonium
ions, or aminium ion radicals, are in remarkable
contrast to those just mentioned in that they have
not been observed to produce any epoxides. Dr. M.
J. Shapiro has also found that Barton's cation ra-
dical $(p\text{-}BrC_6H_4)_3N^{+\cdot}$ converts biadamantylidene into
dioxetane and not into epoxide. It is striking in
this connection that the cationoid radical complex
$(CH_3)_2N\cdot ZnCl_2$ (39) reacts, also thermally, with
oxygen and olefins to yield, not peroxides, but
epoxides, as inferred from the structure and configuration of the aminoalcohols isolated.

Photo-oxidation of Sulfides

Dialkl sulfides, like olefins, can undergo
one-O or two-O oxidation, with the obvious difference
that in a sulfone the two oxygen atoms do not remain
bonded to each other. Foote and Peters (5) developed
convincing evidence that singlet oxygen reacts directly with diethyl sulfide to yield a reactive peroxy-
sulfoxide (17) which is capable of converting

$$Et_2S + {}^1O_2 \longrightarrow Et_2\overset{+}{S}\text{-}O\text{-}O^- \longrightarrow Et_2S\overset{\nearrow O}{\searrow O}$$

$$17$$

$$Et_2S + {}^3O_2 \qquad\qquad Et_2SO + R_2SO$$

$$R_2S$$

another sulfide molecule into sulfoxide, or of re-
arranging more slowly into sulfone, in competition
with quenching by dissociation into sulfide and
triplet oxygen. Cleavages attending the photo-
oxidation of benzyl alkyl sulfides (40) could also
be formulated as involving peroxysulfoxides derived
from singlet oxygen. However, the photo-oxidation
of sulfides sensitized by 9,10-dicyanoanthracene
(31) also produced sulfoxides and sulfones. This
oxidation was not inhibited by β-carotene, and
showed a strong reversal of the relative reactivities
of diphenyl and diethyl sulfides compared to those
in singlet oxygen oxidation.
 Another kind of evidence for the involvement
of ion radicals in photo-oxidation comes from the
observation of cleavage of certain sulfides (41)
with formation of unoxidized disulfides. Unlike
the benzyl alkyl sulfides (40), di-t-butyl sulfide
gives on oxidation with several photosensitizers,
in addition to the sulfoxide and sulfone, amounts
of di-t-butyl disulfide varying from traces with
rose bengal or methylene blue in methanol to 97
and 100% in acetone with rose bengal free and bound
on polymer beads, respectively. The yield of the
disulfide may be a measure of the relative rate at
which the cation radical 18 dissociated to the t-
butyl cation and the t-butylthiyl radical in the
various media:

$$(CH_3)_3C\overset{+\cdot}{S}C(CH_3)_3 \rightarrow (CH_3)_3C^+ + (CH_3)_3CS\cdot \rightarrow (CH_3)_3CSSC(CH_3)_3$$

18

The other products of the photo-oxidations are mix-.
tures of di-t-butylsulfoxide and di-t-butylsufone,
these mixtures becoming steadily richer in sulfone
as the reaction progresses. Although singlet oxygen
generated thermally was capable of producing the
sulfoxide and sulfone, the disulfide was absent from
such thermal product mixtures.
 Neither was disulfide produced with sensitizer
in the absence of oxygen. In parallel irradiations
of di-t-butyl sulfide, one under oxygen and the other
under argon, no cleavage or any other reaction is seen
in the experiment without oxygen. These results
show that   oxygen plays an essential part in
generating the cation radicals, even though direct
attack of singlet oxygen on the sulfide (as shown in
the experiments with chemical generation) is no part

of the process. This is further evidence supporting
some such multi-step electron transfer process as
that proposed by Jefford and Boschung in some of
the epoxidation reactions.

Not surprisingly, although superoxide anion is
implicated in the genesis of the cation radicals,
a solution of potassium superoxide-crown ether in
methylene chloride was without effect on di-t-butyl
sulfide alone.

Di-t-butyl sulfide could be converted to disulfide in two hours at room temperature by tris-(p-
bromophenyl)-aminium fluoborate in a stream of oxygen, but neither tetrahydrothiophene nor diphenyl
sulfide underwent cleavage under these conditions.

Conclusions

Photo-oxidation has been shown to proceed, not
only by concerted reactions with singlet oxygen,
but through stepwise mechanisms involving neutral
free radicals and radical ions. In some of the
latter cases singlet oxygen is implicated in the
genesis of the radical ions and their precursors,
and the resulting radical reactions compete with the
direct reaction of the singlet oxygen. Radical and
cation-radical processes appear to be especially
important in photo-epoxidation.

Literature Cited

1. Schenck, G.O., and Schulte-Elte, K., Ann. (1958) 618 185.
2. Bartlett, P.D., and Schaap, A.P., J. Am. Chem. Soc. (1970) 92 3223.
3. Schenck, G.O., and Ziegler, K., Naturwissenshaften (1944) 32 57.
4. Shimizu, N., and Bartlett, P.D., J. Am. Chem. Soc. (1976) 98 4193.
5. Foote, C.S., and Peters, J.W., J. Am. Chem. Soc. (1971) 93 3795.
6. Foote, C.S., and Peters, J.S., 23^d Int. Cong. Pure and Appl. Chem. (1971) 4 129.
7. Kharasch, M.S., and Burt, J.G., J. Org. Chem. (1951) 16 150.
8. Bocke, J., and Runquist, O., J. Org. Chem. (1968) 33 4285.
9. McCapra, F., and Beheshti, I., J. Chem. Soc. Chem. Comm. (1977) 517.
10. Foote, C.S., Accts. Chem. Res. (1968) 1 104.
11. Jefford, C.W.; Boschung, A.; and Rimbault, C.G., Helv. Chim. Acta (1976) 59 2542.
12. Swern, D., "Organic Peroxides," Vol. 2, Chapter 5, pp. 466-475; Wiley-Interscience, New York, 197.
13. Lutz, R.E., and Weiss, J.O., J. Am. Chem. Soc. (1955) 77 1814.
14. Curci, R., and Edwards, J.O., in D. Swern, "Organic Peroxides," Vol. 1, Chapter 4, p. 245.
15. Cvetanovic, R.J., Adv. Photochem. (1963) 1 117-149.
16. The retention index PQ is defined (17) and its use in evaluating rate constant ratios is described by L. K. Montgomery, K. Schueller, and P.D. Bartlett, J. Am. Chem. Soc., 86, 622 (1964). We have recently formulated a more general analysis of stepwise, stereo-equilibrating cycloadditions, containing some derivations not given in the original paper, and describing a few applications. This manuscript, prepared on invitation for a review journal, is available to interested readers on request.
17. Bartlett, P.D., 23^d Intl. Cong. Pure and Appl. Chem. (1971) 4 281.
18. Mayo, F.R., J. Am. Chem. Soc. (1958) 80 2465.
19. Mayo, F.R., and Miller, A.A., J. Am. Chem. Soc. (1958) 80 2480.

20. Bartlett, P.D., and Landis, M.E., J. Am. Chem.
    Soc. (1977) 99 3033.
21. Bartlett, P.D., and Ho, M.S., J. Am. Chem. Soc.
    (1974) 96 627.
22. Jefford, C.W., and Boschung, A., Helv. Chim.
    Acta (1974) 95 3381.
23. Bartlett, P.D., Chem. Soc. Reviews (1976) 5 154.
24. Unpublished work of M. J. Shapiro, N. Shimizu,
    and G.Y. Moltrasio Iglesias.
25. Jefford, C.W., and Boschung, A.F., Tet. Letters
    (1976) p. 4771.
26. Jefford, C.W., and Boschung, A.F., Helv. Chim.
    Acta, December, 1977. We thank Professor
    Jefford for a preprint.
27. Schaap, A.P., and Faler, G.R., J. Am. Chem. Soc.
    (1973) 95 3381.
28. Ho, M.S., Thesis, Harvard University (1974).
29. Dewar, M.J.S.; Griffin, A.C.; Thiel, W.; and
    Turchi, I.J., J. Am. Chem. Soc. (1975) 97 4439.
30. Shapiro, M.J., unpublished work.
31. Eriksen, J.; Foote, C.S.; and Parker, T.L.,
    J. Am. Chem. Soc. (1977) 99 6455.
32. Frimer, A.A.; Bartlett, P.D.; Boschung, A.F.;
    and Jewett, J.G., J. Am. Chem. Soc. (1977) 99
    7977.
33. Jefford, C.W.; Boschung, A.F.; Bolsman, T.A.B.M.;
    Moriarty, R.M.; and Melnick, B., J. Am. Chem.
    Soc. (1976) 98 1017.
34. Srinivasan, V.S.; Podolski, D.; Neckers, D.C.;
    and Westrick, N.J. (1977) Submitted to J. Am.
    Chem. Soc. We thank Prof. Neckers for a preprint.
35. Fridovitch, Accts. Chem. Res. (1972) 5 321.
36. For a review of superoxide chemistry, see E.
    Lee-Ruff, Chem. Soc. Revs. (1977) 6 196.
37. Russell, G.A., in "Radical Ions," ed. by E.T.
    Kaiser and L. Kevan, Chapter 3, Interscience
    Publishers, New York, 1968.
38. Barton, D.H.R.; Haynes, R.K.; Leclerc, G.; Mag-
    nus, P.D.; and Menzies, I.D., J. Chem. Soc.
    (1975) Perkin I, 2055.
39. Michejda, C.J., and Campbell, D.H., J. Am. Chem.
    Soc., (1976) 98 6728.
40. Corey, E.J., and Ouannes, C., Tet. Letters
    (1976) 4263.
41. Shapiro, M.J., and Landis, M.E., unpublished
    work.

RECEIVED December 23, 1977.

Radical Production from the Interaction of Closed-Shell Molecules

Part 7. Molecule-Assisted Homolysis, One-Electron Transfer, and Non-Concerted Cycloaddition Reactions

WILLIAM A. PRYOR

Department of Chemistry, Louisiana State University, Baton Rouge, LA 70803

For some time, my research group has been interested in the processes by which closed-shell, stable molecules interact to produce free radicals (1-6). We have divided these reactions into three mechanistic types: molecule-assisted homolyses (MAH), one-electron transfers, and non-concerted pericyclic reactions. These processes usually occur at surprisingly moderate temperatures and often have an intriguing and challenging kinetic and mechanistic complexity. A large number of reactions of all three mechanistic types are now known.

Many of these processes find practical use. For example, some processes that appear to involve MAH reactions are used in chemical industry: the commercial polymerization of styrene often is self-initiated (7); the autoxidation of acetaldehyde is initiated by ppm levels of ozone (8); and some halogenations are clearly MAH processes (1,9). Electron-transfer reactions find practical use as low-temperature initiation systems: for example, benzoyl peroxide and dimethylaniline produce radicals at temperatures as low as 10°, perhaps by an electron transfer process (3, 4). Radical production from [2+2] cycloaddition reactions also has been clearly demonstrated by a number of workers, using both experimental and theoretical techniques (9a). However, the interception of the intermediate 1,4-diradicals by added radical trapping reagents has been reported in only a few cases (5,10-16).

In the limited space available here, I will not attempt a complete summary or a detailed review of the three classes of reactions; rather, I will present an eclectic report of a few reactions on which my own group has done research in the past several years.

First, let me define the three classes of processes which we will consider. Molecule-assisted homolyses generally have a formulation like that given in eqs 1a or 1b, where the homolysis of the A-B bond is assisted by some type of bond-formation process

with molecule(s) C.*   Electron transfer reactions can be formulated as shown in eqs 2a or 2b; these reactions can be followed by

$$A-B + C \longrightarrow A\cdot + BC\cdot \qquad (1a)$$

$$A-B + 2C \longrightarrow AC\cdot + BC\cdot \qquad (1b)$$

$$A-B + C \longrightarrow A-B\overset{+}{\cdot} + C\overset{-}{\cdot} \qquad (2a)$$

$$AB + C: \longrightarrow AB\overset{-}{\cdot} + C\overset{+}{\cdot} \qquad (2b)$$

the scission of the $A-B\overset{+}{\cdot}$ or $A-B\overset{-}{\cdot}$ bond. We shall limit our discussion of pericyclic reactions to [2+2] cycloadditions which lead
to scavengable radicals; these processes can be described as in
eqs 3 and 4, where S is a reactive molecule or radical that can
trap the 1,4-diradical in competition with the ring closure
reaction, eq 4a, leading to cyclobutanes.

$$\| + \| \longrightarrow \cdot\sqcap\cdot \qquad (3)$$

$$(\cdot M_2\cdot)$$

$$\cdot M_2 \cdot \Big< \begin{array}{c} \longrightarrow \square \qquad (4a) \\ \\ \overset{S}{\longrightarrow} \cdot SM_2\cdot \qquad (4b) \end{array}$$

Molecule-Assisted Homolyses

Introduction. The first systematic discussion of this field,
to my knowledge, was by Semenov (17), in his book published in
English translation in 1958. I discussed the principles of MAH
reactions and reviewed a number of examples in my book in 1966
(1), and Benson (18) has considered the application of his thermochemical techniques to such processes. In 1974, Harmony presented
a relatively complete survey of MAH reactions, including 150 references (9). Her review does not cover non-concerted cycloadditions, nor does she treat one-electron transfer reactions in
depth. Her chapter also gives an unsatisfactory picture of

---

*In the textbook "Free Radicals", ref. 1, I used the expression
"molecule-induced homolysis" and the abbreviation MIH, a phrase-
ology retained by Harmony (9). More recently, I have used
"molecule-assisted homolysis" to avoid confusion of MAH processes
with the induced decomposition of initiators caused by radicals,
a propagation step, rather than an initiation process:

$$AB + R\cdot \longrightarrow A\cdot + RB$$

radical production from the self-initiated polymerization of various vinyl monomers ("thermal polymerizations"), but we have recently reviewed that area in detail (7). Nevertheless, her review does provide, for the first time, a compilation of most of the key references on MAH processes in non-polymer systems.

An accelerated homolysis is defined as one that proceeds at an accelerated rate over that expected for a simple unimolecular homolysis. If a compound A-B undergoes unimolecular homolysis, eq 5, the rate constant can be predicted from the Arrhenius equation, eq 6, where BDE(A-B) is the bond dissociation energy of the A-B bond (18). However, if AB undergoes an assisted homolysis

$$A-B \longrightarrow A\cdot + B\cdot \qquad (5)$$

$$k = A \exp(-E_a/RT) = 10^{16}\exp[-BDE(A-B)/RT] \text{ sec}^{-1} \qquad (6)$$

with another molecule (or molecules) C, eq 7, then the heat of reaction for this process can only be calculated if the stoichiometry of the reaction is known. For example, if the process can be described by eq 1a, then the heat of reaction equals the BDE of the A-B bond minus that of the B-C bond; thus, the overall

$$AB + nC \longrightarrow \text{Radicals} \qquad (7)$$

endothermicity of the process is reduced. In general the activation energy also is reduced and an acceleration in the rate of the reaction is observed.

MAH Transfer of a Hydrogen Atom. The assisted transfer of a hydrogen atom, often from a C-H bond, is perhaps the most fascinating of these processes to an organic chemist. Assisted C-H bond homolyses were postulated very early to rationalize complex radical chain reactions. For example, in 1947 Hinshelwood (19) proposed that the primordial initiation process in hydrocarbon autoxidations is the spontaneous hydrogen abstraction reaction of oxygen, eq 8. This process remains controversial to this day (20, 21). Since oxygen is not a closed-shell molecule, but a ground

$$RH + O_2 \longrightarrow R\cdot + HO_2\cdot \qquad (8)$$

state triplet, eq 8 actually does not meet the formal definition of an MAH process. However, it is usually discussed with other examples, since it has a superficial similarity (and perhaps also because it has some of the same exasperatingly inaccessible experimental features).

A hydrogen atom transfer from carbon in an MAH process was proposed by Flory in 1937 to rationalize the spontaneous polymerization of styrene (7). He suggested that 1,4-diradicals are formed via eq 9 and are converted to monoradicals by the transfer reaction shown in eq 10. However, Hammond, Kopecky, and our group have used kinetic isotope effects to rule out reaction 10 as the predominant process that converts diradicals to monoradicals in styrene ($\underline{7}$).

$$2PhCH=CH_2 \longrightarrow Ph\overset{\bullet}{C}H-CH_2-CH_2-\overset{\bullet}{C}HPh \qquad (9)$$

$$(\cdot M_2 \cdot)$$

$$\cdot M_2 \cdot + M \longrightarrow PhCH=CH-CH_2-\overset{\bullet}{C}HPh + Ph\overset{\bullet}{C}H-CH_3 \qquad (10)$$

$$(HM_2 \cdot) \qquad\qquad (HM \cdot)$$

Another mechanism for the spontaneous initiation reaction in styrene was proposed by Mayo in 1961 ($\underline{7}$). It involves the reactions shown in eqs 11-12.

$$2PhCH=CH_2 \xrightarrow[\text{dimerization}]{\text{Diels-Alder}} \qquad\qquad (11)$$

(M)

(AH)

$$AH + M \xrightarrow{MAH} \qquad\qquad + HM\cdot \qquad (12)$$

(A·)

$$A\cdot \text{ (or } HM\cdot) + M \longrightarrow M_n^{\bullet} \qquad (13)$$

(Growing polymeric chain radical)

Recently, the occurrence of the MAH process, eq 12, has been probed by two methods: (i) Buchholz and Kirchner ($\underline{22}$) and Pryor and Patsiga ($\underline{6a,20}$) used the uv absorption of AH to follow its rate of appearance and measure its steady state concentration. Buchholz and Kirchner obtain a steady state concentration for AH of about 0.6×10^{-4} $\underline{M}$ at $64°$. (ii) We had previously published a computer simulation ($\underline{23}$) of the thermal polymerization of styrene in which we assumed that eqs 11-12 were the only initiation mechanism; this simulation predicts the steady state concentration of AH to be 5×10^{-4} $\underline{M}$ at $60°$. This certainly is in acceptable agreement with the later experimental measurement by Buchholz and Kirchner. In addition, our simulation gives the chain transfer constant for AH, i.e., k_{14}/k_{15}, to be about 1. Thus, AH is a remarkably reactive hydrocarbon toward radicals.

(For example, the transfer constant of Ph_3CH is only 3×10^{-4} at $60°$ (23a).)

$$AH + M_n^{\bullet} \longrightarrow A^{\bullet} + M_n\text{-}H \tag{14}$$

$$M_n^{\bullet} + M \longrightarrow M_{n+1}^{\bullet} \tag{15}$$

A large number of MAH transfers of hydrogen atoms have been inferred, and in the space available here we can do no more than list a few to indicate the diversity and wide occurrence of the process. For example, reaction 16 was postulated by Semenov to rationalize gas phase cracking rates (17).

$$C_2H_5\text{-}H + CH_2=CH_2 \longrightarrow 2C_2H_5^{\bullet} \tag{16}$$

Many of the halogens give MAH reactions, but the mechanism usually involves attack on the halogen molecule by π-electrons from an unsaturated molecule as a halogen atom adds to the unsaturated linkage:

$$R_2C=CR_2 + X_2 \longrightarrow R_2\overset{\displaystyle X}{\underset{\displaystyle |}{C}}\text{-}\overset{\bullet}{C}R_2 + X^{\bullet}$$

However, some halogenations involve MAH H-atom transfers. The first example discovered, and perhaps the most clearly established occurs in the fluorination of organic materials (27). Here, the driving force is clearly the very large exothermicity of processes like eq 17; e.g., for propane, eq 17 would be 3 kcal/mole exothermic.

$$R\text{-}H + F_2 \longrightarrow R^{\bullet} + HF + F^{\bullet} \tag{17}$$

A variety of halogenation agents evidence MAH processes. Many of the reaction systems are quite complex, and it is not always clear whether C-H bond breaking is involved or not. For example, Walling and his coworkers have studied *tert*-butyl hypochlorite in detail. This reagent reacts with a variety of organic compounds, including ethers, aldehydes and alcohols in MAH processes (28).

We have already mentioned the reaction of ozone with aldehydes, and the fact that an MAH process is used in commerce to initiate the autoxidation of acetaldehyde to produce peracetic acid. Ozone reacts with a wide variety of organic materials to produce radicals; these reactions are fascinating because they occur at very low temperatures and with almost every class of organic compound (29-33). Even alkanes react--in this case to give alcohols, eq 18.

$$RH + O_3 \longrightarrow [R^{\bullet} \cdot O_3H] \longrightarrow ROH + O_2 \tag{18}$$

Although the process is partially stereospecific, radicals are postulated to be involved (34,34a,35). We will return to this subject when we discuss reaction 19, in which the MAH of an O-H bond occurs.

$$ROOH + O_3 \longrightarrow ROO\cdot + HO\cdot + O_2 \qquad (19)$$

Radicals are produced in the low temperature reactions of dimethylaminomalononitrile, but the mechanism is unknown (36). An MAH scission of a C-H bond, eq 20, can be suggested (36), although this reaction would be strongly endothermic and probably too slow.

$$\underset{\displaystyle CN-\underset{\displaystyle |}{C}H-CN}{\overset{\displaystyle NMe_2}{|}} + H-\underset{\displaystyle |}{\overset{\displaystyle NMe_2}{C}}(CN)_2 \longrightarrow \underset{\displaystyle CN-\underset{\displaystyle |}{C}H\cdot}{\overset{\displaystyle NMe_2}{|}} + HNC + \underset{\displaystyle \cdot\overset{\displaystyle |}{C}(CN)_2}{\overset{\displaystyle NMe_2}{|}} \qquad (20)$$

Assisted transfers of a hydrogen atom from two other atoms—sulfur and oxygen—might be mentioned to demonstrate the variety of MAH processes. Hiatt and Bartlett (24) have shown that ethyl thioglycolate and styrene react to produce radicals. At high thiol/styrene ratios, the rate of radical production is too fast to be accounted for by the thermal reaction of styrene with itself, and we (2,7) have suggested that the reaction is a direct MAH of an S-H bond, eq 21.

$$RSH + PhCH=CH_2 \longrightarrow RS\cdot + Ph\overset{\bullet}{C}H-CH_3 \qquad (21)$$

Styrene also reacts with hydroperoxides to produce radicals (25, 26). *A priori*, one might have thought that a hydrogen atom transfer would be involved, eq 22; however, isotope effect data of

$$ROOH + PhCH=CH_2 \longrightarrow ROO\cdot + Ph\overset{\bullet}{C}H-CH_3 \qquad (22)$$

Walling and Heaton do not support this formulation. This reaction system has some of the most complex and baffling kinetic and mechanistic difficulties of any MAH system yet studied.

Demonstration of MAH Hydrogen Atom Transfers from C-H Bonds.

The Self-Initiated Polymerization of Styrene and the Chemistry of

Methylenecyclohexadiene (Isotoluene)

Introduction. The brief review given above indicates that MAH reactions in general, and assisted homolysis of C-H bonds in particular, can no longer be regarded *terra incognita*. Yet, few systems are well understood or have mechanisms that are known with certainty. Our love affair with this field began some twenty years ago with studies of the thermal polymerization of styrene.

This is a particularly fascinating process. Styrene is an unusually stable molecule (it has only vinylic and aromatic hydrogens) and yet it initiates its own polymerization at a well-
defined and reproducible rate. (The rate is about 1%/hr at 90°.)
The initiation reaction has been convincingly shown to involve
styrene itself, and not some adventitious impurity (7).
     Currently the most popular mechanism to rationalize the self-
initiated polymerization of styrene, eqs 11-13, involves the formation of the Diels-Alder adduct, AH, from two styrene molecules,
and the MAH reaction of this dimer with a third styrene molecule.
This novel and very elegant mechanism was originally suggested by
Frank Mayo, and derived from his observation of phenyltetralin and
phenylnaphthalene among the oligomers. It is clear that AH is the
source of the phenyltetralin-type products. However, while a
critical review (7) of the evidence establishes that AH is present
during polymerization, there is no conclusive evidence that it is
involved in eq 12, the MAH step.
     No synthesis or isolation of AH has been reported, although
our recent work (6) suggests that AH could be prepared *in situ*,
at least. However, we have reported the synthesis of an analogue
of AH which was chosen so as to have a lower driving force for
H-atom transfer (and consequently to be more easily isolable
than AH) and still be sufficiently reactive so as to be able to
initiate the polymerization of olefins by an MAH mechanism. The
adduct, BH, formed from 2-vinylthiophene and 4-phenyl-1,2,4-
triazoline-3,5-dione, appears to have this property. Solutions
(0.01 to 0.1 $\underline{M}$) of BH appear to initiate the polymerization of
styrene and also of methyl acrylate, a monomer which does not
initiate its own polymerization, eq 23.

$$\text{(BH)} \qquad\qquad \text{(B·)} \qquad\qquad (23)$$

(BH)   (B·)

     MAH Reactions of Methylenecyclohexadiene (Isotoluene). In a
quest for a simpler and perhaps more reactive model for AH, my
group (6,6a), and Kopecky and Lau (37) independently and simultaneously, have studied the reactions of methylenecyclohexadiene
(MCH). Initially, both we and Kopecky utilized the synthesis
invented by Bailey (38), but that method can provide only dilute
solutions of MCH and requires repetitive and time-consuming

preparative glpc separations. Therefore, both groups developed
new syntheses. Ours utilizes the pyrolysis of compound I.

+ CO   (24)

(I)   (MCH)

   Surprisingly, MCH is relatively stable in acid free, degassed
heptane or benzene; however, MCH does rearrange to form toluene
with a half life of about 50 hrs at 80°.* However, MCH decomposes
in styrene solution (ca. 10^{-3} $\underline{M}$) with a first order rate constant
of 4.3×10^{-4} sec^{-1} at 60°, corresponding to a half life of 27
minutes. This accelerated disappearance of MCH in styrene could
be due to several processes: (1) an MAH initiation reaction;
(2) a Diels-Alder reaction of MCH with styrene or, less likely at
the concentrations studied, with itself; (3) an ene reaction of
MCH with styrene or itself; (4) chain transfer of MCH with the
polystyryl radical produced from the self-initiated polymerization
of styrene.
   Fully expecting that MCH would initiate the polymerization of
styrene, we investigated solutions from 0.001 to 0.01 $\underline{M}$ at 60°.
To our disappointment and considerable surprise, MCH does not
cause an increase in the observed rate of thermal polymerization.
Kopecky and Lau have reached the same conclusion.
   As I have said, the half life of MCH is about 27 minutes in
styrene at 60°, and only a few percent of the styrene is converted
to polymer in the first hour of reaction. Therefore, we were
concerned that the transfer and ene reactions of MCH might be
depleting its concentration so rapidly that there was none left
to undergo the MAH reaction. That is, we were concerned that we
had not used sufficiently high initial concentrations of MCH to
observe initiation by it. However, as I will show below, this is
not the case. Before we consider the possible reactions of MCH
and the fraction of MCH undergoing each possible pathway, let's
examine the possible reactions of AH itself.

---

*The rate appears to be slower in the presence of hydroquinone,
suggesting a radical chain mechanism for the rearrangement of MCH
to toluene (6b). (In the presence of oxygen, the product of the
reaction is benzyl hydroperoxide (6b).) However, it is difficult
to find an inhibitor that produces a radical that does not abstract hydrogen from the super-reactive MCH. (For example, DPPH
and galvinoxyl destroy MCH, and of course styrene cannot be used
as an inhibitor.) Thus, it is difficult to prove that the rearrangement involves a radical chain.

Figure 1 shows the reactions of AH, and it can be seen that major processes using up AH are: the formation of trimer, the MAH process, and chain transfer. The rate constants for all three of these reactions can be estimated in the following way (6,6a). If it is assumed that all of the trimeric product A-Sty is produced by an ene reaction (eq f in Figure 1A) rather than by radical recombination, reaction h, then the rate constant for the ene reaction of AH can be calculated from the rate of appearance of the trimer A-Sty measured by Buchholz and Kirchner (22)* The rate constant for transfer of AH can be calculated from its transfer constant, obtained from our computer simulation (23), and the known value of k_p for styrene. And finally, the rate constant of the MAH reaction of AH can be calculated from the rate at which radicals are formed in styrene (calculated from the observed rate of thermal polymerization), assuming that all radicals come from this postulated MAH reaction. The steady state concentration of AH was measured by Kirchner, and is 6 x 10^{-5} M at about 60°. Thus, knowing the rate constants for the ene, MAH, and transfer reactions of AH, and the steady-state concentration of AH, the fraction of AH that undergoes each of these three reactions can be calculated. Table I shows these rate constants and also shows the percent of AH that reacts by each of these paths. The data are quite surprising: 94% of AH undergoes an ene reaction, and only about 4% undergoes an MAH process. Transfer is even less important, despite the large transfer constant. Table I also gives the moles of AH that pass through each pathway in the first hour of reaction; as can be seen, 7 x 10^{-7} moles of free radicals are produced.

For MCH, a similar conclusion is reached (6,6a). Kopecky and Lau showed that almost all of the MCH is converted to the ene (or radical recombination) product. Thus, we can set the rate constant for the ene reaction as approximately equal to the total rate constant (measured by uv) for the disappearance of MCH. Again, the transfer constant is known (see below), from which the rate constant for transfer can be calculated. If we <u>assume</u> that the rate constants for the MAH reactions of MCH and AH are equal,

---

*It is not likely that all A-Sty trimer arises from an ene reaction. (I.e., from reaction f in Fig. 1A.) As I have pointed out (7), phenyltetralin probably arises from the disproportionation reaction of caged radicals, shown as eq i in Fig. 1A. (Any A· that diffuses into free solution would be expected to add to styrene and not to abstract hydrogen to give phenyltetralin.) If cage disproportionation occurs, then cage combination also must occur. (See eq h in Fig. 1A.) However, the conclusion reached here is not dependent on whether trimer arises from the combination of radicals within a cage or an ene process, since neither reaction produces free radicals that can initiate the polymerization of styrene.

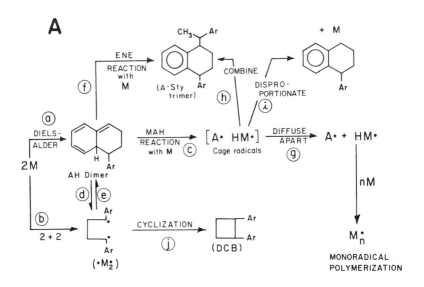

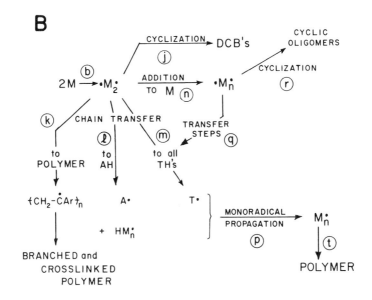

Figure 1. (A) This chart outlines the Diels–Alder dimerization of 2 styrene monomers (or similar vinyl aromatics) to form AH and the subsequent possible reactions of AH to give oligomers or to produce free radicals. The [2 + 2] cycloaddition of 2 monomer units to form the 1,4-diradical ·M₂· also is shown. (B) The reactions of 1,4-diradical that convert it to oligomers such as dicyclo-butanes (DCB) or to monoradicals. (See Refs. 5, 6, and 7.)

then the data in Table I are obtained: virtually all of the MCH disappears by the ene reaction and less than 1% disappears by transfer or by an MAH process. However, by doing a point-by-point integration of the amount of MCH that goes through each reaction, the number of moles of MCH that would have undergone the MAH reaction in 1 hour is calculated to be 6 x 10^{-5}. Thus, if we <u>assume</u> that the rate constants for the MAH reactions of MCH and AH are identical, then 85-fold more moles of radicals would have been produced in 1 hour from a solution initially 0.01 M in MCH than are produced by the steady state (6 x 10^{-5} M) concentration of AH. If this were true, the rate of thermal polymerization (proportional to the square root of the rate of initiation) would be observed to increase by about 9-fold in the MCH solution. Since we could easily observe an increase in $R_{p,th}$ of 2-fold, the rate constant for the MAH reaction of MCH must be <u>at least</u> 21-times smaller than that for AH.

Thus, the calculations shown in Table I can be summarized as follows. If we <u>assume</u> that MCH and AH undergo an MAH reaction with the same rate constant, then an initially 0.01 M solution of MCH (the most concentrated we studied) would produce 85-times more radicals in 1 hour than does the steady state concentration of AH, despite the fact that only a very small fraction of MCH undergoes the MAH reaction. Since this would produce a rate of polymerization 9-times greater than the thermal rate, our initial assumption that MCH and AH undergo the MAH reaction with the same rate constant must be in error. In fact, since we could quite easily detect a 2-fold increase in the rate of polymerization, the rate of radical production from MCH must be at least 21-fold slower than for AH.

We also measured the transfer constant of MCH, and it is about 10 at 60°. Thus, MCH is extraordinarily reactive toward radicals; 10 is not only the world's record for a transfer constant for a hydrocarbon, but it is one of the largest transfer constants known. Only thiols and transition metal compounds have transfer constants this large or larger. For comparison, the transfer constant of triphenylmethane (in styrene at 60°) is about 10^5 smaller than that of MCH! Despite this strikingly large transfer constant, MCH undergoes the MAH reaction an order of magnitude (or more) more slowly than does AH. Surely this is extremely surprising and is hard to reconcile with the Diels-Alder mechanism for the initiation of styrene.

There are two possibilities at this point. The most conservative is that MCH is a poor model for AH; this allows the postulated MAH reaction of AH to be retained as the initiation process in styrene. The least conservative is that the entire Diels-Alder mechanism for styrene's initiation is wrong!

It certainly does seem possible that MCH is a poor model for AH, and that MCH might undergo a more nearly concerted ene reaction than does AH, thus giving a smaller yield of free radicals. <u>Figure 2</u> outlines the reactions of an MCH-like molecule.

Table I. A Comparison of the Rate Constants for Reaction and Moles Reacted in 1 Hour for AH and MAH in Styrene at 60° (6)

Process	$[AH]_{ss}$ = 6 x 10⁻⁵ [a]			$[MCH]_0$ = 0.01 [b]		
	k	Moles/hr [c]	%	k	Moles/hr [c]	%
Ene	8.5 x 10⁻⁶ [d]	1.6 x 10⁻⁵	94.	--	--	80-100 [h]
Transfer	145 [e]	2 x 10⁻⁸	0.1	10³ [i]	4 x 10⁻⁵	0.5
MAH	3.7 x 10⁻⁷ [f]	7 x 10⁻⁷	4.	[3.7 x 10⁻⁷] [j]	6 x 10⁻⁵	0.7 [k]
Total	9.0 x 10⁻⁶ [g]	1.7 x 10⁻⁵	100.	5.3 x 10⁻⁵ [i]	8.2 x 10⁻⁵	100.

(a) Steady state AH concentration (22). (b) The half life of MCH is 0.5 hr (6). (c) Moles reacted by this path in first hour by point-to-point integration. (d) from ref (22). (e) from ref (23). (f) Calculated from the rate of radical appearance indicated by the observed rate of thermal polymerization. (g) from the measured rate of AH appearance (22) and assuming that $-d[AH]/dt = +d[AH]/dt$ at the steady state. (h) Ref (37). (i) Ref (6). (j) Assumed as identical to the AH value. (k) Not in agreement with experiment. Clearly, the value of k_{MAH} must be smaller for MCH than for AH.

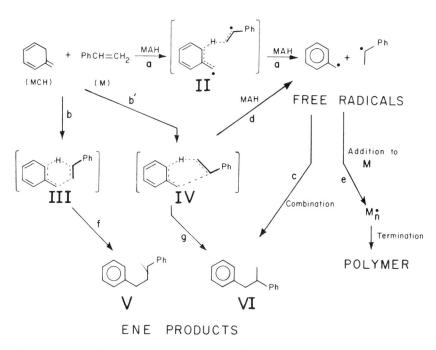

Figure 2.   Possible reactions of methylenecyclohexadiene (MCH) and the formation of ene products. Compound VI can be produced either by a concerted ene reaction or by a process (Equations 2a and 2c) involving diradicals. The possible leakage of radicals from the ene reaction's extended transition state (II) to initiation polymerization is shown (6, 6a).

Ene products V and VI can be produced by reactions b and b'. In
addition, one of the ene products, VI, can be formed by the recombination of free radicals, reactions a,c. (Kopecky and Lau have
reported that V and VI are produced in approximately a 1:3 ratio
from MCH.) For a more hindered molecule such as AH, models indicate that an extended transition state such as II may be relatively more preferable to III or IV, thus explaining the larger yield
of radicals relative to ene products. Alternatively, it is possible that a transition state like IV may have a more distorted
shape for AH, and that more radicals "leak out" of the ene reaction via eq d. In either case, the faster rate of radical production from AH relative to MCH is rationalized as due to the greater
hindrance in the AH-styrene reaction relative to the MCH-styrene
ene process.

Thus, the faster production of radicals from AH relative to
MCH could be rationalized in terms of an ene reaction that involves a transition state with varying amounts of radical character and leading to varying yields of scavengable free radicals.
As the argument has been presented above, this rationalization
does not appear to be impossible. However, Figure 3 shows a tabu-
lation of all of the model compounds synthesized to date in an attempt to model the behavior of AH. Although the second and third
compounds in the list are reported to initiate polymerization, MCH
does not. Furthermore, as shown in Figure 3, a compound that
would appear to give an even more hindered ene reaction transition
state than AH also does not initiate the polymerization of styrene
(private communication from D. Aue). This is hard to reconcile
with a rationalization of the lack of MAH reaction by methylenecyclohexadiene based on the hindrance of its ene reaction.

There are other problems with the Diels-Alder mechanism. For
example, it appears from UV evidence that AH builds up to its
steady-state concentration so slowly that an induction period
should be observed in the rate of polymerization. However, none
has been reported, although an induction period in chain transfer
by AH is easily observed (6a).

Thus, I suggest that the Diels-Alder mechanism can continue
to be accepted as the mechanism for initiation as long as it is
clearly recognized that it is not conclusively established and
that it may be incorrect. If the Diels-Alder mechanism is wrong,
it is difficult to suggest a superior mechanism to take its place.
The only mechanism that has been suggested and not yet disproven
is one involving 1,4-diradicals (5,7), and it too has its difficulties (5). The only conclusive experiment appears to be the
synthesis and testing of AH itself, and we are attempting to do
that.

The Reaction of Ozone with *tert*-Butyl Hydroperoxide. An MAH

Hydrogen Atom Transfer from Oxygen

   Introduction. As we have remarked above, ozone reacts with

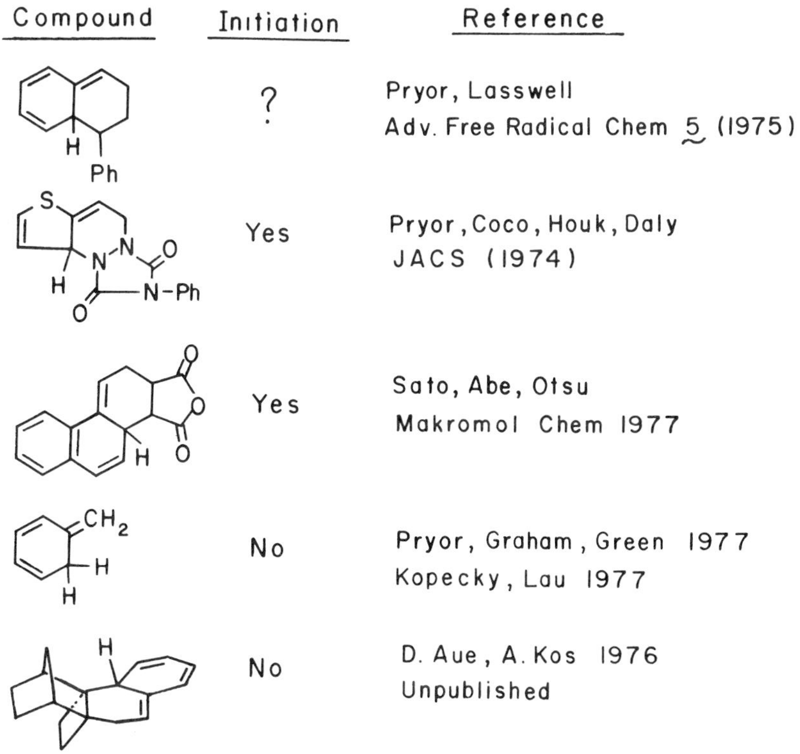

Compound	Initiation	Reference
	?	Pryor, Lasswell Adv. Free Radical Chem 5 (1975)
	Yes	Pryor, Coco, Houk, Daly JACS (1974)
	Yes	Sato, Abe, Otsu Makromol Chem 1977
	No	Pryor, Graham, Green 1977 Kopecky, Lau 1977
	No	D. Aue, A. Kos 1976 Unpublished

Figure 3.   Compounds synthesized and tested as models for AH. The Diels–Alder adduct of styrene, AH itself, is shown at the top of the figure; it is postulated to initiate the polymerization of styrene. Of the models tested, two initiate and two do not.

virtually every type of organic molecule to form radicals (29,30, 31). Our interest in ozone was sparked by indications that the pathology caused by ozone in smog is partially due to the autoxidation of polyunsaturated fatty acids (PUFA) in lung lipids initiated by ozone (32,51). A priori, several mechanisms can be envisioned for radical production from O_3-substrate reactions.

(i) The simplest radical-producing reaction is the unimolecular homolysis of ozone, eq 25. The heat of this reaction is

$$O_3 \longrightarrow O_2 + O \qquad (25)$$

25 kcal/mole (assuming O_2 is formed in its ground state), predicting a half life for ozone of 10^4 hrs at $-20°$, 10^2 at $0°$, and 10 min at $37°C$ (18,39). Thus, eq 25 is too slow to be an important initiation process at the low temperatures at which ozonolyses are usually conducted, but it could play an important role at $37°C$ in biological systems where ozone toxicity is observed.

(ii) Perhaps the next simplest radical-producing reaction would be an electron-transfer, eq 26. Bailey has suggested that

$$X + O_3 \longrightarrow X \cdot^+ + O_3 \cdot^- \qquad (26)$$

an initial electron transfer is responsible for the radical production observed at $-78°$ in the reactions of ozone with trimesitylvinyl alcohol (40). If $O_3\cdot^-$ were produced, it might form reactive radicals. The rate constant for dissociation of $O_3\cdot^-$ in aqueous solution at $25°$, eq 27, is 10^3 sec^{-1} (41). Thus, HO·

$$O_3 \cdot^- + H_2O \longrightarrow O_2 + HO\cdot + HO^- \qquad (27)$$

could be the radical that initiates autoxidation (and pathology in bio-systems) in some of the reactions of ozone (42). This is reminiscent of the suggestion that the pathology caused by superoxide, $O_2\cdot^-$, is mediated by HO· radicals (43,43a,43b) generated by by the reaction of superoxide with H_2O_2, eq 28.*

$$O_2 \cdot^- + H_2O_2 \longrightarrow O_2 + \cdot HO\cdot + HO^- \qquad (28)$$

---

*Reaction 28 is too slow to be important in biological systems (43b). However, we have recently shown that the reaction of the superoxide anion radical with organic hydroperoxides, eq i, is fast (45b). Since PUFA forms hydroperoxides in vivo both by an enzymatic path and by autoxidation, it appears that reaction i could be responsible for the HO· radicals observed in biological systems.

$$O_2 \cdot^- + ROOH \longrightarrow O_2 + RO\cdot + HO^- \qquad (i)$$

(iii) Ozone could undergo a concerted 1,3-dipolar insertion into a C-H bond or O-H bond, eq 29a (44,44a). When R-H is

$$R-H + \overset{\delta+}{O}=O-\overset{\delta-}{O} \longrightarrow RO_3H \longrightarrow ROO\cdot + HO\cdot \qquad (29a)$$

benzaldehyde, $PhCO-OOOH$ is an intermediate and an insertion mechanism appears reasonable (35a). However, when R-H is an alkane, an MAH hydrogen abstraction mechanism is usually written, eq 29b, although the reaction is stereospecific (34,34a,35).

$$R-H + O_3 \longrightarrow [R\cdot \quad HO_3\cdot] \longrightarrow ROH + O_2 \qquad (29b)$$

(iv) Ozone reacts with olefins some 10^6 faster than with alkanes (44). Thus, in systems that contain olefinic unsaturation, the fastest reaction is addition of ozone to give the trioxide, eq 30. Even at 10^{-3} M olefin, this addition is so fast that ozone homolysis, eq 25, cannot compete at $37°$ C or below. The trioxide then undergoes rapid dissociation and recombination to form the Criegee ozonide, eqs 31-32.

$$O_3 + R_2C=CR_2 \longrightarrow R_2C\overset{\overset{\displaystyle O-O-O}{|\quad\quad|}}{\underline{\quad\quad}}CR_2 \qquad (30)$$

(VII)

$$\text{(VII)} \longrightarrow [R_2C\overset{\overset{\displaystyle\cdot O}{|}}{\underline{\quad}} \overset{\overset{\displaystyle OO\cdot}{|}}{CR_2}] \longrightarrow R_2C=O + [R_2\overset{\cdot}{C}-OO\cdot \longleftrightarrow R_2\overset{+}{C}-OO^-] \qquad (31)$$

(VIII)   (IX)

$$R_2C=O + \text{(IX)} \longrightarrow R_2C\overset{\overset{\displaystyle O-O}{|\quad\,|}}{\underline{\quad}}O-\overset{}{C}R_2 \qquad (32)$$

Criegee ozonide

To the extent that the zwitterion-diradical, IX, has free radical character, it can be the source of initiating radicals:

$$\text{(IX)} + R'H \longrightarrow R_2\overset{\cdot}{C}OOH + R'\cdot$$

However, it appears doubtful that IX has sufficient lifetime and/or radical character to be the initiating species in PUFA autoxidations induced by ozone.

(v) We have shown that the Criegee ozonide does not homolyze to initiate autoxidation in PUFA-ozone-air systems (32,45). In contrast, the trioxide (VII) does dissociate at temperatures that are sufficiently low so as to rationalize radical production in olefins induced by ozone (46,47,48), but it is doubtful that the diradical produced in this dissociation, VIII, has sufficient lifetime to initiate PUFA autoxidation. In the gas phase, diradical VIII may undergo a "back-bite" reaction (51a):

$$
\begin{array}{cc}
\overset{\text{O}\cdot}{\underset{|}{\text{RHC}}}\text{---}\overset{\text{OO}\cdot}{\underset{|}{\text{CHR}}} \longrightarrow & \overset{\text{O}}{\underset{\|}{\text{RC}}}\text{---}\overset{\text{OOH}}{\underset{|}{\text{CHR}}}
\end{array}
\qquad (33a)
$$

In analogy with this process, we have suggested (45a) that the trioxide formed from ozone addition to one of the double bonds in a polyunsaturated fatty acid can undergo the similar reaction, eq 33b, in which an allylic hydrogen atom is abstracted. Eq 33b

$$
\text{(VII)} \longrightarrow \overset{\cdot\text{OO}}{\underset{|}{\text{RCH=CH-CH}_2}}\text{---}\overset{\text{O}\cdot}{\underset{|}{\text{CH}}}\text{---}\text{CHR} \longrightarrow \overset{\text{HOO}}{\underset{|}{\text{RCH-CH-CH}}}\text{---}\overset{\text{O}\cdot}{\underset{|}{\text{CH-CHR}}} \qquad (33b)
$$

generates a diradical that may well have sufficient lifetime to initiate autoxidation.

(vi) Finally, radicals can be produced by MAH reactions of ozone with O-H bonds. In particular, the reaction of ozone with hydroperoxides might involve an MAH process, eq 34. We began

$$
\text{ROOH} + O_3 \longrightarrow \text{ROO}\cdot + \text{HO}_3\cdot \qquad (34)
$$

investigating the reactions of ozone with hydroperoxides for several reasons. In the first place, radicals are produced at very low temperatures (47,48,49,50). Secondly, lipid hydroperoxides are produced during the autoxidation of PUFA by ozone-containing smog, and we wished to know if ozone reacts with these lipid hydroperoxides to produce radicals (32,51).

Analysis of the reaction of ozone with $tert$-butyl hydroperoxide in CHCl$_3$, CCl$_4$, or CFCl$_3$ as solvents at temperatures from 24 to $-60°$ showed that the products are $tert$-butyl alcohol, acetone, water, and di-$tert$-butyl peroxide (44,44a). The reaction scheme, shown in eqs 35-43, rationalizes these products as resulting from free radical reactions of ozone and the hydroperoxide, where T = $tert$-C$_4$H$_9$.

$$
\text{TOOH} + O_3 \xrightarrow{\;i\;} \text{TOO}\cdot + \text{HO}\cdot + O_2 \qquad (35)
$$

$$
\text{HO}\cdot + \text{TOOH} \xrightarrow{\;i'\;} \text{H}_2\text{O} + \text{TOO}\cdot \qquad (36)
$$

$$
2\text{TOO}\cdot \underset{K}{\overset{\;}{\rightleftharpoons}} [\text{TO}_4\text{T}] \xrightarrow{\;a\;} [2\text{TO}\cdot + O_2] \xrightarrow{\;f\;} \text{TOOT} + O_2 \qquad (37)
$$

$$
\xrightarrow{\;(1-f)\;} 2\text{TO}\cdot + O_2 \qquad (38)
$$

$$
\text{TO}\cdot + \text{TOOH} \xrightarrow{\;p\;} \text{TOH} + \text{TOO}\cdot \qquad (39)
$$

$$TO\cdot \xrightarrow{\beta} CH_3COCH_3 + CH_3\cdot \tag{40}$$

$$CH_3\cdot + O_2 \xrightarrow{o} CH_3OO\cdot \tag{41}$$

$$CH_3OO\cdot + TOO\cdot \xrightarrow{t'} \text{Non-radical products} \tag{42}$$

$$TOO\cdot + O_3 \xrightarrow{x} [TO_5\cdot] \longrightarrow TO\cdot + 2O_2 \tag{43}$$

Under conditions where the production of acetone is insignificant (e.g., in $CFCl_3$ at $-4°$), kinetic analysis of eqs 35-43 gives the rate laws shown in eqs 44 and 45.

$$Rate(-O_3) = k_i(TOOH)(O_3) + k_x\left(\frac{k_i}{k_t}\right)^{\frac{1}{2}}(TOOH)^{\frac{1}{2}}(O_3)^{3/2} \tag{44}$$

$$Rate(-TOOH) = \frac{2k_i}{f}(TOOH)(O_3) + k_x\left(\frac{k_i}{k_t}\right)^{\frac{1}{2}}(TOOH)^{\frac{1}{2}}(O_3)^{3/2} \tag{45}$$

Eq 44 can be rearranged to give eq 46. Thus, a plot of the quantity of the left side of eq 46 versus $\{[TOOH]/[O_3]\}^{1/2}$ should give a straight line with slope equal to k_i and intercept $k_x(k_i/k_t)^{1/2}$. We have performed the analysis in $CFCl_3$ at $-4°C$ both for TOOH and for the deuterated hydroperoxide, TOOD.

$$\frac{\text{Initial rate}(-O_3)}{(O_3)^{3/2}(TOOH)^{\frac{1}{2}}} = k_i\frac{(TOOH)^{\frac{1}{2}}}{(O_3)^{\frac{1}{2}}} + k_x\left(\frac{k_i}{k_t}\right)^{\frac{1}{2}} \tag{46}$$

Our data give $k_i = 5.4$ M^{-1} sec^{-1} for *tert*-butyl hydroperoxide, and $k_H/k_D = 2.8$ for *tert*-BuOOD at $-4°$. This primary isotope effect establishes the mechanism of eq 35 as an assisted scission of an O-H bond, rather than a rate limiting electron transfer followed by a rapid proton transfer, eqs 35a,b. In addition, the heat of

$$ROOH + O_3 \longrightarrow ROOH^{\overset{+}{\cdot}} + O_3^{\overset{-}{\cdot}} \tag{35a}$$

$$ROOH^{\overset{+}{\cdot}} + O_3^{\overset{-}{\cdot}} \longrightarrow ROO\cdot + HO_3\cdot \tag{35b}$$

reaction of eq 35a can be shown to be too large for the electron transfer process to be facile (44).

Although the kinetic law for the reaction of ozone with *tert*-butyl hydroperoxide is complex with a first order and a chain term, under conditions where the hydroperoxide is present in 10-30 mole excess, 50-80% of the reaction goes by the first order process. (That is, calculations show that most of the reaction is due to the first term in eqs 44 and 45.) Under these conditions, approximate values of k_i can be obtained by simply measuring initial rates. Data obtained in this way give an apparent activation energy for the reaction of ozone and *tert*-butyl hydroperoxide of 7 kcal/mole and a log A value of 7. Reaction 35 can be calculated to be 14 kcal/mole endothermic (44,44a), and the heats of reaction and activation energies for eqs 36-43 have been measured or can be calculated using Benson's methods. Thus, the activation energies for the two terms in the kinetic law, eq 46, are calculated to be about 14 and (5 + 14/2 - 8/2) = 8 kcal/mole, respectively. Thus, depending on the kinetic chain lengths, the observed activation energy for this mechanism is predicted to be between 14 and 8 kcal/mole, in excellent agreement with the observed value of 7 kcal/mole.

The heat of reaction for a concerted 1,3-insertion by ozone into the O-H bond also can be calculated using Benson's estimate of the group term for an oxygen atom attached to two other oxygens (18). This calculation shows that the direct insertion to form *tert*-butyl hydropentoxide as an intermediate is far too endothermic to be allowed by our approximate activation energy of 7 kcal/mole (44,44a). The calculated heats of reaction for the insertion process are shown over the arrows in eq 35c.

$$\text{TOOH} + \text{O}_3 \xrightarrow{\Delta H = +25} \text{TO}_5\text{H} \xrightarrow{\Delta H = -11} \text{TOO}\cdot + \text{O}_2 + \text{HO}\cdot \qquad (35c)$$

By the same approximate technique that was used to obtain activation energies, the rates for reaction of ozone were determined for a variety of types of organic species. Table II gives data of this type. It can be seen that the relative rates of reaction of ozone with alkenes: ROOH: alkanes is approximately in the ratios of 10^6: 10^3: 10^1.

1,4-Diradicals from [2+2] Cycloaddition Reactions

1,4-Diradicals. Orbital symmetry arguments suggest the [2+2] cycloaddition should not be concerted, and experiments of various types suggest that these reactions do, indeed, involve diradicals (9a,53-55). The arguments are primarily based on the lack of stereospecificity in cycloaddition reactions. However, recently, several reports have been published in which 1,4-diradicals have been trapped by added reagents. Table III presents these data. Most of the diradicals are produced photochemically,

Table II.   Approximate Second Order Rate Constants for the
Reaction of Ozone with a Series of Organic Compounds
in CCl_4 at $24°$ (44,57)

Substrate	$k(M^{-1}sec^{-1})$	k_{rel}
2,3-Dimethylbutane	0.2	(1)
Cyclohexane	0.02	0.1
tert-Butyl alcohol	0.02	0.1
Cyclopentanol	1.3	6.
Benzene	0.05	0.2
tert-Butyl hydroperoxide	37.	185.
1-Pentene	8×10^4	4×10^6

Table III.   Trapping of 1,4-Diradicals by Scavengers

Diradical	Scavenger	Diradical half life (sec)	Ref
$C_6F_5\dot{C}H-CH_2CH_2-\dot{C}HC_6F_5$	$C_6F_5CH=CH_2$	–	(56)
$\underset{\overset{\displaystyle OH}{\|}}{Ph\dot{C}}-CH_2CH_2-\dot{C}H-OMe$	n-BUSH-d$_1$	$10^{-6} - 10^{-7}$	(10)(11)
$\underset{\overset{\displaystyle OH}{\|}}{Ph\dot{C}}-CH_2CH_2-\dot{C}Me_2$	O_2	–	(12)
$\underset{\overset{\displaystyle OH}{\|}}{Me\dot{C}}-CH_2CH_2-\dot{C}H_2$	HBr (gas)	$10^{-4.9}$	(14)
$\underset{\overset{\displaystyle OH}{\|}}{Ph\dot{C}}-CH_2CH_2-\dot{C}Me_2$	Methyl methacrylate	–	(13)
	$(t-Bu)_2NO\cdot$	–	(15)
$\underset{\overset{\displaystyle OH}{\|}}{Ph\dot{C}}-CH_2CH_2-\dot{C}R_2$	$(t-Bu)_2C=Se$	$10^{-6.4}$	(16)

and thus could be triplet series that must undergo intersystem
crossing prior to ring closure to cyclobutanes. However, the
recent report by Bottini in which a diradical produced by the
thermal cycloaddition of 1-methyl-3,4-cyclohexadiene is trapped
by a nitroxide suggests that sufficiently reactive trapping
reagents can scavenge even singlet diradicals.

      In addition to data of the type shown in Table III, a consid-
erable literature exists for polymer systems; these latter data
are much less well-known to chemists interested primarily in
cycloaddition phenomena. The data are of two types. (*i*) Compounds that decompose to produce diradicals can, albeit somewhat
inefficiently, initiate the polymerization of vinyl monomers.
Compounds X-XIV are examples of diradical initiators that have

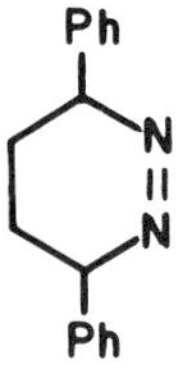

X

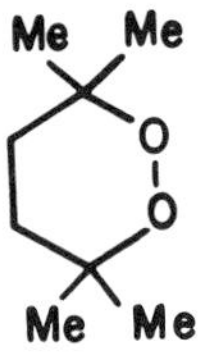

XI

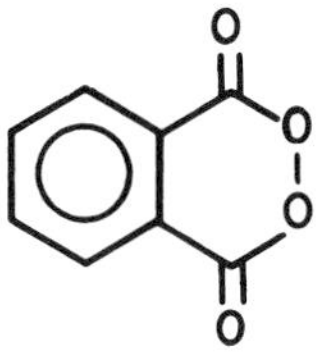

XII

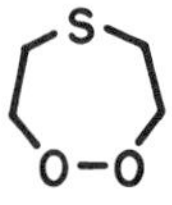

XIII

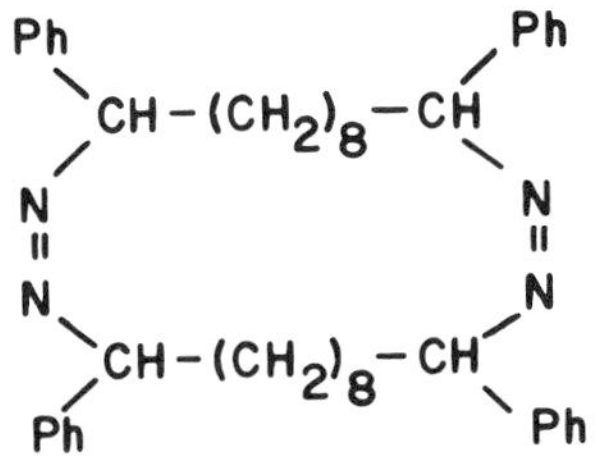

XIV

been studied in this way (7). (*ii*) When olefins are heated in the presence of scavengers, the scavengers disappear. In the case of styrene, the scavengers often disappear faster than monoradicals are formed, and it has been suggested that the scavengers react with diradicals, with AH (Figure 1A), or with monomer directly. The field is complex, and we have reviewed all of the data in some detail (7).

Study of Pentafluorostyrene (PFS). A Particularly Likely

Candidate for Initiation of Polymerization by 1,4-Diradicals? As was described above, my research group has been involved for some time with efforts to establish the Diels-Alder mechanism for initiation in the thermal polymerization of styrene suggested by Mayo. However, we also have argued that some of the initiation in styrene might be due to 1,4-diradicals. (See Figure 1B.)

It seems clear that all monomers do not undergo thermal polymerization by the same mechanism. For example, Table IV lists monomers that are susceptible to thermal spontaneous polymerization; at least some of those listed do not appear capable of forming a Diels-Alder adduct analogous to AH that could initiate by an MAH process. Furthermore, as is shown in Figure 1, the yields of dimers and trimers are indicative of the presence of oligomers that might be involved in initiation processes; and, as shown in Table V, the relative yields of dimers and trimers differ enormously when different monomers are compared.

In this regard, pentafluorostyrene (PFS) is particularly interesting. It undergoes a spontaneous polymerization at a rate that is slower than that of styrene, but still quite appreciable. There is some evidence that PFS might form diradicals (9a), but it does not appear to form a Diels-Alder adduct analogous to styrene's AH. Therefore, we have attempted to find evidence that the initiating species in the thermal polymerization of PFS is a 1,4-diradical (5). The stumbling block in any such diradical mechanism, and the factor that has mitigated against its general acceptance, is the mechanism by which the diradicals are converted to monoradicals (6,7). Since evidence of several sorts suggests that the polymer (at least in the styrene case) is formed via growth of monoradicals, and since it is clear that 1,4-diradicals would undergo rapid ring closure to form cyclobutanes (DCB's in Fig. 1B), a very rapid process must convert the diradicals to monoradicals if diradicals are to initiate polymerization.

We at first thought that this process for PFS might be that originally suggested for styrene by Flory, eq 10. Therefore, we investigated the polymerization of PFS-β,β-d_2. In this case, like that for styrene, no primary isotope effect is observed (5). Thus, if radicals are produced in PFS by eq 47, they are not converted to monoradicals by eq 48 to an appreciable extent.

Table IV. Monomers that Undergo Thermal Polymerization (7)

Monomer	$R_{P,th}$ at $90°$ (%/hr)
Styrene	1
2-Bromostyrene	11
2-Fluorostyrene	2
PFS	0.03
3,4-Dichlorostyrene	7
2,4-Dichlorostyrene	20
2,5-Dichlorostyrene	15
2,6-Dichlorostyrene	0.01
Methyl methacrylate	0.01

Table V. Yields of Dimers and Trimers from Styrene and 2-Vinyl-
thiophene (7)

	μ moles/100 g monomer		
		Dimers	Trimers of
	Polymer	$[4+2]^a$ $\quad$ $[2+2]^b$	A-Sty Type
Styrene, $85°$	15	70 $\qquad$ 35	830
2-Vinylthiophene, $100°$	20	900 $\qquad$ --	0

(a) Derivatives of Diels-Alder dimers; e.g., phenyltetralin for
styrene. (b) 1,2-Diarylcyclobutanes.

$$2 \text{ PFS-d}_2 \longrightarrow C_6F_5-\overset{\bullet}{\text{C}}\text{H}-\text{CD}_2-\text{CD}_2-\overset{\bullet}{\text{C}}\text{H}-C_6F_5 \qquad (47)$$

$$(\text{XV})$$

$$(\text{XV}) + \text{PFS-d}_2 \longrightarrow C_6F_5-\text{CH}=\text{CD}-\text{CD}_2-\overset{\bullet}{\text{C}}\text{H}-C_6F_5 \qquad (48)$$

$$+ C_6F_5-\overset{\bullet}{\text{C}}\text{H}-\text{CD}_3$$

In thinking about another mechanism by which diradicals might be converted to monoradicals in PFS, we became intrigued by the fact that very pure PFS precipitates at very low conversions and appears to be crosslinked. This suggests that the polymer is attacked by growing mono- or diradicals to cause branching and ultimately cross links, eq 49. Some of the oligomers that have

$$\cdot M_n^{\bullet} + (-\overset{\overset{\displaystyle C_6F_5}{|}}{\text{CH}}-\text{CH}_2-)_n \longrightarrow HM_n^{\bullet} + (-\overset{\overset{\displaystyle C_6F_5}{|}}{\underset{\bullet}{\text{C}}}-\text{CH}_2-)_n \qquad (49)$$

benzylic hydrogens could also undergo a transfer reaction of the same type; eq 50 shows this process for cyclobutanes.

$$\cdot M_n^{\bullet} + \boxed{\begin{array}{c} \text{Ar} \\ \text{Ar} \end{array}} \longrightarrow HM_n^{\bullet} + \boxed{\begin{array}{c} \overset{\bullet}{}\text{Ar} \\ \text{Ar} \end{array}} \qquad (50)$$

If chain transfer is the mechanism by which diradicals are converted to monoradicals, then the addition of transfer agents should cause an increase in the rate of polymerization. This experiment was tried very early in the study of styrene's thermal polymerization with negative results (7); in fact, it was one of the coffin nails in the diradical-initiation mechanism for styrene. However, for PFS, transfer agents such as toluene do increase the rate of polymerization (5). Table VI gives data on the effect of transfer agents on the rate of polymerization for both styrene and PFS (5). For styrene it can be seen that even very high concentrations of transfer agents do not affect the rate other than by diluting the concentration of monomer, and hence, reducing the rate of polymerization by a solvent effect. In contrast, for PFS, molecules with benzylic hydrogens such as toluene and diphenylmethane cause a dramatic increase in the rate of polymerization.

Figure 1B illustrates the possible fates of diradicals formed in styrene and PFS systems. They can cyclize to form cyclobutanes (eq j) or macrocyclic oligomers (eq ℓ), they can react with polymer to yield polymer-radicals and HM· (eq k), or they can react with endogenous transfer agents that have benzylic hydrogens (TH's) to yield monoradicals (eq m). Possible TH molecules

include the cyclobutanes and all other oligomers shown in Figure
1. Thus, our best suggestion at present for the mechanism of conversion of diradicals to monoradicals in PFS is that reaction m in
Figure 1, the transfer process, is the critical monoradical-form-
ing reaction (5). Although the arguments are not overwhelming, it
does appear possible that the transfer reaction m could compete
with the cyclization reaction j or r (5).

Table VI.   Rates of Polymerization of PFS and Styrene in the
            Presence of Transfer Agents with Benzylic Hydrogens

Monomer	Transfer agent	Wht %	Rate %/hr	$R_{P,obs}/R_{P,calc}$ [c]
PFS[a]	None	--	0.17	(1)
	PhCH$_3$	8.3	0.24	1.5
	PhCH$_3$	8.4	0.28	1.8
	Ph$_2$CH$_2$	9.6	0.44	2.9
Styrene[b]	None	--	0.10	(1)
	PhH	43	0.05	0.9
	PhCH$_3$	47	0.05	0.9
	PhCH$_3$	43	0.05	0.9
	Ph$_2$CH$_2$	46	0.06	1.1

(a) 100°.   (b) 60°.   (c) The $R_{P,calc}$ values are obtained by
assuming that the transfer agent acts only as an inert diluent,
with no effect on the rate, and that polymerization is first
order in monomer.

Abstract

    A general review is presented of the mechanisms for reactions
in which radicals are formed at accelerated rates by the interaction of closed-shell molecules. We have grouped these mechanisms into three classes: molecule-assisted homolysis, electron
transfer, and non-concerted cycloaddition reactions.
    The spontaneous thermal polymerization of styrene is postu-
lated to involve an initiation step in which the reactive Diels-
Alder dimer of styrene (AH in Figure 1A) undergoes a molecule-
assisted homolysis reaction with another styrene molecule. The
evidence for this process is reviewed, and it is concluded that
it is not at all conclusive. Furthermore, methylenecyclohexadiene

(MCH, eq 24) was synthesized, and, although it appears to be an excellent model for AH, it does not initiate the polymerization of styrene. The implications of this and similar experiments on the mechanism of the spontaneous initiation of styrene are discussed.

Reaction between *tert*-butyl hydroperoxide and ozone produces radicals at a rapid rate even at temperatures as low as -20°C. We have studied the kinetics of this reaction in $CFCl_3$ at -4°C and have obtained the kinetic isotope effect and an approximate activation energy. Using these data, both an electron-transfer reaction and a concerted 1,3-dipolar insertion reaction can be eliminated as mechanistic possibilities for the peroxide-ozone reaction. We conclude the mechanism involves a molecule-assisted homolysis:

$$TOOH + O_3 \longrightarrow TOO\cdot + O_2 + HO\cdot$$

The spontaneous polymerization of pentafluorostyrene (PFS) has been studied. It appears unlikely that this monomer undergoes an initiation by the Diels-Alder mechanism. Rather, we suggest it undergoes initiation by a mechanism involving 1,4-diradicals. Measurements of the kinetic isotope effects on the polymerization of PFS-β,β-d_2 prove that the 1,4-diradical is not converted to monoradicals by donating a hydrogen atom to another molecule of PFS. Instead, it appears that the 1,4-diradicals undergo chain transfer reactions with oligomers that possess benzylic hydrogens. (See Figure 1B.) If this is correct, then the capture of thermally produced 1,4-diradicals by transfer agents is sufficiently fast to compete with the closure of the diradicals to form cyclobutanes.

Acknowledgement

The work described in this report was done by a series of talented graduate students and post-doctoral fellows, and I want to acknowledge their considerable contributions. Without them, this chapter would not have been written. I particularly want to acknowledge the contributions of Professor Michael Kurz and Drs. W. David Graham, John G. Green, Masashi Iino, and Daniel Church.

This research was supported by grants from the National Science Foundation, the National Institutes of Health, and Dow Chemical Company. I express my sincere thanks to those agencies for their continuing support.

Literature Cited

1. Pryor, W. A., "Free Radicals", pp 119-126, 184-186, 290. McGraw-Hill, New York, 1966.
2. Pryor, W. A., Coco, J. H., Daly, W. H., and Houk, K. N., *J. Amer. Chem. Soc.* (1974) 96, 5591.
3. Pryor, W. A., and Hendrickson, W. H., Jr., *J. Amer. Chem. Soc.* (1975) 97, 1580.
4. Pryor, W. A., and Hendrickson, W. H., Jr., *J. Amer. Chem. Soc.* (1975) 97, 1582.
5. Pryor, W. A., Iino, M., and Newkome, G. R., *J. Amer. Chem. Soc.* (1977) 99, 6003.
6. Pryor, W. A., Graham, W. D., and Green, J. G., *J. Org. Chem.* (1977), in press.
6a. Graham, W. D., Green, J. G., and Pryor, W. A., to be submitted.
6b. Pryor, W. A., and Graham, W. D., *J. Org. Chem.* (1978), in press.
7. Pryor, W. A., and Lasswell, L. D., "Diels-Alder and 1,4-Diradical Intermediates in the Spontaneous Polymerization of Vinyl Monomers" in "Advances in Free Radical Chemistry," Vol V, pp 95-27-94. Academic Press, New York, 1975.
8. Briner, E., and Wegner, P., *Helv. Chim. Acta* (1943) 26, 30.
9. Harmony, J. A. K., "Molecule-Induced Radical Formation" in "Methods in Free Radical Chemistry", E. S. Huyser, ed., Vol 5, pp 101- 176. Marcel Dekker, Inc., New York, 1974.
9a. Bartlett, P. D., *Science* (1968) 159, 833.
10. Wagner, P. J., and Zepp, R. G., *J. Amer. Chem. Soc.* (1972) 94, 287.
11. Wagner, P. J., and Liu, K.-C., *J. Amer. Chem. Soc.* (1974) 96, 5952.
12. Grotewold, J., Previtali, C. M., Soria, D., and Scaiano, J. C., *J. Chem. Soc., Chem. Comm.* (1973) 207.
13. Hamity, M., and Scaiano, J. C., *J. Photochem.* (1975) 4, 229.
14. O'Neal, H. E., Miller, R. G., and Gunderson, E., *J. Amer. Chem. Soc.* (1974) 96, 3351.
15. Bottini, A. T., Cabral, L. J., and Dev, V., *Tetrahedron Letters*, in press, and private communication.
16. Scaiano, J. C., *J. Amer. Chem. Soc.* (1977) 99, 1494.
17. Semenov, N. N., "Some Problems of Chemical Kinetics and Reactivity," Vol I pp 260-271. translated by J. E. S. Bradley, Pergamon Press, New York, 1958.
18. Benson, S. W., "Thermochemical Kinetics," 2nd Ed., pp 98, 221-225, 239. Wiley-Interscience, New York, 1976.
19. Cullis, C. F., and Hinshelwood, C. N., *Disc. Faraday Soc.* (1947) 2, 117. Also see p. 262 in ref. 17.
20. Pryor, W. A., and Patsiga, R. A., *Spectroscopy Letters* (1969) 2, 61-68; 353-355.
21. See p. 239 in ref. 18.
22. Buchholz, K., and Kirchner, K., *Makromolec. Chem.* (1976), 17, 935.

23. Pryor, W. A., and Coco, J. H., *Macromolecules* (1970) 3, 500.
23a. Brandrup, J., and Immergut, E. H., "Polymer Handbook", 2nd edition, Wiley-Interscience, New York, 1975.
24. Hiatt, R. R., and Bartlett, P. D., *J. Amer. Chem. Soc.* (1959) 81, 1149.
25. Benson, S. W., *J. Chem. Phys.* (1964) 40, 1007.
26. Walling, C., and Heaton, L., *J. Amer. Chem. Soc.* (1965) 87, 38.
27. Miller, W. T., Koch, S. D., and McLafferty, F. W., *J. Amer. Chem. Soc.* (1956) 78, 4992.
28. Walling, C., and Mintz, M. J., *J. Amer. Chem. Soc.* (1967) 89, 1515.
29. Diaper, D. G. M., *Oxid. Combustion Rev.* (1973) 6, 145.
30. Bailey, P. S., *Chem. Rev.* (1958) 58, 925.
31. Bailey, P. S., "Ozone Reactions with Organic Compounds," American Chemical Society, Washington, D. C., 1972.
32. Pryor, W. A., Stanley, J. P., Blair, E., and Cullen, G. B., *Arch. Environ. Health* (1976) 31, 201.
33. Mayo, F. R., Ed., *Adv. Chem. Series* (1968) 77.
34. Hamilton, G. A., Ribner, B. S., and Hellman, T. M., *Adv. Chem. Series* (1968) 77, 15.
34a. Hellman, T. M., and Hamilton, G. A., *J. Amer. Chem. Soc.* (1974) 96, 1530.
35. Whiting, M. C., Bolt, A. J. N., and Parish, J. H., *Adv. Chem. Series* (1968) 77, 4.
35a. White, H. M., and Bailey, P. S., *J. Org. Chem.* (1965) 30, 3037.
36. deVries, L., *J. Amer. Chem. Soc.* (1977) 99, 1982.
37. Kopecky, K. R., and Lau, M. P., submitted for publication.
38. Bailey, W. J., and Baylouny, R. A., *J. Org. Chem.* (1962) 27, 3476.
39. Benson, S. W., *Adv. Chem. Series* (1968) 77, 74.
40. Bailey, P. S., Ward, J. W., Potts, F. E., Chang, Y., and Hornish, R. E., *J. Amer. Chem. Soc.* (1974) 96, 7228.
41. Gall, B. L., and Dorfman, L. M., *J. Amer. Chem. Soc.* (1969) 91, 2199.
42. Willson, R. L., *Chem. Ind.* (1977) 183 (March 5 issue).
43. Fong, K., McCay, P. B., Poyer, J. L., Misra, H. P., and Keele, B. B., *Chem.-Biol. Interactions* (1976) 15, 77.
43a. Beauchamp, C., and Fridovich, I. J., *Biol. Chem.* (1970) 245, 4641.
43b. Rigo, A., Stevanoto, R., Finazzi-Agro, A., and Rotilio, G., *FEBS Letters* (1977) 80, 130.
44. Kurz, M. E., and Pryor, W. A., to be submitted to *J. Amer. Chem. Soc.*
44a. Pryor, W. A., and Kurz, M. E., to be submitted to *Tetrahedron Letters*.
45. Pryor, W. A., and Burguieres, G., to be submitted.
45a. Pryor, W. A., *Photochem. Photobiol.*, in press.
45b. Pryor, W. A., and Thomas, M., (1978), to be submitted.

46. Ref. 18, p 267.
47. Bartlett, P. D., and Gunther, P., J. Amer. Chem. Soc. (1966) 88, 3288.
48. Bartlett, P. D., and Lahav, M., Israel J. Chem. (1972) 10, 101.
49. Barnard, D., McSweeney, G. P., and Smith, J. F., Tetrahedron Letters (1960) 1.
50. Taube, H., and Bray, W. C., J. Amer. Chem. Soc. (1940) 62, 3357.
51. Pryor, W. A., Environ. Health Perspectives (1976) 16, 180.
51a. O'Neal, H. E., and Blumstein, C., J. Chem. Kinetics (1973) 5, 397.
52. Howard, J. A., "Free Radicals", edited by J. K. Kochi, Vol II, p 32, Wiley-Interscience, New York, 1973.
53. Bartlett, P. D., Quart. Revs. (1970) 24, 473.
54. Bartlett, P. D., Pure Applied Chem. (1971) 4, 281.
55. Huisgen, R., Accounts Chem. Res. (1977) in press.
56. Pryor, W. A., Polymer Preprints (1971) 12, (No. 1), 49.
57. Williamson, D. G., and Cvetanovic, R. J., J. Amer. Chem. Soc. (1970) 92, 2949.

RECEIVED December 23, 1977.

4

Organic Peroxides as Free Radical Generators: Some Competitions

R. HIATT

Department of Chemistry, Brock University, St. Catharines, Ontario, Canada

In the decomposition of peroxides many reactions compete with unimolecular homolysis and more than a few of these are thoroughly appreciated in a general, qualitative sense. Some quantitative information is available, but usually applies to conditions where the nonhomolysis reaction(s) predominate. Precise predictions for real-world situations fail through a superabundance of ignorance--a circumstance for which we offer no immediate remedy.

Three kinds of competing reactions are considered here: (1) electrocyclic decompositions; (2) free radical-induced decompositions; (3) decompositions by nucleophiles. Our results of the past few years, mainly for hydroperoxides and alkyl peroxides, are reviewed and some current specific problems are mentioned.

1. Electrocyclic Decompositions of Peroxides

A family of reactions involving concerted transfer of hydrogen from carbon to a hetero-atom may be discerned (1). Despite the dissimilarity of products, there are sufficient unifying

I. Peroxides

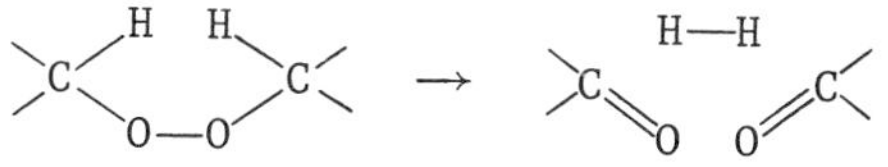

(1)

II. Peroxyesters

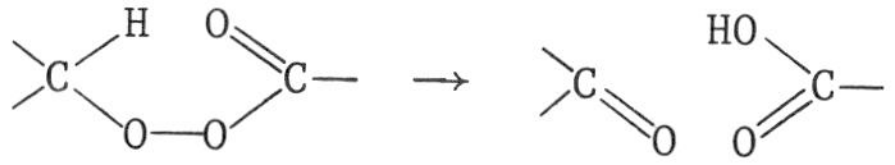

(2)

III. Tetroxides

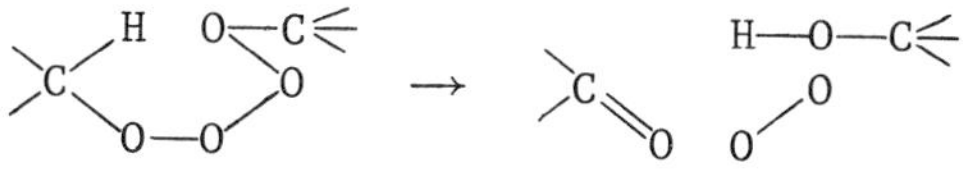

(3)

IV. Acylperoxy hemiacetals

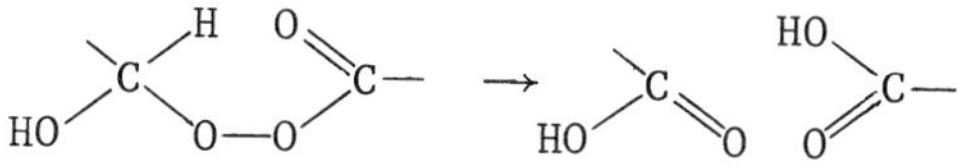

(4)

© 0-8412-0421-7/78/47-069-063$05.00/0

features to justify the generic classification: all are thermally
allowed and photolytically nonallowed by orbital symmetry rules.
(Photolysis yields only alkoxy radicals in cases I and II.   For
III and IV the experiments would be much more difficult.)   All
proceed quite readily in nonpolar solvents but show a marked
aversion to the vapor phase, in so far as known.[*]   In some instances there are competing equilibria; e.g.,

$$\text{products} \longleftarrow \underset{\text{products}}{\rlap{\hspace{2em}}}\quad \rightleftharpoons \quad \longrightarrow \text{products} \quad (6)$$

$$\text{products} \longleftarrow \quad \rightleftharpoons \quad \quad (7)$$

For all cases, the competing homolysis yields oxy radical pairs
capable of self-disproportionation, a complicating feature for
both product analyses and solvent-cage considerations.
      Historically the reaction can be traced to Blank and Finken-
beiner (2) who found that dilute alkaline solutions of H_2O_2 and
formaldehyde yielded H_2 quantitatively at room temperature.

$$H_2O_2 + CH_2O \xrightarrow{OH^-} \quad \rightarrow \quad H_2 + 2\ HCO_2^- \quad (8)$$

The yield of H_2 obtained by heating hydroxymethylene peroxide in
solution is also virtually quantitative, but very low upon vapor
phase pyrolysis (3).
      Yields of H_2 are only 40-75% in the thermal decomposition
(80-110°C), of peroxyhemiacetals (4), but whether this reflects

----

[*]For type III's the evidence is circumstantial. Gas phase
kineticists evidently produce alkoxy radicals from peroxy radical
species which in solution would yield only nonradical products.
However for reactions such as

$$2\ CH_3O_2\cdot \xrightarrow[25°C]{vapor} \begin{array}{l} 2\ CH_3O\cdot + O_2 \\ \\ CH_3OH + CH_2O + O_2 \end{array} \quad (5)$$

the possibility of the competing electrocyclic route has not been
considered and the efficiency of alkoxy radical production is
unknown. For acylperoxyhemiacetals (type IV), there is as yet
no reliable information.

$$iPrCH_2O_2H + iPrCHO \overset{K}{\rightleftharpoons} \begin{array}{c} iPr \quad H \quad H \quad iPr \\ \diagdown C \diagup \quad \diagdown C \diagup \\ H \diagup \quad \diagdown O—O \diagup \quad \diagdown OH \end{array}$$

$$\overset{\Delta}{\underset{\diagdown}{\diagup}} \begin{array}{l} H_2 + iPrCHO + PrCO_2H \\ iPrCH_2O\cdot + iPrCH(OH)O\cdot \end{array} \qquad (9)$$

a more competitive homolysis or a partly dissociated peroxide is not clear.*

Table I reflects some of our own attempts to discover the

Table I. H_2 from Pyrolyses of Alkyl Peroxides (R_2O_2) in Solution

R	T°C	Solvent	% $H_2{}^{\underline{a}}$	$10^5 k_H{}^{\underline{b}}$	E_{aH} kcal/mol
s-Bu	130.0	toluene	22	1.35	32.2
Ph$_2$CH	130.0	toluene	95	19.3	28.0
i-Pr	140.0	toluene	7.5	5.0	32.1
i-Pr	140.0	i-PrOH	8.4	28.5	
i-Pr	140.0	MeOH	30.5	67.2	17.2
i-Pr	140.0	H$_2$O	41.5	296	

$\underline{a}$Based on total decomposition. $\underline{b}$For H_2 formation, sec^{-1}.

factors governing the competition between homolysis and electro-

$$R_2C \begin{array}{c} H \quad H \\ \diagdown \quad \diagup \\ \diagup \quad \diagdown \\ O—O \end{array} CR_2 \xrightarrow{} \overset{k_H}{\underset{k_r}{\diagup\diagdown}} \begin{array}{l} H_2 + 2\ R_2CO \\ 2\ R_2CHO\cdot \end{array} \qquad (10)$$

cyclic decomposition in sec-alkyl peroxides. Not unexpectedly a weaker C–H bond results in more rapid H_2 formation at the expense of O–O scission. (The α-hydroxyl groups, (*vide supra*), presumably exerts the same influence.)

---

* Certainly some radicals are produced since induced decomposition is extensive in the absence of radical traps. However, measurements (5), (6) of the equilibrium constant at lower temperatures indicate that K can be no greater than 3–4 at 100°C. The variation in H_2 yields with initial concentration can be computer-modelled with fair success, assuming an induced term for free hydroperoxide decomposition.

The phase effect has been more intriguing. Early attempts
(7), (8) to demonstrate an effect of solvent polarity failed;
neither k_H nor k_r is highly sensitive and the effects are qualitatively in the same direction. Recent work (9) using H_2O and
alcohols as solvents (Table I) has shown a dependence. Log k_H
correlates lineally with Kosower's 2-values (10). The acceleration in polar media may be due to the products being more polar
than the starting peroxide or to dimunition of dipole-dipole
interaction in the cisoid transition state--or both. In any case
there is now an experimental basis for believing that the electrocyclic decomposition would proceed less rapidly in the vapor phase
than in toluene or benzene.

The problem lies in the homolytic decomposition which also
shows a real, if less pronounced, (and less readily explained) acceleration in more polar media. One correlation of log k_r with
Z is not very good, but is excellent with the Hildebrand δ-values
(11). Thus homolysis should be less rapid in the vapor phase
than in toluene or benzene also, and should not completely over-
whelm the electrocyclic decomposition--as it does. In fact,
translation of alkyl peroxides from nonpolar solvents to the vapor
phase has no measurable effect on their rates of homolysis, as the
results of many workers attest (12), (13).

Of course, this new quandary is based on straight line men-
tality. Ideally a plot of log k_r vs. solvent "polarity" should
curve upwards and have as slope assymptotically approaching zero
at the low end of the scale. If we really understood the effect
of medium on peroxide homolysis the correct solvent parameter
probably would be obvious.

2. Free Radical-Induced Decomposition of Allylic Peroxides and Hydroperoxides

Alkyl peroxides may be attacked by free radicals at three
sites: the O-O bond, a β C-H bond, or an α C-H bond, (if available). (The last mode has not been unambiguously demonstrated,
except for Me_2O_2 in the vapor phase, where little else can happen.) Ordinarily it is not difficult to eliminate any induced
component in a thermal decompositions; e.g., using toluene as
solvent does the job for sec-alkyl peroxides.

We have found allylic peroxides, i.e., allyl tert-butyl to
be more susceptible (14). Heating 0.1-0.2 molar solutions in
toluene at 130°C gives an estimated 14-15% induced decomposition,
while DBPO-peroxide-toluene mixtures at 60°C yield similar results.

At the higher temperature at least half of the induced
reaction results from benzyl radical addition followed by s_Hi.

$$PhCH_2 \cdot \quad C{=}C{-}CO_2t\text{-}Bu \;\longrightarrow\; PhCH_2{-}C{-}\overset{\bullet}{C}{-}CO_2t\text{-}Bu$$
$$\downarrow \quad\;\; O \tag{11}$$
$$PhCH_2{-}C{-}C\overset{/\!\!\!\diagdown}{-}C + t\text{-}BuO\cdot$$

At 60°C no epoxide was produced, although bibenzyl was, and the yield of t-BuOH was nearly quantitative. We conclude that the reaction here was

$$t\text{-BuO} \cdot + \; C{=}C{-}\underset{\underset{H}{|}}{C}{-}O_2 t\text{-Bu} \;\rightarrow\; C{=}C{-}\overset{\bullet}{C}{-}O_2 t\text{-Bu} + t\text{-BuOH}$$
$$\downarrow$$
$$C{=}C{-}C{=}O + t\text{-BuO} \cdot \tag{12}$$

From the average peroxide concentration (0.14 M), the amount decomposed (0.0415 mmol), and the number of t-BuO·s involved (0.66 mmol), the rate constant for t-BuO· abstraction from $C{=}C{-}\overset{H}{\underset{|}{C}}{-}O{-}/H$ is calculated to be 5.5 times as great as for abstraction from PhCH$_3$. This seems about right,[*] and suggests that direct t-BuO· abstraction will also be a factor in the higher temperature self-induced decomposition.

Pyrolyses of allylic hydroperoxides have revealed a novel aspect of induced decomposition associated with addition to the double bond (16). Here there are peroxy radicals in the system as well, produced either from RO$_2$H by H-abstraction or from solvent radicals + O$_2$. Addition of peroxys leads to the sequence

$$RO_2 \cdot + \; C{=}C{-}C{-}O_2 H \;\rightarrow\; RO_2{-}C{-}\overset{\bullet}{C}{-}C{-}O_2 H \tag{13}$$

$$RO_2{-}C{-}\overset{\bullet}{C}{-}C{-}O_2 H \;\underset{b}{\overset{a}{<}}\; \begin{array}{l} RO \cdot + \; C\overset{O}{\overbrace{\;\;\;}}C{-}CO_2 H \\[4pt] RO_2{-}C{-}C\overset{O}{\overbrace{\;\;\;}}C + \cdot OH \end{array} \tag{14}$$

$$RO_2{-}C{-}C\overset{O}{\overbrace{\;\;\;}}C \;\rightarrow\; RO \cdot + \cdot OC{-}C\overset{O}{\overbrace{\;\;\;}}C \tag{15}$$

Both branches are interesting: reaction 14a converts an inactive peroxy radical to alkoxy while converting the original hydroperoxide to a new one; reaction 14b + 15 converts a peroxy radical to three alkoxy's.

As yet we have not been able to isolate the presumed intermediates. Low temperature, extraneous radical-induced decompositions of 2,3 dimethyl-2-hydroperoxy-3-butene (TMEH) (16) gave some addition, and epoxide formation, when the radical was CH$_3$·. With Ph$_3$C, however, the only induced product was Ph$_3$COH from s$_{H2}$ displacement. Clearly the propensities for alkyl radicals to attack at the several available sites require elaboration.

Thermal decompositions of allyl hydroperoxide have proved much more complicated than those of TMEH. For one thing, it is

---

[*] Comparisons of t-BuO· and t-BuO$_2$· reactivity vs. PhCH$_3$ and Ph$_3$CH, and t-BuO$_2$· vs. PhCH$_3$ and (PhCH$_2$)$_2$O (15) form the basis of this supposition.

primary and subject to electrocyclic decomposition. H_2 is a major product and the kinetics are pseudo zero-order in RO_2H. Secondly, the allylperoxy radical loses O_2:

$$H_2C\!\!=\!\!CH\!-\!CH_2O_2\cdot \; \underset{}{\overset{170°C}{\rightleftharpoons}} \; H_2C\!\!=\!\!CH\!-\!CH_2\cdot + O_2 \qquad (16)$$

Products of allyl radical coupling are found, as well as solvent oxidation products, (PhCHO and PhCH$_2$OH).*

Thirdly, a problem with all allylic peroxides, but particularly severe here, the product α,β-unsaturated aldehyde (or ketone) polymerizes, incorporating other interesting products as well as some undecomposed hydroperoxide. The polymerization of acrolein by itself is gloriously complicated, and the nonvolatile residue from thermal decomposition of allyl hydroperoxide in toluene has so far defied analysis. Since the polymer constitutes the major, if not the only, chain termination product, some knowledge of its structure is necessary in order to obtain a free radical count and determine the efficacy of reaction 15.

3. Nucleophilic Reductions of Hydroperoxides

The extra oxygen of a hydroperoxide is readily removed by phosphines, arsenes and stibines (17). Other nucleophiles, e.g., sulfides and sulfoxides, amines, and olefines, accomplish this feat, though less rapidly (Table II).

Table II. t-BuO$_2$H + Nucleophiles; Alone and with Vanadium
Catalysis

N: (solvent)	$10^4 \; k_2^{\underline{a}}$ (°C)		k_3, relative$^{\underline{b}}$ (°C)
Ph$_3$P (EtOH)	7200.	(35)	4×10^5 (25)
Ph$_3$Sb (EtOH)	5700.	(35)	1×10^6 (25)
Ph$_3$As (EtOH)	93.	(35)	2×10^5 (35)
Thioxane (H$_2$O)	1.35	(25)$^{\underline{c}}$	–
Styrene (PhH)	0.0004	(70)$^{\underline{d}}$	6×10^3 (60)
Aniline (PhH)	–		1 (66)

$^{\underline{a}}$M^{-1}sec^{-1}. $^{\underline{b}}$Apparent third order rate constant with catalytic amounts of VO(Acac)$_2$ relative to aniline (k$_3$ = 0.24 M^{-2}sec^{-1}). $^{\underline{c}}$Reference 18. $^{\underline{d}}$Reference 19.

---

*The TMEH results are puzzling in contrast. Under identical conditions this hydroperoxide does not rearrange, as would be expected if eq. 16 were operative, nor are PhCHO and PhCH$_2$OH among the products.

$$R\diagdown_O\diagup^O\diagdown_H + N: \longrightarrow ROH + N{\to}O \tag{17}$$

Competing with this reaction is the one-electron reduction giving
MAH or MIRF (20). The 2-electron--but not 1-electron--reductions

$$RO_2H + N: \longrightarrow Free Radicals \tag{18}$$

are enormously accelerated by certain metal ions or complexes;
e.g., Mo^{VI}, V^V, some Fe porphyrins enzymes (Table II). Whether
this says anything about MAH depends upon the mechanism of the
metal catalyzed reaction.

The simplest view of the latter is that metal ion complexation reduces electron density on the peroxidic oxygen, facilita-
ting nucleophilic displacement--just as H^+ catalyzes displacement
by I^-. (If this is the case, MAH and 2-electron reduction can
hardly share N^+—OH as a common intermediate.)

However there is some evidence for the involvement of metal
ion-peroxo intermediates; $MoO(O_2)_2\cdot HMPA$ has been isolated from
the reaction of MoO_3 with H_2O_2 and shown to epoxide olefins
selectively (21), (22). Thus a more comprehensive sequence may be

$$RO_2H + M \underset{}{\overset{K}{\rightleftharpoons}} RO_2H\cdots M \xrightarrow[N:]{k_a} [NOH^+ ROM^-] \longrightarrow ROH + N{\to}O \tag{19}$$

$$\Big\downarrow k_b$$

$$M—O + ROH$$

$$M—O + N: \overset{K'}{\rightleftharpoons} M—O\cdots N \xrightarrow{k_c} M + N{\to}O \tag{20}$$

The gross kinetics of the hydroperoxide-metal ion-nucleophile
reaction can not distinguish between the two pathways. However,
comparison of the apparent 3^d-order rate constants with the
nucleophilicity of N--for a constant RO_2H—M system--might be
useful. For very good and very poor nucleophiles, the rate will
depend on nucleophilicity, with k_a and $K'k_c$, respectively, rate
controlling. Mediocre nucleophiles should not show this dependence since k_b will govern the rate.

Our own randomly collected data, Table II, show no evidence
of this postulated triphasic curve. The leveling out in the
P, As, Sb series might be the expected plateau, but could equally
well indicate near diffusion control. More work is required.
Additional work with $MoO(O_2)_2\cdot HMPA$ using a greater variety of
nucleophiles, under conditions directly comparable to RO_2H—Mo
date should also prove interesting.

70

Literature Cited

1. Hiatt, R., Glover, L., and Mosher, H. S., J. Am. Chem. Soc. (1975) 97, 5234.
2. Blank, O. and Finkenbeiner, H., Ber. (1898) 31, 2979.
3. Jenkins, A. D. and Styles, D. W., J. Chem. Soc., 1953, 2337.
4. Durham, L. J. and Mosher, H. S., J. Am. Chem. Soc. (1962) 82, 4537; (1964) 84, 2811.
5. Antonovskii, V. L. and Terent'ev, V. A., Zh. Organ. Khim. (Eng. Trans.) (1967) 3, 232, 972.
6. Sauer, M. C. V. and Edwards, J. O., J. Phys. Chem. (1971) 75, 3377.
7. Hiatt, R. and Szilagyi, S., Can. J. Chem. (1970) 48, 615.
8. Hiatt, R. and Thankachan, C., Can. J. Chem. (1974) 52, 4090.
9. Hiatt, R. and Rahimi, P., Int. J. Chemical Kinetics (in press).
10. Kosower, E. M., J. Am. Chem. Soc. (1968) 80, 3253.
11. Hildebrand, J. H., Prausnitz, J. M., and Scott, R. L., "Regular and Related Solutions," Van Nostrand-Reinhold, New York, 1970.
12. Hiatt, R., "Organic Peroxides," Vol.III, p. 26, D. Swern, Ed. Wiley, New York, 1972.
13. Yip, C. K. and Pritchard, H. O., Can. J. Chem. (1971) 49, 2290.
14. Hiatt, R. and Nair, V. G. K., unpublished work.
15. Ingold, K. U., "Free Radicals," Vol. I, Ch. 2, J. K. Kochi, Ed., Wiley, New York, 1973.
16. Hiatt, R. and McCarrick, T., J. Am. Chem. Soc. (1975) 97, 5234.
17. Hiatt, R., McColeman, C., and Howe, G. R., Can. J. Chem. (1975) 53, 559.
18. Dankleff, M. A. P., Curci, R., Edwards, J. O., and Pynn, Y-Y., J. Am. Chem. Soc. (1968) 90, 3209.
19. Walling, C. and Heaton, L., J. Am. Chem. Soc. (1965) 87, 38.
20. Harmony, J. A. K., "Methods in Free Radical Chemistry," Vol. V, Ch. 2, E. S. Huyser, Ed., Marcel Dekker, New York, 1974.
21. Mimoun, H., Seree de Roch, I., and Sajus, L., Tetrahedron (1970) 26, 37.
22. Arakawa, H., Moro-oka, Y., and Ozaki, A., Bull. Chem. Soc. Japan (1974) 47, 2958.

RECEIVED December 23, 1977.

5

The Anchimeric Acceleration of *tert*-Butyl Perbenzoate Homolysis: Reactions on the Borderline between Radical and Ionic Chemistry

J. C. MARTIN

Department of Chemistry, Roger Adams Laboratory, University of Illinois, Urbana, IL 61801

Substantial accelerations of radical formation by neighboring group participation in the homolysis of oxygen-oxygen bonds has been shown in several suitably substituted peresters and diacyl peroxides (1 - 8).

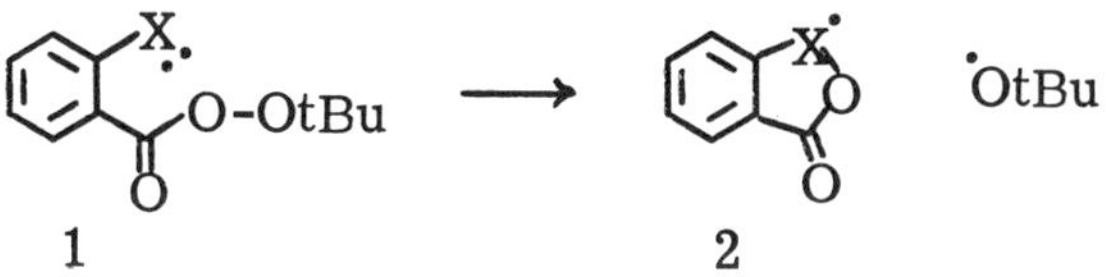

1   2

Ortho-substituted *tert*-butyl perbenzoates, 1, may show anchimerically accelerated bond homolyses leading to bridged radicals of type 2 plus *tert*-butoxy radicals.

Table I. Decomposition of 1 in Chlorobenzene

X	k_{rel} (60°)	ΔH^{*}(kcal/mol)	ΔS^{*}(eu)	Ref
SPh	2.78 x 10⁴	23.0	-3.4	2
SCH₃	1.81 x 10⁴	22.6	-5.5	2
SOPh	72.7	29.4	4.5	9
CH=CPh₂	67.0	26.3	-5.0	5
I	54.1	28.1	-0.7	2
C(CH₃)₃	3.8	34.2	12.5	2
SO₂CH₃	0.4ᵃ	38.0	9.5	9
C≡CPh	1.8	35.4	15.0	10
H	1.0	34.1	10.0	4

ᵃRelative to H at 120°C.

The formation of a stable bridged radical of type 2 requires the substituent to accommodate one additional electron beyond its ground state complement and one might expect the degree of transition state stabilization from the formation of the X-O bond to reflect the ability of the substituent to provide a low energy orbital for this additional electron. We shall discuss reasons for expecting the parallel between the stability of 2 and transition state stabilization to be only approximate.

Mechanism of Reaction. The relative rates of Table I show that the largest accelerations so far observed result from participation of sulfide (sulfenyl) sulfur (2). The neighboring phenyl sulfinyl group shows a much smaller accelerating influence (9) while a sulfonyl substituent shows no acceleration (10). Neighboring iodine (2) and neighboring 2,2-diphenylvinyl (5) substituents provide substantial accelerations, while the very bulky o-tert-butyl (2) substituent shows no acceleration. This rules out explanations for the acceleration based on steric influences of the ortho substituent.

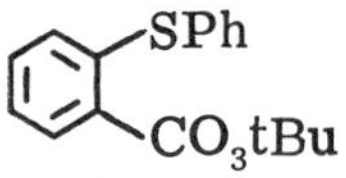

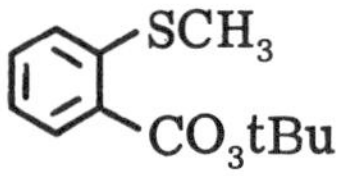

More work has been done (1-4, 6, 8) on systems with neighboring sulfenyl sulfur than on any of the other systems listed in Table I. The decompositions of 3 and 4 occur with very similar rates, both showing acceleration of more than 10^4 at 60° (2). We can therefore be reasonably sure that the sulfur of the phenylthio substituent, rather than the phenyl, is responsible for the observed acceleration. The products of decomposition of 3 observed in the original work of Bentrude (1, 2) are adequately explained by the mechanism originally proposed, that of Scheme I.

Scheme I

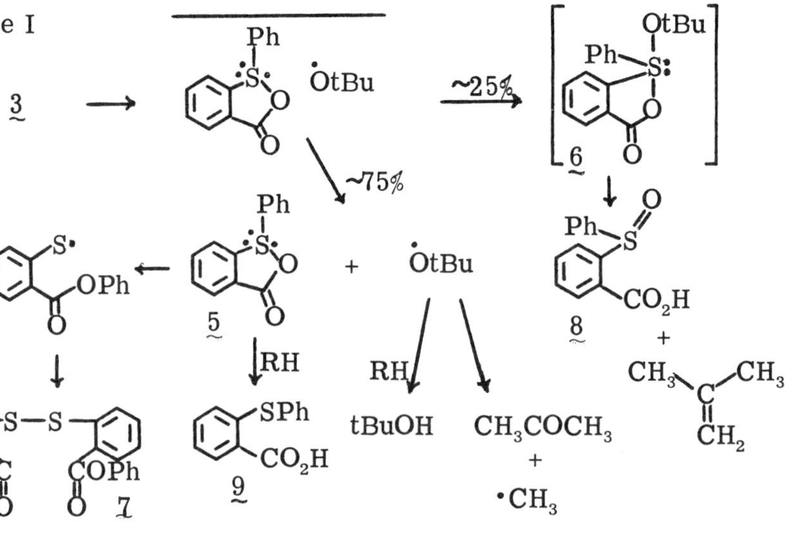

The initially formed cage radical pair was postulated to give free radicals to the extent of ca. 75% in chlorobenzene solution at 80°C, with free radical 5 giving 9 by hydrogen abstraction or 7 by S to O migration of the phenyl group. Under various condi‑ tions scavenging of reactive radicals produced in the reactions of Scheme I could be accomplished (using galvinoxyl as the scav‑ enger) to the extent of 30-98% of the radicals expected.

A minor pathway (ca. 25%) for the production of products from the radical pair was suggested to involve cage recombina‑ tion to yield the tetracoordinate sulfur(IV) species 6. Although no precedents were at that time available for such species the postulate that 6 could serve as a reactive intermediate, precur‑ sor to 8 and isobutylene, provided an appealing economy of speculative mechanistic detail.

The suggested intermediacy of 6 led to the idea that a re‑ lated species, lacking access to the pictured decomposition path‑ way in which a tert-butoxy ligand to sulfur loses a methyl proton to yield sulfoxide and isobutylene, might be stable enough to al‑ low direct observation or isolation.

This idea was vindicated in the work of Arhart (11, 12) who succeeded in oxidative introduction of hexafluorocumyloxy lig‑ ands (OR$_F$) to the sulfur of diphenyl sulfide to give sulfurane 10, a crystalline material stable indefinitely at room temperature.

$$Ph_2S + KOR_F \xrightarrow[CCl_4]{Br_2} \begin{matrix} & OR_F \\ & | \\ Ph\diagdown \!\!\!\!S\colon \\ Ph\diagup \!\!\!\! \\ & | \\ & OR_F \end{matrix} + KBr$$

(where OR$_F$ = OC(CF$_3$)$_2$Ph)

10

-80°C
tBuOH

H$_2$O

$$Ph_2S{=}O + 2R_FOH + (CH_3)_2C{=}CH_2 \qquad Ph_2S{=}O + 2R_FOH$$

Sulfurane 10, a ketal analog of a sulfoxide, is very sensitive to moisture and reacts very rapidly at -80°C with tert-butyl alcohol to give isobutylene (13). Mechanisms studies (14) pointing to the intermediacy of a tert-butoxysulfurane 10a in the reaction of 10 with tert-butyl alcohol are consistent with the postulated inter‑ mediacy of 6 in Scheme I.

$$\begin{matrix} & OtBu \\ & | \\ Ph\diagdown \!\!\!\!S \\ Ph\diagup \!\!\!\! | \\ & OR_F \end{matrix}$$

10a

The Five-membered Ring Effect. The remarkable stabi‑ lizing effect of a five-membered ring linking an apical with an equatorial position of a trigonal bipyramidal (TBP) molecule has long been recognized in phosphorane chemistry (15) and has more recently been shown to be important in sulfuranes as well, since these sulfur(IV) species may be considered to have TBP geometry if the sulfur lone pair is considered to be one of the equatorial ligands. Sulfuranes stabilized by this five-membered ring effect (e.g., 11 and 12) are orders of magnitude less reactive than acyclic analogs 10 and 10a.

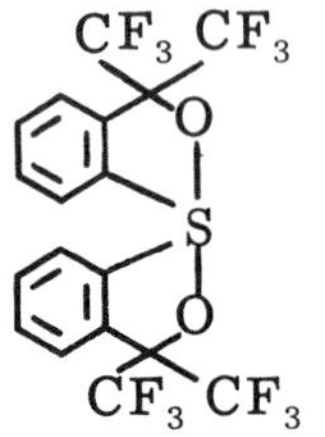

11 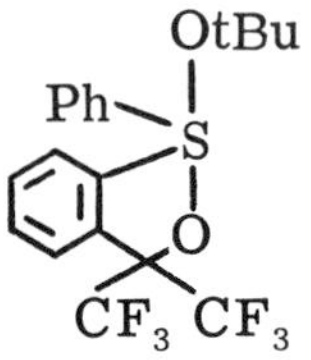   12

Characterization of the Sulfurane Intermediate. The great stability of these cyclic sulfuranes suggested the possibility that the postulated intermediate sulfurane (6) of Scheme I, a close analog of 12, might in fact have sufficient stability to allow its isolation and characterization. This was found by Livant (8) to be the case. The reaction of Scheme II produced 6 in high yield.

Scheme II

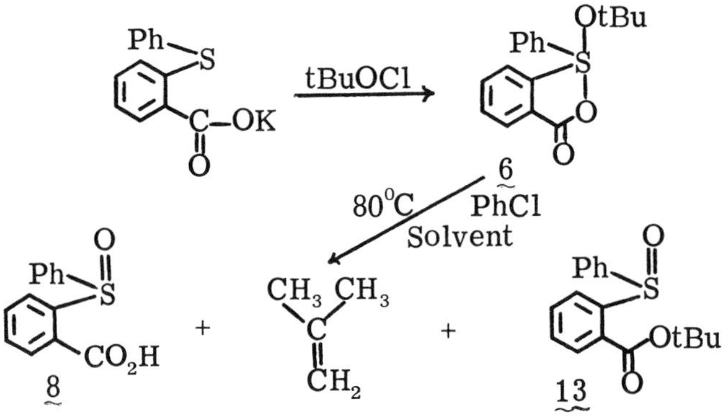

Sulfurane 6 was found to be a stable crystalline compound decomposing, by a reaction showing some autocatalysis by acid produced in the reaction, to give isobutylene and 8 in keeping with the postulate of Scheme I. An additional product, tert-butyl ester 13 was characterized in this reaction. It is sufficiently unstable that its conversion to 8 and isobutylene would probably have been complete after the long reaction times used in the earlier product studies of the decomposition of perester 3.

The failure to observe any radical products ($\underset{\sim}{9}$, $\underset{\sim}{7}$ or acetone) in the decomposition of $\underset{\sim}{6}$ allows us to rule out a mechanism for the anchimerically accelerated homolysis of $\underset{\sim}{3}$ which begins with a biphilic insertion of sulfenyl sulfur into the O-O bond of $\underset{\sim}{3}$ to yield $\underset{\sim}{6}$ directly with $\underset{\sim}{6}$ serving as the precursor to radical products. ~Analogous biphilic insertion reactions have been used to produce sulfuranes from the reaction between sulfoxylates and dioxetanes (18).

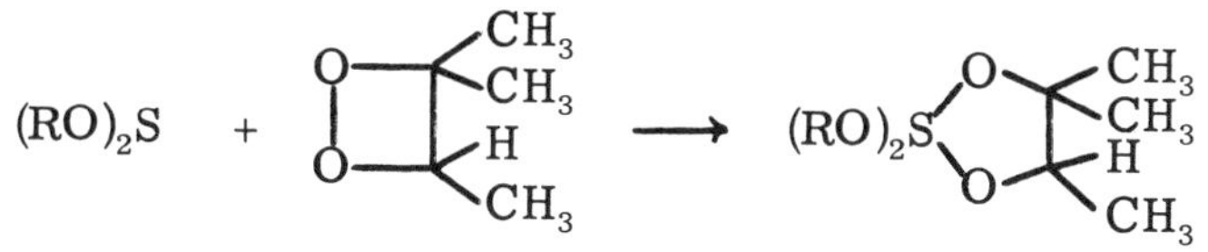

Since $\underset{\sim}{6}$ does not produce free radicals we may rule out the mechanism~of Scheme III for this anchimerically accelerated bond homolysis.

Scheme III

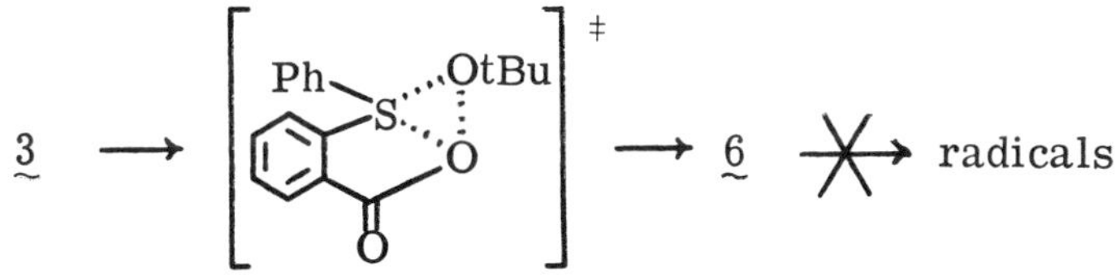

A reexamination of the thermolysis of perester $\underset{\sim}{3}$ at $70^{0}C$ in chlorobenzene, using ^1H NMR to monitor the reaction~throughout its course, showed evidence for the intermediacy of $\underset{\sim}{6}$. Indeed the finding that its stability is comparable to that of the perester requires an appreciable concentration of $\underset{\sim}{6}$ to be built up at intermediate times if it is an intermediate in the reaction. Maximum concentrations of $\underset{\sim}{6}$ approximately 19% of the original perester concentration were observed during the thermolysis of $\underset{\sim}{3}$ after about 12 minutes at $70^{0}C$.

The Transition State. The polar nature of transition states for anchimerically accelerated bond homolyses of peresters of type $\underset{\sim}{1}$ was established very early (2, 3, 4). For example, the dependence on substituent electronic effects of the rate of radical production from $\underset{\sim}{3}$ was probed by kinetic studies of $\underset{\sim}{14}$. The dependence on Hammett σ values for substituent X is described by a ρ value of -1.3.

The polar character of the transition state for the decomposition of 3 is also manifested in a remarkably large dependence on solvent polarity. The fraction of radicals scavengeable varies with solvent in the range from 30 to 98% at 40°C, but not in a manner related to solvent polarity. The rate of decomposition of 3 in a variety of solvents is directly correlated with measures of solvent polarity such as the Kosower Z values (19) with a sensitivity to solvent polarity about 64% of that observed for the ionization of p-methoxyneophyl tosylate (20). It also shows a kinetic salt effect with a rate dependence of radical formation on the concentration of lithium perchlorate in tetrahydrofuran which provides strong evidence for appreciable transition state polarization.

The pronounced charge polarization in the transition state for bond homolysis is not limited to cases involving neighboring sulfenyl sulfur nor to peresters. One of the more thoroughly studied cases is that of a series of diaroyl peroxides showing accelerated bond homolysis from neighboring group participation of an olefinic double bond in the ortho position (5).

The rate of radical production from peroxides 16 in chlorobenzene solvent depends on substituents X and Y in a manner correlated with Hammett σ constants with ρ_X = -1.2 and ρ_Y = +0.6. The development of positive charge in the vinyl substituent and negative charge in the leaving carboxyl group is in keeping with the transition state charge distribution pictured for resonance structure 17b in the reactions examined in this study by Koenig (5).

The closely related perester, 18, also shows a strongly accelerated rate of decomposition (Table I). It gives products from bridged radical 19 in which oxygen 18 tracer experiments indicate 88% retention of the label in the product carbonyl oxygen (21). This evidence for the maintenance of bond integrity in the intermediate radical 19 also points to bridging in the polar transition state leading to this radical.

The Nature of the First-formed Intermediate. Despite the very polar nature of the transition state, the products of the decomposition of 3 (or 14) are reliably free-radical. The high energy intermediate which is first encountered along the reaction pathway for decomposition of a radical initiation is a geminate radical pair, within the solvent cage. In the case of 3, the large degree of charge polarization in the transition state leading to this intermediate leads one to question whether the cage pair is

better described as a radical pair, resembling transition state
resonance structures 15a (or 15c), or an ion pair resembling
15b. If the latter were the case it should be possible to generate
the same species by reaction of a sulfonium species with tert-
butoxide anion. The ionic sulfonium triflate 20 was treated with
potassium tert-butoxide in chlorobenzene at 80°C, the same conditions used for the product studies for the decomposition of 3.
The products included isobutylene and sulfoxide acid 8, the products expected for the reaction via 6. None of the free radical
products, specifically acetone, was observed. We may conclude
that there is no crossover between the two sets of intermediates,
the radical pair from the perester decomposition and the ion pair
formed in the reaction of Scheme IV (8), either directly or by
formation of perester by combination of the ion pair to form an
O-O bond.

Scheme IV

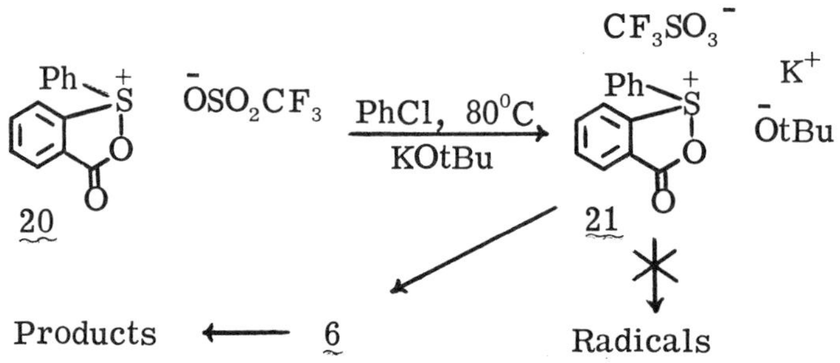

The absence of radical products in the reaction of Scheme
IV, and the formation of only a limited amount of product in the
perester decomposition (Scheme III) which could have arrived by
a non-radical path (via 6), could be explained in at least two
ways. The two pair species, one from perester and the other
from the pairing of ions, could differ in that the ion sources
contribute potassium and triflate ions which could be included
in an ion aggregate, 21, possibly more likely to give ionic products than is the unperturbed sulfonium--tert-butoxide pair.
Further work will be required to confirm or deny this explanation. The associated uncertainty remains a defect in the experiment.

Another more interesting explanation of the failure to see
extensive crossover of the two reaction paths is based on a possible difference in geometry between the sulfonium ion and the
corresponding sulfuranyl radical derived from it by one-electron
reduction. If this difference in geometry is sufficiently large a
Franck-Condon factor could slow the electron transfer sufficiently to allow diffusion of radicals from the cage to compete with
conversion to the ion pair.

While we can place considerable confidence in a prediction
of strongly pyramidal geometry for the cyclic acyloxysulfonium

ion of 20, the preferred geometry of sulfuranyl radical, 5, is less obvious.

Geometry and Structure of the Sulfuranyl Radical. Evidence relating to the preferred geometry of the transition state leading to sulfuranyl radical from perester 3 has been sought in kinetic studies of model peresters of constrained geometry. For example, perester 22 differs from 3 in having the S-aryl substituent a part of a six-membered ring. The further perturbation of the system introduced upon going from 3 to 22 is the introduction of an electron-withdrawing acyl substituent on both aromatic rings in 22. After allowing for the rate retarding effect, which can be estimated from the known ρ = -1.3, the observed rate of decomposition of 22 is slower by a factor of ca. 10^3 (at 50°C) than the rate expected from the operation of this electronic effect alone (3). This factor was attributed to a steric effect in the study by Fisher (6).

22

Even if we concede that the transition state for the decomposition of 22 is constrained away from the geometry which is optimum for bridging to the neighboring sulfur it is clear that extrapolation from this to speculation on the geometry of the product radical is dangerous.

Evidence that these displacements by neighboring sulfur occur on the acyloxy oxygen, rather than the tert-butoxy oxygen, can be adduced from the observation of only slightly attenuated (ΔH^* = 23.7 kcal/mole, ΔS^* = -5.5 eu) anchimeric acceleration in perester 23. Participation at the tert-butoxy oxygen in this molecule would lead to a seven-membered ring in the transition state. Attack at the acyloxy oxygen as envisioned in the mechanism we have adopted leads to the six-membered ring in the transition state, more in keeping with the observed high level of anchimeric acceleration.

23

Literature models may be cited in support of three different
bonding modes for bridged radical 5: (a) π-sulfuranyl radicals
with a 3-center, 4-electron (3c-4e) bond and a π-delocalized
unpaired electron, (b) σ-sulfuranyl radicals with a 3-center,
3-electron (3c-3e) bond, and (c) radicals containing a 2-center,
3-electron (2c-3e) bond.

(a) The direct observation of a number of sulfuranyl radicals, e.g., dialkoxyarylsulfuranyl radical 24 (22), by ESR provides evidence that π-sulfuranyl radicals may in some cases
adopt a T-shaped geometry as shown for 24. The unpaired electron is delocalized over the aromatic π-system in this radical.
The O-S-O bond linking the two apical alkoxy ligands to sulfur is
therefore most likely a hypervalent (17) 3c-4e bond. On the
basis of several analogies, this predicts a near colinearity of
the two O-S axes. Radical 24 was formed by the reaction of
tert-butoxy radicals with diphenyl disulfide.

tBuO–Ṡ–OtBu

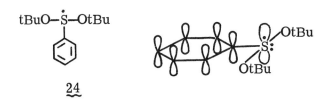

24

While sulfuranyl radicals derived from perester homolysis
with neighboring sulfur participation have not been directly observed by ESR, results of CIDNP studies of the decomposition
of 4 have been interpreted in terms of a π-sulfuranyl radical
intermediate 25 (23).

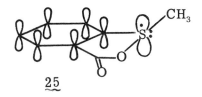

25

The interpretation of observed (23) nuclear spin polarization
of CH₃ protons in the earlier noted (2) products of decomposition
of this perester was based on results of CNDO/2 SCF-MO calculations without minimizing energy with respect to geometry.
Other bonding schemes must therefore still be considered possibilities.

(b) An alternative bonding scheme consistent with a geometry similar to that postulated in 25 would have the hypervalent
colinear three-center bond occupied by three rather than four

electrons. This scheme, which places two electron pairs on sulfur orthogonal to the apical three-center bond, has analogies in phosphoranyl radicals for which such a bonding scheme has been postulated from ESR studies (24, 25). In this bonding model the odd electron is in a 3c-3e bond along the O-S-C axis of 25, orthogonal to the π-system, hence we call such a radical a σ-sulfuranyl radical.

(c) The third model, which also has a three-electron bond orthogonal to the π-system, might be expected to differ in geometry from that with the O-S-CH$_3$ axis colinear. Analogies for a 2c-3e bonding scheme have been established from ESR studies of radical anion 26 (26). The one-electron reduction of the corresponding disulfide to give 26 introduces the electron into an orbital in the molecular plane which is approximately described as the antibonding combination of sulfur 3p atomic orbitals (26).

$$\left[\begin{array}{c} \mathrm{S-S} \\ \text{(naphthalene ring)} \end{array} \right]^{\cdot\,-}$$

26

Similar bonding schemes may be invoked for the dialkylated analogs of 26, radical cations 27 (27), which shows equivalent hyperfine coupling to all twelve protons, and the more stable 28 (28), a possible beneficiary of a five-membered ring stabilizing effect similar to that noted earlier for closed shell sulfuranes.

$$\left[\mathrm{Me_2S\text{-}SMe_2} \right]^{+}_{\cdot}$$

27

$$\left[\begin{array}{c} \mathrm{S} \\ \mathrm{S} \end{array} \right]^{+}_{\cdot}$$

28

Species 27 and 28 are generated by one-electron oxidation of the dialkyl sulfide. The interaction between the resulting sulfide cation radical and another sulfide sulfur is overall a bonding interaction since the antibonding orbital is only singly occupied.

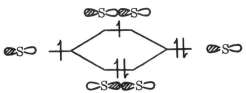

The interaction of a sulfide cation radical with any nucleophile might be expected to result in some net bonding by such an interaction. Radical 5 for example could be imagined to have an S-O bond formed by interaction of the doubly occupied oxygen 2p orbital of the carboxylate ion interacting intramolecularly with the singly occupied sulfur 3p orbital of the adjacent sulfide sulfur of 29.

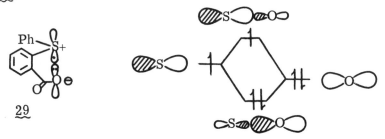

29

Polarity of the S-O Bond in 5. The MO scheme shown for the 2c-3e S-O bond in 29 (one representation of 5) shows the doubly occupied lower MO to have a heavier contribution from the lower energy oxygen 2p atomic orbital than from the higher energy sulfur 3p orbital. The S-O bond in such a species would therefore be polar with negative charge on oxygen and positive charge on sulfur.

An argument can be made for a similar charge polarization in the S-O bonds described by bonding schemes a (3c-4e) and b (3c-3e) as well. In both cases a three-center bond joins two apical ligands to sulfur. It has been shown that such bonds to sulfur, at least in the 3c-4e case, are highly polarizable (29, 30). Specifically to the point in the present case, it has been shown (29) that the carbonyl stretching frequency in sulfuranes of type 30 are very sensitive to the electron-withdrawing power of apical substituent X.

30

The change in $\nu_{C=O}$ as X is changed from OSO_2CF_3 (1832 cm^{-1}) to Cl (1740 cm^{-1}) to O-t-Bu (1640 cm^{-1}) suggests an appreciable increase in carboxylate anion character as the electron-withdrawing character of X decreases. One might expect a hypervalent radical of type 25 (or its 3c-3e analog) with a carbon-centered ligand at one apical position and a carboxylate oxygen at the other to be very dipolar.

One might therefore expect bridged radical 5 to be strongly dipolar no matter which bonding mode (a, b, or c) is properly involved to describe the S-O bond. It is possible that this dipolar character, reflecting a major contribution from 15c to the description of the transition state resonance hybrid, is largely responsible for the solvent dependence of rate of decompositions

observed for perester **3**. The bridged radical product may be
represented for any of the three structural hypotheses (a, b, or
c) by the geometrically ambiguous resonance structures 31a and
31b.

31a   31b

Relationship to Electron Transfer Mechanisms. The intermolecular reactions of dialkyl or alkyl aryl sulfides with tert-
butyl perbenzoates have been shown by Pryor and Hendrickson
(31) to give only a small amount (0.7 - 3.5%) of scavengeable
radical product. The rates of disappearance of the perbenzoates
are strongly accelerated by addition of the sulfides. The mechanism which has been postulated involves an electron transfer
from sulfur to the peroxide and kinetic isotope effect data have
been presented in support of this mechanistic postulate (32).
The interpretation of deuterium kinetic isotope effects, for substitution of D for H on a carbon bonded to the sulfur, invoked
the change in C-H stretching force constants upon removing an
electron from the delocalized HOMO of the sulfide, a molecular
orbital with significant contributions from the hydrogen 1s orbitals.
    It is clear from the arguments presented earlier in this
paper that a continuum of transition state structures may obtain
for such reactions with the free sulfide cation radical at one extreme and a variable amount of covalent bonding to a nucleophile,
either intramolecularly or intermolecularly. For example, the
relative contributions of the two resonance structures written
for **31** can vary continuously with substitution or conditions.
    The much larger fraction of radical products from the de-
composition of **3**, relative to the intermolecular analogs reported
by Pryor (31, 32), can be explained as a manifestation of the
five-membered ring effect discussed earlier. It is clear that a
five-membered ring containing sulfonium sulfur is, because of
angle strain and other factors, very much disfavored in comparison with such a ring containing sulfuranyl sulfur (16, 17,
33). The ionic pathway for decomposition of 3 leads to such a
sulfonium species and is therefore disfavored, because of the
five-membered ring effect, relative to the homolytic pathway
leading to 5 which has a different, more favorable geometry
about sulfur.

Simultaneous Participation of Multiple Neighboring Groups.
The strength of a hypervalent 3c-4e bond increases with electronegativity of the apical ligands. In a sulfuranyl radical such as

25 (3c-4e, bonding mode a), or its σ-sulfuranyl radical analog (3c-3e, bonding mode b), one of the apical positions is occupied by an oxygen but the other is occupied by a more electropositive carbon. A stronger bond might be expected in the transition state if both apical ligands were oxygen-centered (17). A search for a radical initiator which might decompose through such a transition state led us to examine the possibility that three neighboring groups might, in a suitable molecule, be simultaneously involved in transition state bond-making. Perester 32, studied by Chau (7), provided the first example of such a radical initiator. It decomposes to give stable sulfurane 34 and tert-butoxy radicals.

Table II. Perester Decomposition in Chlorobenzene

Compound	k_{rel}, 0°C	ΔH*(kcal/mol)	ΔS*(eu)	Ref
tert-butyl perbenzoate	1.	33.6	8.2	34
3	1.1×10^6	22.3	-5.7	2
32	8.9×10^7	17.6	-13.9	7
35	3.2×10^2	28.1	-0.66	2
36	3.9×10^4	22.4	-11.9	7
38	2.3×10^5	25.5	2.5	37

The data of Table II show that the decomposition of bis-perester 32 is anchimerically accelerated by a factor of 10^8 at 0°C, a temperature within the range at which rate measurements were made for 32. This is larger than the anchimeric acceleration noted for monoperester 3 by a factor of about 10^2. Evidence for the simultaneous breaking of two O-O bonds and formation of two S-O bonds, in the manner suggested in the drawing for transition state 33, is even more compellingly discernible in the activation parameters listed in Table II. The remarkable lowering of ΔH* for the decomposition of 32 (17.6 kcal/mol) compared to tert-butyl perbenzoate (33.6 kcal/mol) is accomplished only at the cost of achieving the very precisely ordered geometry of transition state 33. This lowers ΔS* from +8.2 eu (for the unsubstituted perbenzoate) to -13.9 eu (for 32). The monoperester, decomposing by a process forming one S-O bond simultaneously with cleavage of a single O-O bond via transition state 15, is intermediate in both ΔH^* and ΔS^*. The rate accelerations observed for 3 and 32 would be larger still were it not for these partially compensating entropy effects. The entropy effects are clearly consistent with the postulated transition state structures.

The bonds represented by dotted lines in transition state 33 could be described as a single extended five-center six-electron (5c-6e) bond. Molecular orbitals constructed from oxygen 2p and sulfur 3p orbitals are pictured below for both transition state and product.

The description of 3c-4e bonds, the sort which Musher (35) has called hypervalent, in terms of linear combinations of p_z orbitals of the three colinear atomic centers is pictured for the product sulfurane (34) at the right below. The 5c-6e bond postulated for transition state 33 is illustrated in similar terms in the MO diagram to the left. Arduengo and Burgess (36) have discussed related multicenter bonds.

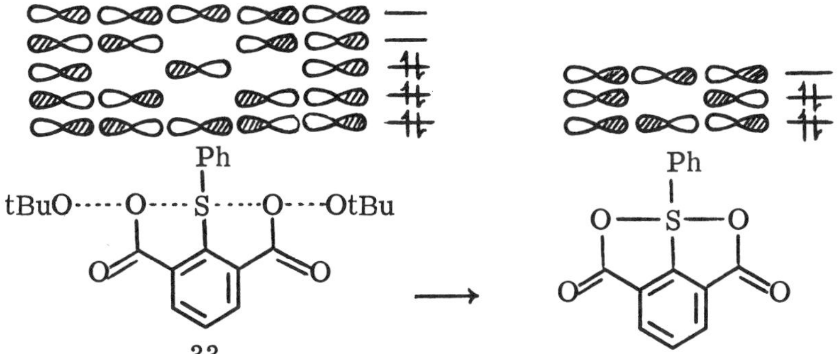

A very similar set of observations for 35 and 36, the iodo analogs of 3 and 32, provides further confirmation for this hypothesis. The simultaneous participation of three neighboring groups in the decomposition of 36, to give 37 and tert-butoxy radicals, is evidenced in a rate of decomposition 10^2 faster for 36 than for 35. This effect is amplified in ΔH^* values (Table II).

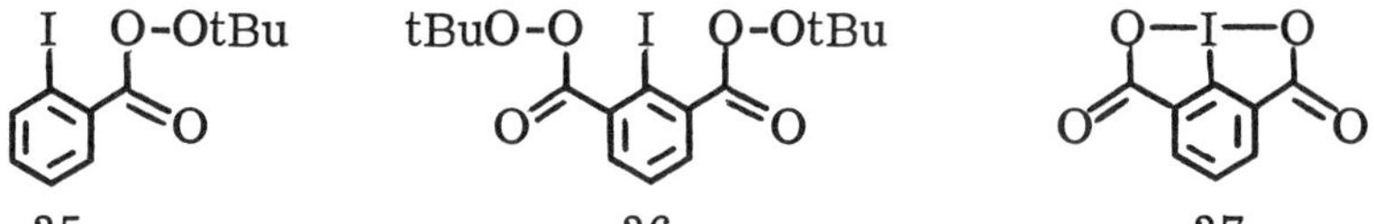

The simultaneous three-neighboring-group participation (S3NGP) mechanism favored for these molecules, via transition states such as 33, is in theory extendable to larger numbers of neighboring groups which might simultaneously engage in bond-making in a transition state. (Note that we have refrained from designating one group as a "reaction center" and another as a "neighboring group" but have simply counted all groups involved in bonding changes in a transition state.)

A search for an example of the S5NGP mechanism led Chau (37) to synthesize tetraperester 38. The pictured S5NGP process leading directly to persulfurane 39 appears not to be the major pathway for the decomposition of 38.

From a consideration of the kinetic data for 38 (Table II), and spectroscopic evidence for the intermediacy of 40 and 41 in the reaction, we suggest the major pathways for decomposition to be the two S3NGP processes possible for the tetraperester, leading first to 40 and 41 (37).

The conclusion that 39 was the eventual product of decomposition of 38 was clouded by our failure to find methods for purification which would allow us to obtain satisfactory elemental analyses for 39. The more recent complete characterization by Lam (38) of a related persulfurane (42) provides a model which makes our assignment of structure for 39 much firmer.

Comparison of the product from thermolysis of 38, which was assigned structure 39, with the more rigorously characterized 42 makes it seem likely that 39 is indeed the proper structure of this product. The possibility of anchimeric acceleration by an S5NGP process remains open for this reaction, perhaps providing a minor pathway for decomposition. This prompts us to continue our search for a more favorable case for the observation of such a mechanism.

Acknowledgment. The research upon which this review is based was supported in part by grants from the National Cancer Institute (CA-13963) and the National Science Foundation (MPS75-17742).

Literature Cited

1. Martin, J. C. and Bentrude, W. G., Chem. Ind. (London) (1959), 192.
2. Bentrude, W. G. and Martin, J. C., J. Am. Chem. Soc. (1962), 84, 1561.
3. Martin, J. C., Tuleen, D. L., and Bentrude, W. G., Tetrahedron Lett. (1962), 229.
4. Tuleen, D. L., Bentrude, W. G., and Martin, J. C., J. Am. Chem. Soc. (1963), 85, 1938.
5. Koenig, T. W. and Martin, J. C., J. Org. Chem. (1964), 29, 1520.
6. Fisher, T. H. and Martin, J. C., J. Am. Chem. Soc. (1966), 88, 3382.
7. Martin, J. C. and Chau, M. M., J. Am. Chem. Soc. (1974), 96, 3319.
8. Livant, P. and Martin, J. C., J. Am. Chem. Soc. (1976), 98, 7851.
9. Unpublished work of R. Arhart.
10. Unpublished work of Der-shing Huang.
11. Martin, J. C. and Arhart, R. J., J. Am. Chem. Soc. (1971), 93, 2339, 2341.
12. Arhart, R. J. and Martin, J. C., J. Am. Chem. Soc. (1972), 94, 4997.
13. Arhart, R. J. and Martin, J. C., J. Am. Chem. Soc. (1972), 94, 5003.
14. Kaplan, L. J. and Martin, J. C., J. Am. Chem. Soc. (1973), 95, 793.
15. Kumamoto, J., Cox, J. R., Jr. and Westheimer, F. H., J. Am. Chem. Soc. (1956), 78, 4858.
16. Martin, J. C. and Perozzi, E. F., J. Am. Chem. Soc. (1974), 96, 3155.
17. Martin, J. C. and Perozzi, E. F., Science (1976), 191, 154.
18. Campbell, B. S., Denney, D. B., Denney, D. Z., and Shih, L. S., J. Am. Chem. Soc. (1975), 97, 3850.

19. Kosower, E. M., J. Am. Chem. Soc. (1958), 80, 3253, 3261.
20. Smith, S. G., Fainberg, A. H., and Winstein, S., J. Am. Chem. Soc. (1961), 83, 618.
21. Martin, J. C. and Koenig, T. W., J. Am. Chem. Soc. (1964), 86, 1771.
22. Gara, W. B., Roberts, B. P., Gilbert, B. C., Kirk, C. M., and Norman, R. O. C., J. Chem. Research (S) (1977), 152.
23. Nakanishi, W., Koike, S., Inoue, M., Ikeda, Y., Iwamura, H., Imahashi, Y., Kihara, H., and Iwai, M., Tetrahedron Lett. (1977), 81.
24. Gillbro, T. and Williams, F., J. Am. Chem. Soc. (1974), 96, 5032.
25. Nishikida, K. and Williams, F., J. Am. Chem. Soc. (1974), 97, 5462.
26. Zweig, A. and Hoffman, A. K., J. Org. Chem. (1965), 30, 3997.
27. Gilbert, B. C., Hodgeman, D. K. C., and Norman, R. O. C., J. Chem. Soc., Perkin Trans., II (1973), 1748.
28. Musker, W. K. and Wolford, T. L., J. Am. Chem. Soc. (1976), 98, 3055.
29. Livant, P. and Martin, J. C., J. Am. Chem. Soc. (1977), 99, 5761.
30. Adzima, L. J. and Martin, J. C., J. Org. Chem. (in press).
31. Pryor, W. A. and Hendrickson, W. H., Jr., J. Am. Chem. Soc. (1975), 97, 1580.
32. Pryor, W. A. and Hendrickson, W. H., Jr., J. Am. Chem. Soc. (1975), 97, 1582.
33. Astrologes, G. W. and Martin, J. C., J. Am. Chem. Soc. (1977), 99, 4390.
34. Blomquist, A. T. and Ferris, A. F., J. Am. Chem. Soc. (1951), 73, 3408.
35. Musher, J. I., Angew. Chem. Int. Ed. Engl. (1969), 8, 54.
36. Arduengo, A. J. and Burgess, E. M., J. Am. Chem. Soc. (1977), 99, 2376.
37. Chau, M. M., and Martin, J. C., J. Am. Chem. Soc., submitted.
38. Lam, W. Y. and Martin, J. C., J. Am. Chem. Soc. (1977), 99, 1659.

RECEIVED December 23, 1977.

6

The Chemistry of Cyclic Peroxides: The Formation and Decomposition of Prostaglandin Endoperoxide Analogs

NED A. PORTER, JOHN R. NIXON, and DENNIS W. GILMORE

Paul M. Gross Chemical Laboratory, Duke University, Durham, NC 27706

Interest in the chemistry and biochemistry of peroxides has been recently stimulated by the fact that the important class of natural products, the prostaglandins, has two members (PGG and PGH) that contain the peroxide linkage. Further, the proposed mechanism for the biosynthesis of these peroxides involves a novel peroxy radical bicyclization. Thus, one mechanism for biosynthesis of PGG, proposed in 1967 (1, 2), involves an autoxidative-type conversion of fatty acid to endoperoxide via a peroxy radical mechanism (Figure 1). For the past three years, our research has focused on aspects of peroxide and free radical chemistry related to the formation and decomposition of monocyclic and bicyclic peroxides analogous to these prostaglandin intermediates, and we present here our results concerning the formation and decomposition of these compounds.

Monocyclic Peroxides

Formation. Although a peroxy radical cyclization mechanism was proposed (1, 2) for prostaglandin biosynthesis, this mode of reaction had received little chemical attention. Early reports suggested that peroxy radical cyclization was an important variant in the autoxidation of polyunsaturated materials such as squalene (3) and cyclododecatriene (4). Products were not fully characterized in these studies, however, due to the difficulties of peroxide isolation and purification.

Perhaps the best documented case of peroxy radical cycliza-tion was the report (5) that α-farnesene undergoes radical autoxi-dation to yield the completely characterized monocyclic peroxide, 1. Autoxidative peroxy radical cyclization has precedent, then, but this autoxidation format is cumbersome for a systematic study.

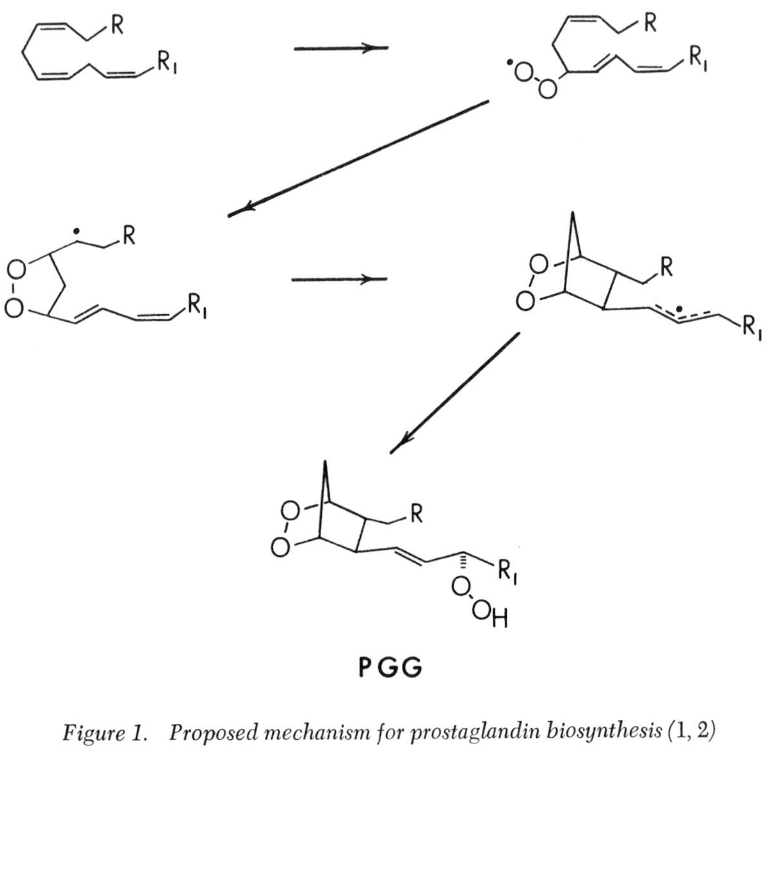

PGG

Figure 1. Proposed mechanism for prostaglandin biosynthesis (1, 2)

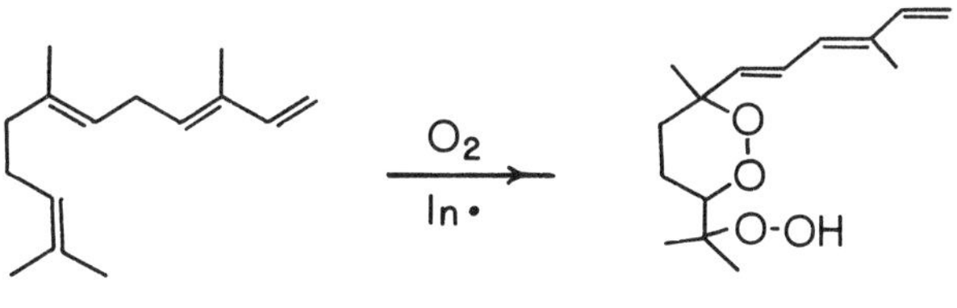

1

We developed a method for the study of peroxy radical cyclization based on the generation of peroxy radicals from unsaturated
hydroperoxides (6, 7). Thus, treatment of the hydroperoxide 2,
with free radical initiators such as DBPO (8) led to the formation
of monocyclic peroxides that could be isolated by liquid chromatography and characterized by standard techniques.

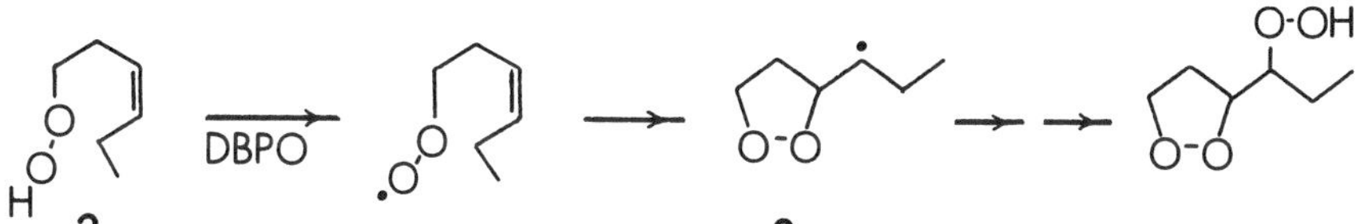

2   **3**

Systematic investigation has shown that the ease of cyclization is [in the terminology proposed by Baldwin (9)] 5 or 6
exocyclic > 6 or 7 endocyclic cyclization.

The mechanism for cyclic peroxide formation presumably
involves formation of β-peroxy alkyl radicals like 3 which can
react with oxygen to yield ultimately cyclic peroxide products.
An alternate pathway available to 3, however, is intramolecular
radical attack on the peroxide linkage (10,11) yielding ultimately
the epoxy-alcohol product. We wished to investigate this S_Hi
pathway systematically and sought other, more controlled methods
for the preparation of radicals like 3.

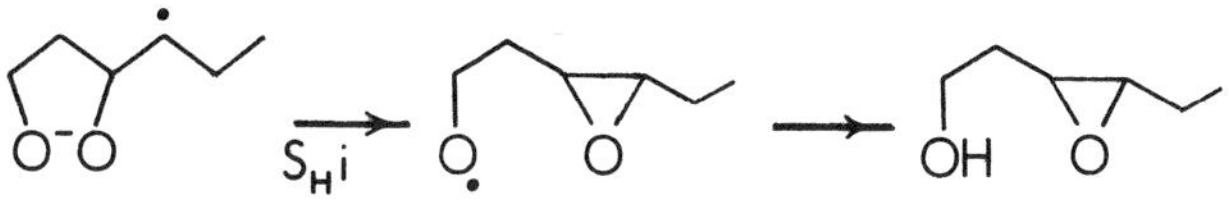

3

Potential methods for generation of radicals like $\underset{\sim}{3}$ involve the use of β-mercurated cyclic peroxides. Alkyl-mercuric compounds react with borohydride to yield the corresponding alkyl radical (12, 13). The mechanism of radical production has been thoroughly investigated (14) and involves an intermediate alkyl-hydrido mercury compound, R-Hg-H. Chain propagation occurs as shown below by radical attack on R-Hg-H.

$$R\bullet + R\text{-}Hg\text{-}H \longrightarrow R\text{-}H + Hg + R\bullet$$

The crucial β-mercurated cyclic peroxide precursors of $\underset{\sim}{3}$ can be prepared by mercuric nitrate initiated cyclization of unsaturated hydroperoxides (15). Yields are excellent and the compounds (as the bromide derivatives) can be purified by high pressure liquid chromatography (hplc). Exocyclic products are formed in cyclizations of $\underset{\sim}{2}$, $\underset{\sim}{4}$, and $\underset{\sim}{6}$ with none of the 6-endo products being detected by nmr or hplc. $\underset{\sim}{5}$ leads to a 3:1 ratio of 6-exo to 7-endo cyclization products. These isomers can be cleanly separated by hplc and quantities of all of the β-mercurated cyclic peroxides are thus readily available.

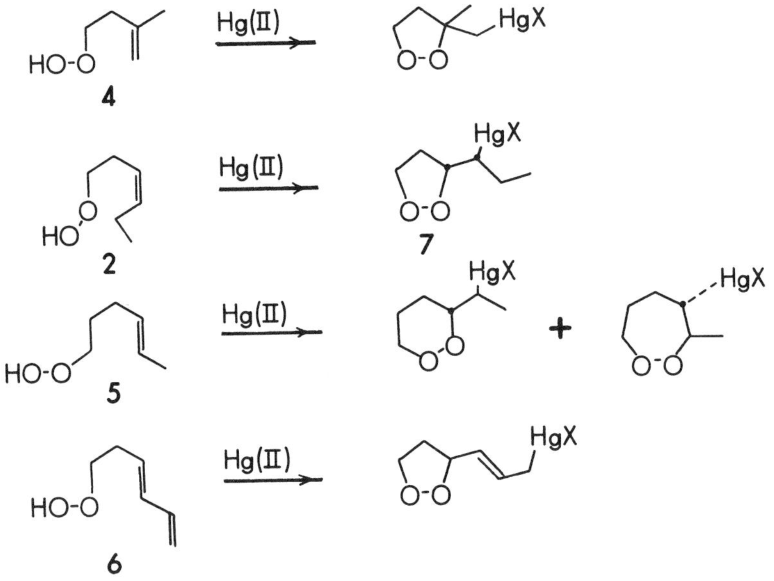

The predictions for cyclization developed by Baldwin (9) are rather vague when applied to these mercury initiated cyclizations. The rules, as stated for nucleophilic attack on 3-membered rings, "seem to lie between those for tetrahedral and trigonal systems, generally preferring exo modes." From our limited observation of peroxide attack on the three-membered ring mercurinium

intermediate, we conclude that <u>exo</u> modes of cyclization are
favored and that 6-<u>endo</u> cyclization is disfavored.

$S_{H}i$ Stereochemistry. Treatment of the β-mercurated cyclic
peroxides with sodium borohydride leads to mixtures of demercu-
rated cyclic peroxides and epoxy-alcohols. Thus, $\underset{\sim}{7}$ reacts to
give the cyclic peroxides and the epoxy-alcohol in a 3:1 ratio.
As noted earlier, these are products expected from a β-peroxy
radical; the cyclic peroxide resulting from H atom abstraction,
the epoxy-alcohol from $S_{H}i$ radical attack on the peroxide linkage.

The relative amount of peroxide product and epoxy-alcohol is
intimately dependent on the structure of the intermediate radical.
In Table I is presented the product distribution of peroxide and
$S_{H}i$ product (epoxy-alcohol) for a series of cyclic peroxide alkyl-
radicals. Note that the exocyclic five-membered ring radicals $\underset{\sim}{8}$
and $\underset{\sim}{9}$ lead predominately to peroxide as does the endocyclic seven-
membered ring radical $\underset{\sim}{11}$. The exocyclic six-membered ring
radical, on the other hand, gives primarily $S_{H}i$ products. The
amount of peroxide and epoxy-alcohol produced does vary somewhat
depending on how borohydride addition is carried out. The product
compositions reported in Table I could be reproducibly obtained,
however, by quickly syringing the borohydride reducing agent into
a mixed phase CH_2Cl_2/H_2O solution of the β-mercurated cyclic
peroxide at 0°. In addition, by reducing mixtures of the various
β-mercurated peroxides, the product ratios presented could be
confirmed.

We suggest that the $S_{H}i$ reaction reported here offers the
possibility of studying the geometric requirements of $S_{H}2$ attack
on the peroxide linkage. The data suggest that the critical
geometric parameter for the $S_{H}i$ reaction is the dihedral angle,
φ, about the OC bond between the attacking radical and the leaving oxygen. We assume that for maximum $S_{H}i$ reactivity, this
dihedral angle must be 180°. A view down the O–C bond of a space-
filling model of a six-membered ring peroxide with a –CH_2• center
attached α to the peroxide linkage is presented in Figure 2. The
critical dihedral angle about the O–C bond is 180°, or nearly so,
for this six-membered ring radical. For the analogous radical
derived from a five-membered ring, the O–C dihedral angle is
substantially less than 180°, having a maximum of approximately
165° in the most favorable conformation for $S_{H}i$ attack. Thus, the

Table   I.   Product distribution from borohydride reduction of
             β mercurated cyclic peroxides.

	peroxide	$S_H i$
8	75	25
9	90	10
10	10	90
11	100	−

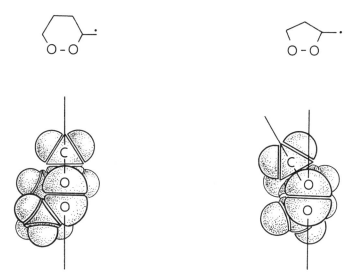

Figure 2.   Geometric requirements of S_Hi radical attack on the cyclic peroxide bond

five-membered ring peroxides lead to less S_Hi product than the
six-membered ring peroxides since the five-membered ring radical
cannot attain the required transition state geometry without undue
strain. It is impossible for radical 11 to assume the required
conformation since the radical center is located within the ring
(endocyclic) rather than on an exocyclic carbon chain. As a
consequence, no S_Hi reaction from 11 is detected and only cyclic
peroxide product is isolated.

Little is known about the stereochemical requirements of S_H2
radical attack on first row elements. An S_N2-type transition
state has been proposed (16, 17) for radical attack on cyclopropane carbon and strong evidence has been presented that suggests
that halogen atom attack on cyclopropane carbon occurs with
stereochemical inversion. Our observation of alkyl radical attack
on the peroxide linkage suggests that backside displacement is the
rule here also, and it appears that this geometry may be
generally required in S_H2 reactions (18, 19).

Bicyclic Peroxides

Formation. Although bicyclo[2.2.1]endoperoxides are important intermediates in prostaglandin chemistry, they have been
synthetically elusive. Autoxidation of polyunsaturated fatty
acids does lead to mixtures of compounds containing the endoperoxide structure (20, 21). We (22) were also able to isolate
prostaglandin products from the sequence described below. The
lipid hydroperoxide (prepared from soybean lipoxygenase and the
fatty acid (23) was treated with a free radical initiator and the
products, following borohydride reduction, were compared to
authentic prostaglandins (24).

Any of these radical methods suffer from the fact that a serious mixture of diastereomers is formed and alternate routes to the bicyclo[2.2.1]endoperoxides have received considerable attention.

Recently, other methods of endoperoxide synthesis have been reported that make the parent 2,3-dioxabicyclo[2.2.1]heptane system, 12, readily available (25, 26, 27). For example, we have been able to prepare 12 from 3-bromocyclopentanehydroperoxide. The trans bromohydroperoxide is converted to 12 in 30 minutes by treatment with silver acetate. The cis bromohydroperoxide reacts much slower and leads to a mixture of products, including 12.

Decomposition of Endoperoxides. With 12 readily available, its chemistry relative to the prostaglandin endoperoxides can be explored. Salomon and collaborators (26) have reported on aspects of the thermal and base catalyzed decomposition of 12. We have chosen to initially concentrate on the reactions of the endoperoxide with Lewis acids such as Cu^{2+}.

Bartlett, Baumstark, and Landis (28) have shown that dioxetanes are subject to acid catalyzed decomposition (29), and we reasoned that 12 would also be subject to catalytic Cu^{2+} decomposition. In fact, the interaction of endoperoxides and Cu^{2+} has been of interest in prostaglandin chemistry. For example, when enzymatic prostaglandin biosynthesis was allowed to occur with Cu^{2+} or Cu^{2+} and thiol present in the biosynthetic reaction mixture, significantly more PGF (diol) products were obtained than when no Cu^{2+} was added to the reaction mixture (30). Thus, the

PGF (diol):PGE β-hydroxy cyclopentanone ratio obtained in prostaglandin biosynthesis increases from 2:3 with no added reagents to
1:1 with added Cu^{2+}. Addition of Cu^{2+} along with thiol increases
the PGF:PGE ratio even more dramatically to 20:1.

	none	Cu^{2+}	Cu^{2+}/RSH
$\dfrac{F}{E}$	$\dfrac{2}{3}$	$\dfrac{1}{1}$	$\dfrac{20}{1}$

        Treatment of 12 in EDTA distilled acetone with Cu^{2+} (7.1 x
10^{-4}M, 25°) leads to consumption of the endoperoxide at a rate of
8 x 10^{-4} min^{-1}. The only products detected are 3-hydroxycyclo-
pentanone and 2-cyclopentenone. 2-Cyclopentenone has been shown
to arise from the β-hydroxycyclopentanone under the conditions of
the reaction and work-up. 3.5 x 10^{-3}M Cu^{2+} leads to consumption
of 12 at a rate of 3.7 x 10^{-3} min^{-1} and this leads to a k_{cat} for
Cu^{2+} catalysis of 1.1 min^{-1} M^{-1}. The endoperoxide decomposes very
slowly at 25° in EDTA purified acetone with no added Cu^{2+}. If
the acetone

is not prepurified by EDTA distillation, then decomposition of the
endoperoxide in this solvent occurs at a significant rate, even
without added Cu^{2+}. This emphasizes the sensitivity of endoperoxides to transition metal impurities and suggests that extreme care
be exercised in pre-purification of solvents to be used in studies
of endoperoxide chemical reactivity.
        The finding that Cu^{2+} catalyzes conversion of 12 to β-hydroxy
cyclopentanone presents an apparent anomoly. Addition of Cu^{2+} to
the PG biosynthetic medium leads to more PGF (diol) products
whereas PGE-type products result from reaction of the endoperoxide
12 with Cu^{2+}. This inconsistency is clarified by consideration of
the fact that the enzymatic mixture contains endogenous thiol
that can react with the added Cu^{2+}, thus modifying the Cu^{2+}

reactivity (31). To test this possibility, we reacted Cu^{2+} with various thiols and then determined by assay with 9,10-dimethyl phenanthroline (DMP) that copper was present as Cu^{1+} (32). Cu^{1+} (as the chloride in methanol) rapidly reduces the endoperoxide 12 to cis-1,3-cyclopentanediol (PGF-type product). In fact, even the Cu^{1+} bis-9,10-dimethylphenanthroline complex reduces 12 slowly to the diol; two Cu^{1+} equivalents being required for every diol equivalent formed.

It thus seems likely that in the biosynthetic studies, the true reactivity of the endoperoxide toward Cu^{2+} is masked by the presence of thiol in the biosynthetic medium. Cu^{2+} acts on the endoperoxide to give PGE products (β-hydroxy cyclopentanone) but this reactivity is never fully recognized since endogenous or exogenous thiol reduces Cu^{2+} to Cu^{1+}. Cu^{1+} then acts to reduce the endoperoxide to PGF-type products.

This study of endoperoxide-Cu^{2+} reactivity demonstrates how the examination of simple endoperoxide model systems can help to clarify the chemistry of the naturally-occurring PGH and PGG molecules and it seems likely that more model studies of this type will provide additional information about the reactivity of prostaglandins, thromboxanes, and prostacyclin.

Acknowledgement

This paper is dedicated to P.D. Bartlett on the occasion of his 70th birthday. The authors acknowledge the initial work in this area carried out by Drs. Max Funk and Ramdas Issac. N.A.P. acknowledges support from an NIH Career Development Award, J.R.N. from an NIH Pharmacology Training Grant, and D.W.G. from the C.R. Hauser Memorial Fellowship. N.A.P. acknowledges discussions with Dr. L.J. Marnett some six years ago that initially interested us in the area of lipid oxidation. Support of the research by NIH and ARO is gratefully acknowledged.

Abstract

Two aspects of the chemistry of cyclic peroxides have been investigated. Alkyl radicals β to the peroxide linkage in cyclic peroxides attack the peroxide linkage in an S_Hi manner giving rise to epoxy-alcohols. The ease of this S_Hi reaction is dependent on the geometric arrangement of the attacking radical center and the peroxide bond. The preferred geometry appears to be an arrangement in which oxygen leaves by backside displacement from the attacking carbon radical. Bicyclic[2.2.1]endoperoxides have been shown to react with Cu^{2+} to yield β-hydroxy cyclopentanone by an acid catalyzed isomerization. Cu^{1+} reduces the endoperoxides to 1,3-cyclopentanediol, and it has been shown that thiol/Cu^{2+} mixtures act as does Cu^{1+} on the endoperoxide.

Literature Cited

1. Samuelsson, B., Granstrom, H., Hamberg, M., Proceedings of the Nobel Symposium, (1967), 2, 31-45.
2. Nugteren, D. H., Beerhuis, R. K., Van Dorp, D. A., ibid., (1967), 2, 45-50.
3. Bateman, R., Quarterly Review, (1954), 8, 164.
4. Van Sickle, D. E., Mayo, F. R., Arluck, R. M., Journal of Organic Chemistry, (1967), 32, 3680.
5. Anet, E. F., Australian Journal of Chemistry, (1969), 22, 2403.
6. Porter, N. A., Funk, M. O., Gilmore, D. W., Isaac, R., Nixon, J. R., Journal of the American Chemical Society, (1976), 98, 6000.
7. Funk, M. O., Isaac, R., Porter, N. A., ibid., (1975), 97, 1281.
8. Bartlett, P. D., Benzing, E. P., Pincock, R. E., ibid., (1960), 82, 1762.
9. Baldwin, J. E., Chemical Communications, (1976), 734.
10. Ingold, K. E., Roberts, B. P., "Free Radical Substitution Reactions," 180, Wiley-Intersciences, New York, London, 1970.
11. Van Sickle, D. E., Mayo, F. R., Gould, E. S., Arluck, R. M., Journal of the American Chemical Society, (1967), 89, 977.
12. Hill, C. L., Whitesides, G. M., ibid., (1974), 96, 870.
13. Jensen, F. R., Miller, J. J., Cristol, S. J., Beckley, R. J., Journal of Organic Chemistry, (1972), 37, 4341.
14. Quirk, R. P., Lea, R. E., Journal of the American Chemical Society, (1975), 98, 5973.
15. Bloodworth, A. J., Loveitt, M. E., Chemical Communications, (1976), 94.
16. Inceremona, J. H., Upton, C. J., Journal of the American Chemical Society, (1972), 94, 301.
17. Shea, K. J., Skell, P. S., ibid, (1973), 95, 6728.
18. Kampmeier, J. A., Jordan, R. B., Liu, M. S., Yamanaka, H., Bishop, D. J., "Abstracts of the 174th American Chemical Society Meetings," (1977), #069, Chicago.
19. Bentrude, W. C., ibid., #160.
20. Nugteren, D. H., Vonkeman, H., Van Dorp, D. A., Recueil, (1967), 86, 1237.

21. Pryor, W. A., Stanley, J. P., Journal of Organic Chemistry, (1975), 40, 3615.
22. Porter, N. A., Funk, M. O., ibid., (1975), 40, 3613.
23. Funk, M. O., Isaac, R., Porter, N. A., Lipids, (1976), 11, 113.
24. Porter, N. A., "Recent Advances in the Chemistry of Prostaglandins and Thromboxanes," Academic Press, New York, 1977.
25. Coughlin, D. J., Salomon, R. G., Journal of the American Chemical Society, (1977), 99, 655.
26. Salomon, R. G., Salomon, M. F., ibid., (1977), 99, 3501.
27. Porter, N. A., Gilmore, D. W., ibid., (1977), 99, 3505.
28. Bartlett, P. D., Baumstark, A. L., Landis, M. E., ibid., (1974), 96, 5557.
29. Bartlett, P. D., Chemical Society Reviews, (1976), 5, 149.
30. Lands, W., Lee, R., Smith, W., Annals of the New York Academy of Sciences, (1971), 180, 107.
31. Chan, J. A., Nagasawa, M., Takeguchi, C., Sih, C. J., Biochemistry, (1975), 14, 2987.
32. Davies, G., Loose, D., Inorganic Chemistry, (1976), 15, 694.

RECEIVED December 23, 1977.

7

Intramolecular Interactions of Acylperoxy Initiators and Free Radicals

THOMAS T. TIDWELL

Department of Chemistry, University of Toronto, Toronto, Ontario, Canada M5S 1A1

In a broad sense radical-induced decomposition may
be considered as the reaction of any substrate promoted
by a free radical, but the more common usage is to re-
strict consideration to those cases in which the molecule undergoing reaction is a free radical initiator
(1,2). Thus radical-induced decomposition may be defined as reaction of a free radical with the primary
source of the radical. This phenomenon is particularly noticeable when a free radical initiator does not
react with first-order kinetics, but is consumed by the
radicals it generates in chain processes. This case
was originally elucidated for the kinetics of benzoyl
peroxide decomposition in solvents such as ethyl ether
(3-5), and was found to occur with radical displacement
(the SH2 reaction) (6) by solvent derived radicals on
the peroxidic oxygen of benzoyl peroxide as shown in
Fig. 1.

Another classic study of benzoyl peroxide involved
the addition of solvent derived radicals to the para
position of the initiator. For example, decomposition
of benzoyl peroxide in cyclohexane led to the formation
of p-cyclohexylbenzoic acid (7). The absence of meta
product led to the suggestion (7,8) that para substitution was enhanced by α-lactone formation concerted with
addition. An alternative route would involve addition
followed by subsequent α-lactone formation (9) or
transannular hydrogen transfer (Fig. 2) (10). In this
case the absence of meta substitution product could be
ascribed to reversibility of the addition, and this is
known to occur (11). A preference for para substitution leading to formation of a radical intermediate
stabilized by an adjacent carboxylate function would
also be expected.

A related phenomenon is molecule induced homolysis.

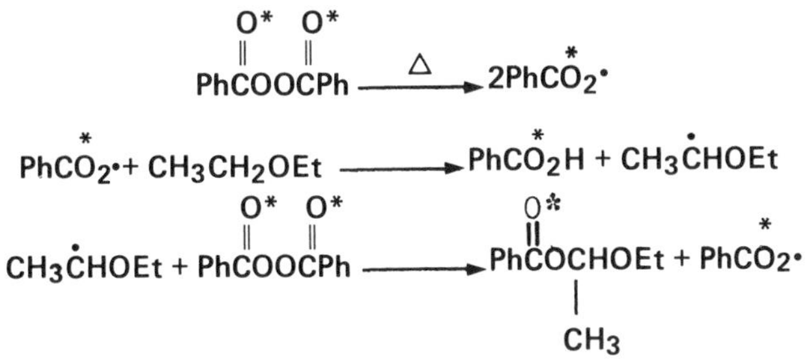

$$\underset{\substack{\| \quad \| \\ \text{PhCOOCPh}}}{\text{O* O*}} \xrightarrow{\Delta} 2\text{Phc} \overset{*}{\text{O}}_2\cdot$$

$$\text{Phc} \overset{*}{\text{O}}_2\cdot + \text{CH}_3\text{CH}_2\text{OEt} \longrightarrow \text{Phc}\overset{*}{\text{O}}_2\text{H} + \text{CH}_3\overset{\cdot}{\text{C}}\text{HOEt}$$

$$\underset{\substack{\| \quad \| \\ \text{CH}_3\overset{\cdot}{\text{C}}\text{HOEt} + \text{PhCOOCPh}}}{\text{O* O*}} \longrightarrow \underset{\substack{\| \\ \text{CH}_3}}{\overset{\text{O*}}{\text{PhCOCHOEt}}} + \text{Phc}\overset{*}{\text{O}}_2\cdot$$

Figure 1. *Radical-induced decomposition*

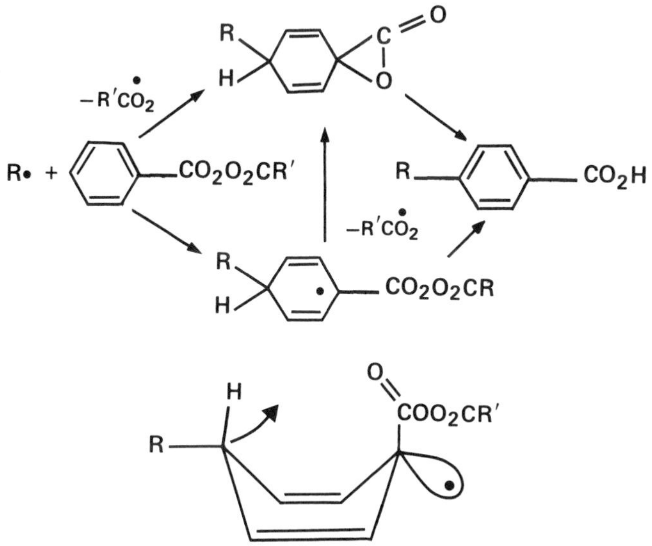

Figure 2. *Transannular hydrogen transfer*

This process, which has been recently reviewed (12),
was defined as "an interaction of two or more non-
radical species which results in the formation of free
radicals", but is not the subject of primary concern
here.

The present work deals with the intramolecular
interaction of free radicals with peroxide initiators.
Our original concept was to examine suitably constructed systems in which it was anticipated that the geometries of the interacting free radical and peroxidic
sites could be accurately defined so that the factors
which affect these interactions could be isolated.
Intramolecular attack of a free-radical on an atomic
center with displacement of a radical constitutes an
SHi process, a subclass of the SH2 reaction, or homolytic substitution (6). The emphasis in our work is on
such displacements on acylperoxy groups, but for a
broader understanding of the phenomena ordinary ester
groups are being examined as well.

The intramolecular reactivity of free radicals
with acylperoxy groups as a function of distance
between the two centers has been delineated by a number
of investigations. These radicals were generated by
homolysis, free radical addition, or atom abstraction
to form the desired system. Radical sites α to
acylperoxy groups lead to α-lactone formation as the
characteristic reaction (eq. 1), whereas β-radicals
lead to elimination (eq. 2-6).

An effort was made to examine the stereochemistry
of β-elimination to give a stilbene system but this
was thwarted since the first step in the reaction in
eq. 7 did not occur to an appreciable extent (20).
However the stereochemistry of formation of 2-butene
from pyrolysis of the isomeric cyclic   dl   and
meso -2,3-dimethylsuccinyl peroxides has been accomplished and found to be the same from either source
(eq. 8) (21). The reaction in equation 2 was stereospecific as shown, and the resulting E-product is that
expected from an anti-elimination from the preferred
conformation of the starting material.

Radicals in γ or δ positions relative to acylperoxy groups also lead to lactone formation (eq. 9-12).
In these lactone-forming reactions homolytic attack
could occur either at carbonyl or peroxidic oxygen
(eq. 13) but apparently this has yet to be determined
for the examples noted in eq. 9-12.

$$t-BuO_3CCMe_2CO_3\underline{t}-Bu \longrightarrow Me_2\overset{\bullet}{C}CO_3\underline{t}-Bu \qquad (1) \quad ^{13,14}$$

$$\downarrow -\underline{t}-Bu\overset{\bullet}{O}$$

Me₂C with a three-membered ring containing C=O and O

$$t-BuO_3CCHPhCHPhCO_3\underline{t}-Bu \longrightarrow Ph\overset{\bullet}{C}HCHPhCO_3\underline{t}-Bu \qquad (2) \quad ^{15}$$

$$\underline{meso}$$

$$\downarrow -\underline{t}-Bu\overset{\bullet}{O},CO_2$$

$$PhCH=CHPh\ (\underline{E})$$

$$\overset{\bullet}{C}H_3 + CH_2=CHCH_2CO_3\underline{t}-Bu \longrightarrow MeCH_2\overset{\bullet}{C}HCH_2CO_3\underline{t}-Bu \qquad (3) \quad ^{16}$$

$$\downarrow -\underline{t}-Bu\overset{\bullet}{O},\ CO_2$$

$$MeCH_2CH=CH_2$$

$$PhC\equiv CCO_3\underline{t}-Bu + R\bullet \longrightarrow Ph\overset{\bullet}{C}=CRCO_3\underline{t}-Bu \qquad (4) \quad ^{17}$$

$$\downarrow -\underline{t}-Bu\overset{\bullet}{O}$$

$$R = PhCH_2 \qquad\qquad PhC\equiv CR$$

$$PhCH=CHCO_3\underline{t}-Bu + R\bullet \longrightarrow Ph\overset{\bullet}{C}HCHRCO_3\underline{t}-Bu \qquad (5) \quad ^{18}$$

$$\downarrow -\underline{t}-BuO^{\bullet},\ CO_2$$

$$PhCH=CHR$$

$$E-\underline{t}-BuO_3CCH=CHCO_3\underline{t}-Bu \xrightarrow{R\bullet} \underline{t}-BuO_3C\overset{\bullet}{C}HCHRCO_3\underline{t}-Bu \qquad (6) \quad ^{19}$$

$$\downarrow$$

$$CH\equiv CH$$

$$\downarrow$$

$$RCH=CHR + CH_2=CHR \longleftarrow - - - - - t-BuO_3CCH=CHR$$

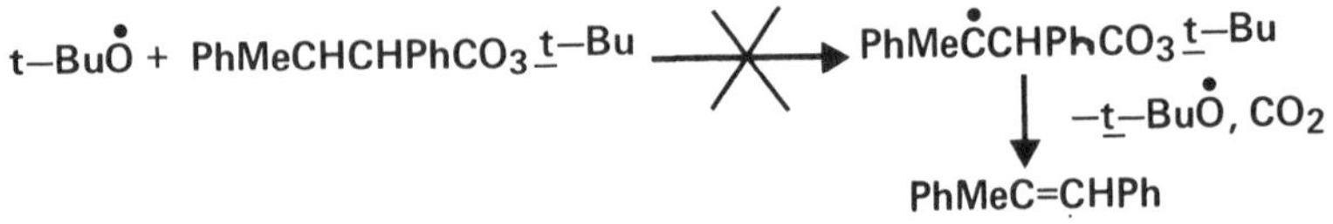

t-BuȮ + PhMeCHCHPhCO₃ t-Bu ──✗──► PhMeĊCHPhCO₃ t-Bu   (7) [20]

$$\downarrow -\underline{t}-Bu\dot{O},\ CO_2$$

PhMeC=CHPh

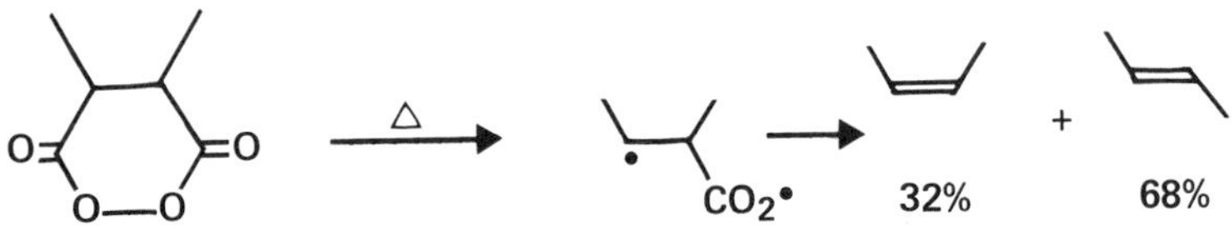

(8) [21]

32%   68%

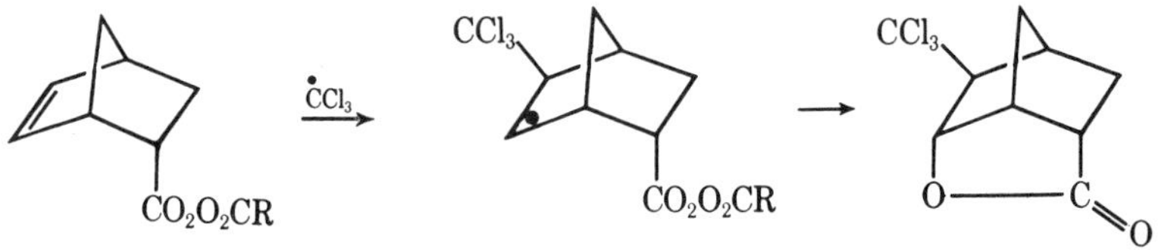

(9) [22]

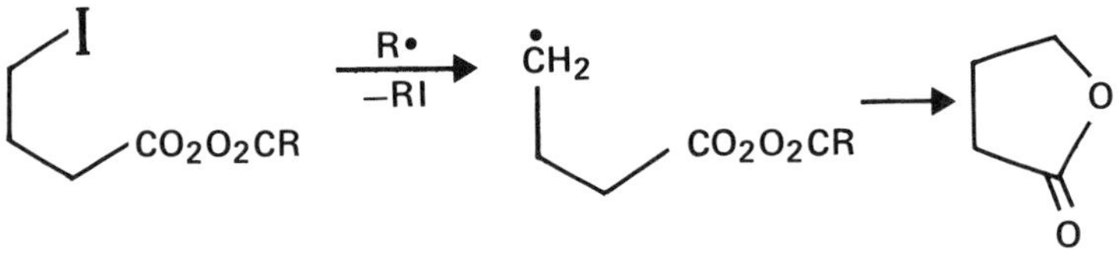

(10) [23]

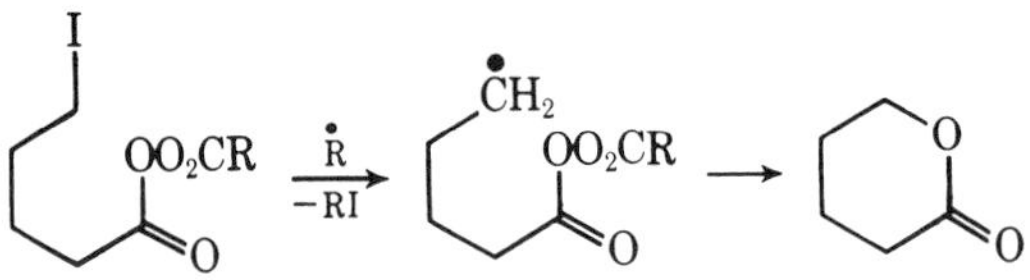

(11) [24]

(12) [25]

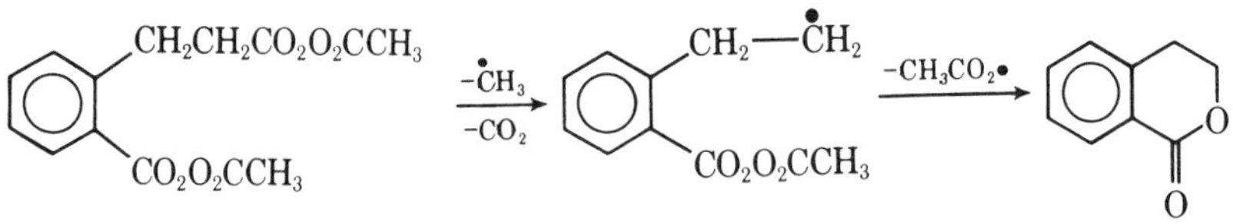

$$
\text{(structures)} \qquad (13)
$$

Some related lactone forming reactions have been
observed in the reactions of acylperoxy alkenes in
which the double bond participates in the homolysis
(eq. 14,15). In the first case (eq. 14), the position
of bond scission has been elucidated and found to occur
at the peroxidic oxygen (27), in agreement with the
results for benzoyl peroxide and tert-butyl perbenzoate
which are attacked on peroxidic oxygen by a variety of
radicals (5,30-32).

Some more compliated cases of reactions of bis-
peroxides have also been reported (eq. 16,17). It was
proposed (34) on the basis of a large negative $\Delta S^{\ddagger}$ that
the reaction in eq. 17 proceeds by the 3-bond path
indicated.

In our own work we were interested in the possibility that radicals delocalized in π-systems might interact with adjacent peroxy groups, for example in the
systems 1 (35) and 2 (1). Induced decomposition paths
can be visualized for both of these peresters; in the

$$
\overset{\bullet}{C}H_2 \!-\!\!\bigcirc\!\!-\! CH_2 CO_3 \underline{t}\text{-Bu} \qquad\qquad \overset{\bullet}{C}H_2 \!-\!\!\bigcirc\!\!-\! CO_3 \underline{t}\text{-Bu}
$$

$$
\underset{\sim}{1} \qquad\qquad\qquad\qquad\qquad\qquad \underset{\sim}{2}
$$

case of 1 the structure is a phenylene analog of a
perester with a β-radical site (eq. 2,3,5) and decomposition to p-xylylene (3) could be contemplated.
Similarly 2 is a phenylene analog of a perester with an
α-radical site and decomposition to an α-lactone (4)
could be contemplated. Radicals 1 and 2 were generated
from the corresponding bisperesters 5 and 6. The

$$
CH_2\!=\!\!\bigcirc\!\!=\!CH_2 \qquad\qquad\qquad CH_2\!=\!\!\bigcirc\!\!\overset{\displaystyle C^{\nearrow O}}{\underset{\displaystyle O}{\diagdown}}
$$

$$
\underset{\sim}{3} \qquad\qquad\qquad\qquad\qquad\qquad \underset{\sim}{4}
$$

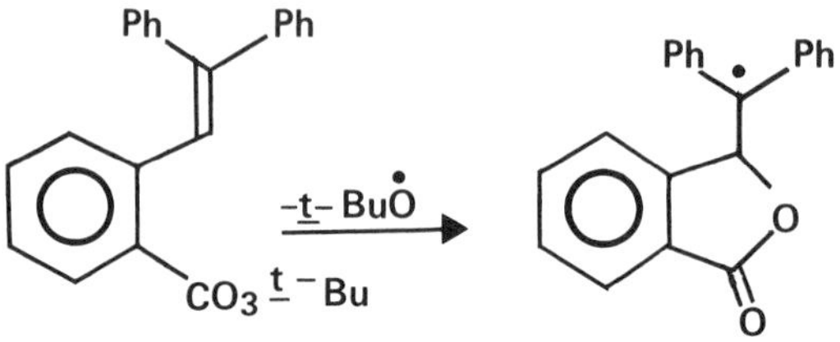
Ph
Ph
CO3 t-Bu
-t-BuO•
Ph
Ph
O
O

(14)[26]

(15)[28,29]

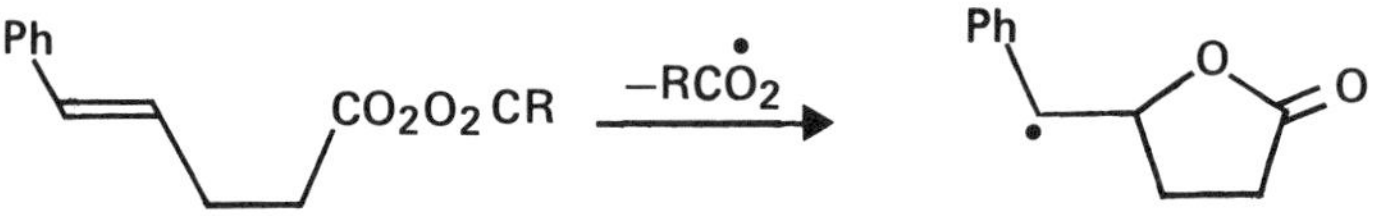
Ph
CO2O2CR
−RCO2•
Ph
O
O

(16)[33]

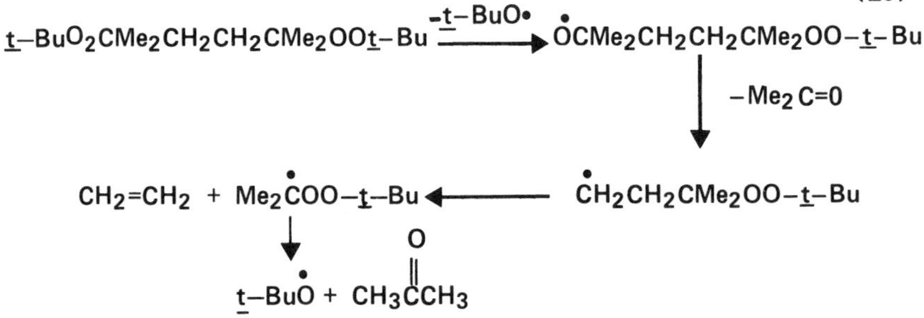
t–BuO2CMe2CH2CH2CMe2OOt–Bu
-t-BuO•
•OCMe2CH2CH2CMe2OO–t–Bu
–Me2C=0
•CH2CH2CMe2OO–t–Bu
CH2=CH2 + Me2•COO–t–Bu
t–BuO• + CH3C(=O)CH3

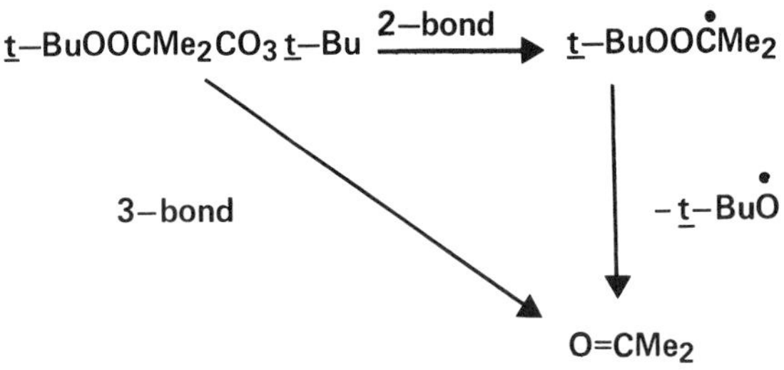
t–BuOOCMe2CO3t–Bu
2–bond
t–BuOO•CMe2
3–bond
–t–BuO•
O=CMe2

(17)[34]

kinetics of decomposition were normal for substituted
phenylperacetates (Table 1). This result showed that
the peresters underwent normal 2-bond cleavage in the
rate determining step with formation of the radicals 1
and 2 as the initial intermediate and excluded the ˜
possibility of synchronous cleavage of more than two
bonds giving the intermediates 3 and 4 directly. However the kinetic studies do not reveal the fate of the
radicals 1 and 2 after they have been formed in the
rate determining step.

Table I. Kinetics of decomposition of substituted
tert-butyl phenylperacetates at 80.0°C in
cumene

Substrate	$k(s^{-1}$ $\times 10^5)$	$\Delta H^{\ddagger}$ (kcal mol^{-1})	$\Delta S^{\ddagger}$ eu
$\underline{t}$-BuO$_3$CCH$_2$⟨◯⟩CH$_2$CO$_3\underline{t}$-Bu $\underset{\sim}{5}$ 35	10.3	26.9	-1.0
$\underline{t}$-BuO$_3$C⟨◯⟩CH$_2$CO$_3\underline{t}$-Bu $\underset{\sim}{6}$ 1	2.30^a	29.2	3.1
⟨◯⟩CH$_2$CO$_3\underline{t}$-Bu $\underset{\sim}{7}$ 35	7.21	27.7	0.9
CH$_2$=CH⟨◯⟩CH$_2$CO$_3\underline{t}$-Bu $\underset{\sim}{8}$ 35	16.2	27.8	2.5
Copolymer of 8 and styrene 35 $\underset{\sim}{}$	23.4	27.2	1.7

^{a}in toluene

Product studies were indicative as to the reaction
pathways of 1 and 2. In the case of 2 the formation of
4 could be excluded as a significant reaction pathway,
as the observed products were all derived from abstraction or radical coupling reactions of 2 (1). The perbenzoate group is thermally stable at the temperature at
which 6 was decomposed and the products indicated that

the peroxide group in 2 remained intact to form perbenzoate containing products. These were analyzed by
hydrogenolysis of the peroxide group and separation of
the derived products.

In the case of 1 products containing perester
groups could not be isolated, and in any event any
peresters derived from 1 by hydrogen abstraction or
radical coupling would be expected to be at least as
reactive as their precursor 5. However the product
distribution was quantitatively what would be expected
from reaction of 1 without formation of 3, and no products uniquely ascribable to 3 were observed. Perhaps
more convincingly radical 9, analogous to 1, could be
formed at 25°C and did not undergo decomposition in the

$$RCH_2\overset{\bullet}{C}H-\!\!\bigcirc\!\!-CH_2CO_3\underline{t}-Bu \qquad\qquad 9$$

time required for reaction with styrene. Thus an argument can be made that 1 is also stable at 70-100°C,
although the evidence is indirect. Using the rate
constants for benzyl radical combination and estimated
diffusion constants for the radicals, average lifetimes
could be estimated as 8×10^{-4}s for 1 at 100°C and 6×10^{-3}
s at 85°C for 2. This is the time which would be re-
quired before dimerization or coupling with solvent
derived radicals occurred.

Once it had been established that 1 was probably
stable in solution, and that 2 was definitely relatively stable it appeared that it would be possible to use
the monomers 8 and 10 in free radical polymerizations
to form polystyrenes with pendant phenylperacetate and

$$CH_2=CH-\!\!\bigcirc\!\!-CH_2CO_3\underline{t}-Bu \qquad\qquad CH_2=CH-\!\!\bigcirc\!\!-CO_3\underline{t}-Bu$$

$$8 \qquad\qquad\qquad\qquad\qquad 10$$

perbenzoate groups. This indeed proved to be the case
(36,37). Both 8 and 10 could be polymerized at 25°C
with initiation by the low temperature initiator di-
tert-butyl peroxyoxalate(DBPO) to give homopolymers
with apparent quantitative retention of the peroxide
functionality, or copolymerized with styrene by the
same initiator. The resulting polymers were then
heated to temperatures at which the pendant peroxide
functions decomposed to give highly insoluble cross-

linked materials. If other monomers such as acryloni-
trile were present graft-polymers were formed (Fig. 3).

Other examples of polymers containing peroxidic
groups are noted in ref. 36. A recent example containing 2 kinds of perester groups and a vinyl group in
shown in figure 4 (38).

Radical 2 (39) and the analogous radical-diacyl
peroxide (39) have also been prepared by abstraction of
benzylic hydrogen and found to survive long enough to
abstract halogen. However under these conditions the
lifetimes of the radical species could not be estimated. The phenolic perester 11 underwent a radical-

induced decomposition on abstraction of the phenolic
hydroxyl, but interestingly the ortho isomer 12 was
resistant to induced decomposition (40,41).

We have also studied the acylic bisperester 13
(42). Qualitative rate data indicate that 13 is about

10 times more reactive than trans-MeCH=CHCH$_2$CO$_3$t-Bu
(14) (43). Inasmuch as 14 gives a primary allylic
radical in the rate determining step whereas 13 could
give a tertiary one the rate difference is not excess-
ive and 13 may be presumed to have the same rate determining step as 14, namely two-bond scission to give an
allylic radical (eq. 18). No lactone was formed from
13. This may seem surprising because other γ-percar-

boxy radicals formed lactones (eq. 9,10). Ring opening

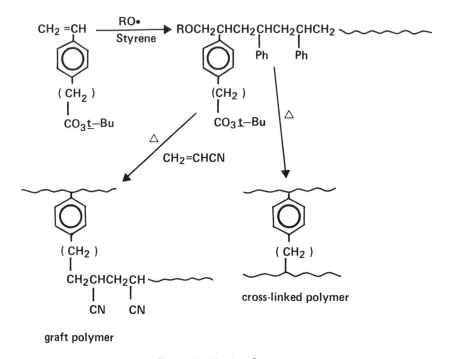

Figure 3.   Graft-polymers

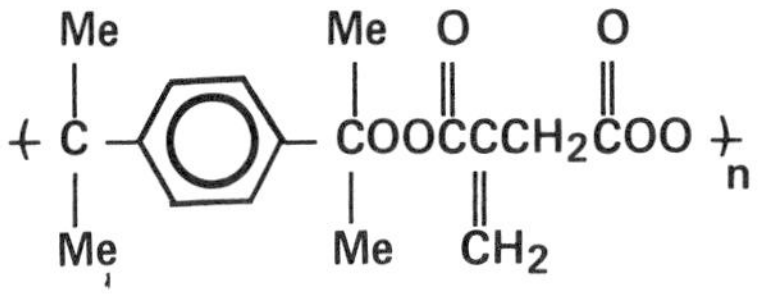

Figure 4.   Polymer containing 2 types
of perester groups

and lactone formation also occurred in the reaction of
trans-di-tert-butylperoxy 3,3-diphenylcyclopropane-1,
2-dicarboxylate (16) as shown in eq. 19 (44). The
yield of 17 was 61% of the isolated material derived
from the carbon skeleton of 16; the remainder was 1,1-
diphenylcyclopropane. Thus lactone formation was quite
efficient in this case, in contrast to 15. However it
is to be expected that the cyclopropyl radical from 16
will be pyramidal (45), and will have a barrier to
inversion (45), so that initially it will have the
structure shown. In this geometry the preferred
direction of ring opening would be to give the cis-
allylic radical, as this gives maximum overlap during
the ring opening (eq. 19). This geometry is favorable
for lactone formation, whereas 15 would be formed in a
trans-geometry unfavorable for lactone formation. A
barrier of 21 kcal/mol for interconversion of allyl
radicals has been reported (46). Lactone formation has
also been observed in a related system (eq. 20) which
does not suffer from geometrical restraints (47).

ortho-Substituted aryl peroxides should be a
source of some interesting interactions between free
radical centers and peroxides. Such substrates have
fixed geometries with the radical site constrained to a
relatively close proximity to the peroxide group. One
example of such a system has already been cited (eq.
12). Another we have examined is o-di-tert-butyl-
diperphthalate (18) (48). This compound had a reactivity relative to tert-butyl perbenzoate of 0.9 and

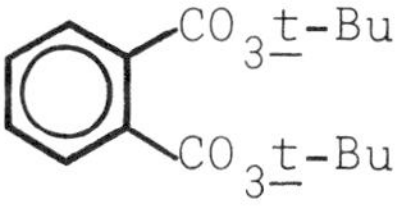

$$\text{(benzene ring with } CO_3\text{t-Bu and } CO_3\text{t-Bu substituents)} \qquad 18$$

similar activation parameters so with confidence may
be assumed to undergo initial one-bond scission to
radical 19. The other observed products from 18 can be
ascribed to arise from 20 and 21, derived from 19 by
hydrogen abstraction, and decarboxylation, respectively.
Two possible induced decomposition pathways of 19 (eq.
21) are not required by the evidence. Phthalic anhy-

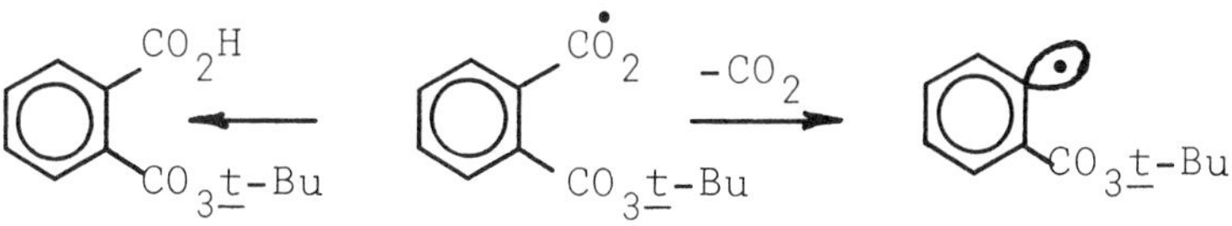

$$\underset{20}{\text{(benzene ring with } CO_2H, CO_3\text{t-Bu)}} \longleftarrow \underset{19}{\text{(benzene ring with } \overset{\bullet}{C}O_2, CO_3\text{t-Bu)}} \overset{-CO_2}{\longrightarrow} \underset{21}{\text{(benzene ring with } CO_3\text{t-Bu)}}$$

dride is formed and there have been claims that this

(19)

(20)

$$\text{(structure)} \xleftarrow{-\underline{t}\text{-BuO}\overset{\bullet}{O}} \underset{\underset{\sim}{1}\underset{\sim}{9}}{} \xrightarrow{-\underline{t}\text{-Bu}\overset{\bullet}{O}} \text{(structure)} \quad (21)$$

occurs by attack on carbon by a carboxylate radical to
give anhydride with expulsion of a peroxy radical (25,
49). However, we found that phthalic anhydride was
formed from 20 under the reaction conditions by a presumably non-radical pathway. Thus the unconventional
SHi displacement of a peroxy radical is not required
to explain the formation of anhydride. Another conceivable SHi reaction of 19 involves displacement of a
tert-butyoxy radical and formation of phthalyl peroxide
(eq. 21) but this product would also be reactive under
the reaction conditions so its presence can neither be
confirmed nor excluded.

   Compound 22, which is an interesting analog of 18,
has recently been reported (50). However the rate of
reaction of 22 in chlorobenzene is reported as 2.8

$$\text{(structure)}\begin{array}{l}\text{CO}_3\underline{t}\text{-Bu}\\\text{CO}_2\text{CH}_2\text{OO}\underline{t}\text{-Bu}\end{array} \qquad \underset{\sim\sim}{22}$$

times that of 18 (in cumene) (48) at 115°, indicating
that 22 reacts by the same rate-determining step as 18,
namely initial l-bond scission to give a carboxylate
radical analogous to 19.

   There has also been a recent report of radical
displacement of an ordinary carboxylate ester (49).
Numerous examples of such reactions involving percarbo-
xylate functions have already been mentioned (cf. Fig.
1, eqs. 1, 9-12) but in the cases where the position
of bond cleavage has been determined all have involved
attack on peroxidic oxygen. The example reported for
an ester would have to involve cleavage of the O-C
bond   (49).

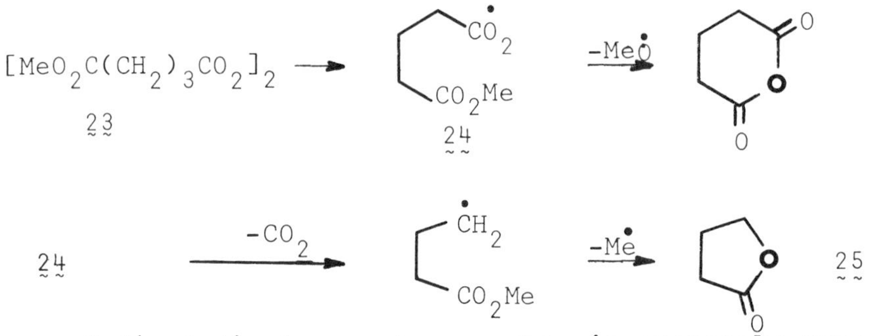
[MeO2C(CH2)3CO2]2
23
CO2
CO2Me
24
-MeO
24
-CO2
CH2
CO2Me
-Me
25

Radical displacement on carbon is extremely rare, excluding examples in which a strained ring is cleaved. One case that has been claimed involves the formation of cyclopropane from the 3-iodo-1-propyl radical (eq. 22). The mechanism in eq. 22 has been criticized (52) on the basis of the high strain generated in the

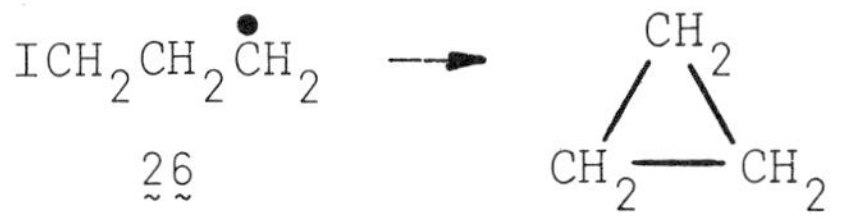
ICH2CH2CH2
26
CH2
CH2
CH2

(22)[51]

reaction, and diradicals were proposed as a more likely source of the cyclopropane (52). The fact that radical 26 would combine would other radicals at near the diffusion controlled rate was felt to eliminate this possibility (51). Radical displacement on hetereo atoms are much more common. Some recent examples involving sulfur include spectroscopic observation of sulfurane 27, presumably formed by SHi attack of sulfur on the peroxidic oxygen (eq. 23) (53), and study of the electron transfer mechanism for some intermolecular

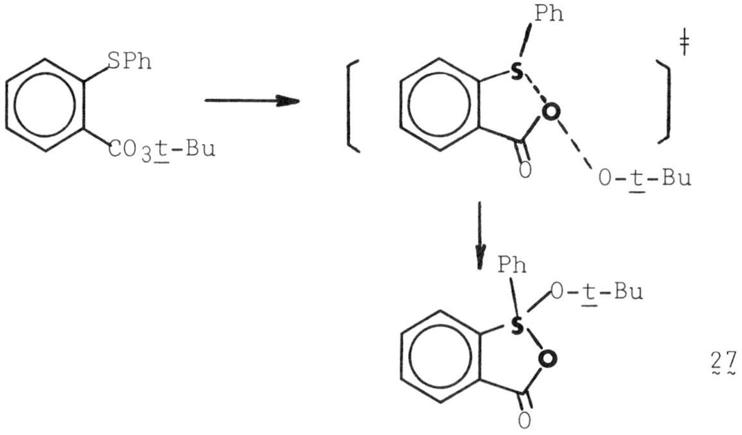
SPh
CO3t-Bu
Ph
S
O
O-t-Bu
‡
Ph
O-t-Bu
27

(23)

reactions of sulfides and peresters ($\underline{54}$).

In the example mentioned above ($\underline{49}$) butyrolactone ($\underline{25}$) was claimed to be isolated to the extent of 35% (mol/mol) from $\underline{23}$ in HOAc. In order to further examine this phenomenon we have studied the decomposition of the peresters $\underline{28}$ and $\underline{29}$ ($\underline{55}$). The kinetic parameters for $\underline{28}$

$$\underline{28} \qquad\qquad \underline{29}$$

(Table II) show the decomposition rate of the perester

Table II. Rates of decomposition of peresters $RCO_3t\text{-Bu}$
    in cumene, extrapolated to 25°C

Perester	k_{rel}	$\Delta H^{\ddagger}$(kcal/mol)	$\Delta S^{\ddagger}$(eu)
$MeCO_3t\text{-Bu}$	0.5	36.9	17.2[a]
$EtCO_3t\text{-Bu}$	3.0	33.3	8.7[b]
$\underline{28}$	1.0	36.1	16.1

[a] In ethylbenzene, ref. $\underline{56}$ [b] Ref. $\underline{57}$

is normal and not affected by the carbomethyoxy group beyond a possible slight inductive retardation. Similarly, the rate of $\underline{29}$ is half that of $PhCH_2CO_3t\text{-Bu}$ ($\underline{7}$), showing that $\underline{29}$ and $\underline{7}$ must react by the same rate-determining step with only a slight inductive role due to the carbomethoxy group.

The products from $\underline{28}$ and $\underline{29}$ both showed the formation of lactones (eq. 24,25). The yield of lactone from $\underline{28}$ decreased with the hydrogen atom donating abil-

$$\underline{28} \xrightarrow{99°} \qquad\qquad\qquad\qquad (24)$$

$$\underline{29} \longrightarrow \qquad\qquad\qquad\qquad (25)$$

ity of the solvent, but in our hands was much less than
that reported (49) for reaction in acetic acid solvent.
The greater yield of lactone from 29 may be attributed
to the fixed geometry of this intermediate which would
encourage cyclization.

Thus SHi reaction is confirmed for 28 and 29. The
question remains as to the mechanism of this process.
Two possibilities to be considered are shown in eq. 26.

$$(26)$$

Formation of an intermediate by attack of carbonyl
oxygen was favored by other authors (49). These
routes should be distinguishable by O-18 labeling and
such experiments are now underway.

Several analogies for this reaction are known. In
an acylic case α-phenethyl benzoate reacts with tri-n-
butyltin hydride by a free radical process (58), but it
has not been demonstrated which oxygen is attacked by
the radical (eq. 27). Radicals are known to attack
intramolecularly on carbonyl oxygen in ketones (eq. 28)
and esters (eq. 29), and attack on carbonyl carbon is
even known (eq. 30).

$$n\text{-Bu}_3\text{Sn}^\bullet + \text{PhCO}_2\text{CHMePh} \longrightarrow \text{PhCO}_2\text{Sn-}n\text{-Bu} + \text{Ph}\overset{\bullet}{\text{C}}\text{HMe} \quad (27)^{58}$$

$$(28)^{59}$$

$$(29)^{60}$$

$$R\cdot \ + \ CH_3\overset{O}{\underset{}{C}}-\overset{O}{\underset{}{C}}CH_3 \ \rightleftharpoons \ CH_3\overset{O}{\underset{R}{C}}-\overset{O}{\underset{}{C}}CH_3 \ \rightleftharpoons \ CH_3\overset{O}{\underset{}{C}}R \ + \ CH_3\overset{\cdot}{C}O \ (30)^{61}$$

In summary the presence of an acylperoxy group and a free radical center in an intermediate may lead to intramolcular induced decomposition by one of several routes, or induced decomposition may not occur. Many interesting examples in which such processes may occur still await study.

Acknowledgements: Financial support by the U.S. Army Research Office-Durham, the Defence Research Board of Canada, and the National Research Council of Canada is gratefully acknowledged. Special thanks is due to my collaborators cited in the footnotes.

Literature Cited

(1) Dalton, A.I., Tidwell, T.T., J. Org. Chem., (1972) 37, 1504.
(2) Suehiro, F., Kaguku Kogaku, (1969) 27, 701.
(3) Nozaki, K., Bartlett, P.D., J. Am. Chem. Soc., (1946) 68, 1686; ibid., (1947) 69, 2299.
(4) Cass, W.E., J. Am. Chem. Soc., (1946) 68, 1976; ibid., (1947) 69, 500.
(5) Denney, D.B., Feig, G., J. Am. Chem. Soc., (1959) 81, 5322.
(6) Ingold, K.U., Roberts, B.P., "Free-Radical Substitution Reactions", Wiley, New York, N.Y. 1971.
(7) Walling, C., Savas, E.S., J. Am. Chem. Soc., (1960) 82, 1738.
(8) Walling, C., Čekovič, Ž., J. Am. Chem. Soc., (1967) 89, 6681.
(9) Cadogan, J.I.G., Hey, D.H., Hibbert, P.G., J. Chem. Soc., (1965) 3939.
(10) DeTar, D.F., Weis, C., J. Am. Chem. Soc., (1957) 78, 4296.
(11) Saltiel, J., Curtis, H.C., J. Am. Chem. Soc., (1971) 93, 2056.
(12) Harmony, J.A.K., Methods in Free Radical Chem. (1974) 5, 101.
(13) Bartlett, P.D., Gortler, L.B., J. Am. Chem. Soc., (1963) 85, 1864.
(14) Gortler, L.B., Saltzman, M.D., J. Org. Chem., (1966) 31, 3821.
(15) Bobroff, L.M., Gortler, L.B., Sahn, D.J., Wiland, H., J. Org. Chem., (1966) 31, 2678.
(16) Ochiai, T.. Yoshida, M., Simamura, O., Bull. Chem. Soc. Jpn., (1976) 49, 2641.

(17) Muramoto, N., Ochiai, T., Yoshida, M.,
Simamura, O., Bull. Chem. Soc. Jpn., (1976) 49, 2518.
(18) Ochiai, T., Usuda, Y., Yoshida, M., Simamura,
O., Bull. Chem. Soc. Jpn., (1976) 49, 2522.
(19) Fedorova, V.A., Darmograi, M.I., Chuchmarev,
S.K., Ukr. Khim. Zh., (1977) 43, 397.
(20) Ochiai, T., Yoshida, M. and Simamura, O., Bull.
Chem. Soc. Jpn., (1976) 49, 2525.
(21) Jones, C.R.; Dervan, P.B.; J. Am. Chem. Soc.,
(1977), 99, 6772.

(22) Hart, H., Chloupek, F.J. J. Am. Chem. Soc.,
(1963) 85, 1155.
(23) Blomstrom, D.C., Herbig, K., Simmons, H.E.,
J. Org. Chem., (1965) 30, 959.
(24) Woolford, R.G., Gedye, R.N. Can. J. Chem.
(1967) 45, 291.
(25) Bylina, G.S., Matveentseva, M.S., Ol'dekop
Yu. A., Zh. Org. Khim., (1975) 11, 2237; Chem. Abstr.
(1976) 84, 43011f.
(26) Koenig, T.W., Martin, J.C. J. Org. Chem.,
(1964) 29, 1520.
(27) Martin, J.C., Koenig, T.W., J. Am. Chem. Soc.
(1964) 86, 1771.
(28) Lamb, R.C., Rogers, F.F. Jr., Dean, G.D., Jr.,
Voight, F.W. Jr., J. Am. Chem. Soc., (1962) 84, 2635.
(29) Lamb, R.C., Spadafino, L.P., Webb, R.G.,
Smith, E.B., McNew, W.E., Pacifici, J.G. J. Org. Chem.
(1966) 31, 147.
(30) Drew, E.H., Martin, J.C., Chem. Ind. (London)
(1959) 925.
(31) Doering, W. von E., Okomoto, K., Krauch, H.
J. Am. Chem. Soc. (1960) 82, 3579.
(32) Rotenberg, K., Neumann, W.P., Avar, G.,
Chem. Ber., (1977), 110, 1628.
(33) Tang, F., Huyser, E.S., J. Org. Chem. (1977)
42, 2160.
(34) Richardson, W.H., Koskinen, W.C. J. Org. Chem.
(1976) 41, 3182.
(35) Lai, L.-F., Tidwell, T.T. J. Am. Chem. Soc.
(1977), 99, 1465.
(36) Dalton, A.I., Tidwell, T.T., J. Polym. Sci.
(1974) 12, 2957.
(37) Lai, L.-F., Tidwell, T.T., unpublished results.
(38) Galibei, V.I., Arkhipova-Kalenchko, E.G.
Zh. Org. Kh., (1977) 13, 227.
(39) Schwartz, M.M., Leffler, J.E. J. Am. Chem.
Soc., (1971) 93, 919.

(40) Huček, A.M., Barbas, J.T., Leffler, J.E.,
J. Am. Chem. Soc., 95, (1973) 4698.
  (41) Barbas, J.T., Leffler, J.E. J. Am. Chem.
Soc., (1975) 97, 7270.
  (42) Dalton, A.I., Tidwell, T.T. unpublished
results; Dalton, A.I., Ph.D. Thesis, University of
South Carolina, (1972).
  (43) Martin, M.M., Sanders, E.B. J. Am. Chem. Soc.
(1967) 89, 3777.
  (44) Swigert, R.D. Ph.D. Thesis, Harvard
University (1964).
  (45) Walborsky, H.M., Collins, P.C., J. Org.
Chem. (1976) 41, 940.
  (46) Crawford, R.J., Hamelin, J., Strehlke, B.,
J. Am. Chem. Soc., (1971) 93, 3810.
  (47) Walborsky, H.M., Chen, J.-C. J. Am. Chem.
Soc., 92, (1970) 7573.
  (48) Cable, D.A., Ernst, J.A., Tidwell, T.T.,
J. Org. Chem. (1972) 37, 3420.
  (49) Nikishin, G.I., Starostin, E.K. and Golovin,
B.A. Izv. Akad. Nauk. SSSR, (1973) 825.
  (50) Kuznetsov, V.I., Ivanchev, S.S. Zh. Org.
Khim. (1977) 13, 718.
  (51) Drury, R.F., Kaplan, L. J. Am. Chem. Soc.
(1973) 95, 2217.
  (52) Ref. 6, pp. 82-83.
  (53) Livant, P., and Martin, J.C. J. Am. Chem.
Soc. (1976) 98, 7851.
  (54) Pryor, W.A., Hendrickson, W.H. J. Am. Chem.
Soc. (1975) 97, 1580, 1582.
  (55) Rynard, C., Thankachan, C., Tidwell, T.T.
unpublished results.
  (56) Rüchardt, C., Bock, H., Chem. Ber. (1971)
104, 577.
  (57) Pryor, W.A., Smith, K., Int. J. Chem. Kinet.
(1971) 3, 387.
  (58) Khoo, L.E., Lee, H.H., Tetrahedron Lett.
(1968) 4351.
  (59) Menapace, L.W., Kuivila, H.G., J. Am. Chem.
Soc. (1964), 86, 3047.
  (60) Tanner, D.D., Law, F.C.P. J. Am. Chem. Soc.
(1969), 91, 7535.
  (61) Bentrude, W.G., Darnall, K.R. J. Am. Chem.
Soc. (1968) 90, 3588.

RECEIVED December 23, 1977.

8

Decomposition of Diazenes and Diazene *N*-Oxides. Consideration of Three-Electron Stabilization.

FREDERICK D. GREENE, JAMES D. BURRINGTON,
and ABRAHAM M. KARKOWSKY

Department of Chemistry, Massachusetts Institute of Technology,
Cambridge, MA 02139

The carbon-nitrogen bond is a strong link of average bond energy 75 kcal/mole. There are, of course, many examples in which homolytic cleavage of this bond may be effected under relatively mild conditions. Diazenes (azo compounds) comprise a well-documented example (1). Decomposition of diazenes affords dinitrogen and radicals. The question of whether the rate-

$$Ra-N{\diagdown}_{N-R_b} \longrightarrow Ra\cdot \quad N_2 \quad \cdot R_b \qquad (eq.\ 1)$$

$$\longrightarrow Ra-N_2\cdot \qquad \cdot R_b \qquad (eq.\ 2)$$

determining step involves synchronous two-bond cleavage (eq. 1) or one-bond cleavage (eq. 2) has received a great deal of study (1, 2). For alkyldiazenes (including aryl-substituted alkyl) the evidence strongly favors cleavage of both carbon-nitrogen bonds in the rate-determining step. A convincing line of evidence is the dependence of rate on substituent [e.g. Table 1 (3)].

Table 1
Rate of Decomposition of Diazenes, $Ra-N{\equiv}N-R_b$
rel k, predicted, 200°C

Ra	Rb	Synchron (eq. 1)	Stepwise (eq. 2)	rel k, Observed
1-Norbornyl	1-Norbornyl			1
1-Norbornyl	t-Butyl			6.6×10^4
t-Butyl	t-Butyl	$(6.6\times10^4)^2$	$2(6.6\times10^4)$	7×10^7

Indeed, these findings might appear to be a logical consequence of a consideration of the energetics of the reaction. The enthalpy change, $\Delta H°$, for eq. 1 is approximately +25 kcal/mole for R = alkyl, e.g. t-butyl (4); the enthalpy of activation is +42 kcal/mole (5). In comparison with the average bond energy of 75 kcal/mole for the carbon-nitrogen bond, the considerably lower activation energy (+42) can be achieved by some synchronous

cleavage of both carbon-nitrogen bonds with the attendant recoup-
ing of the energy cost by partial formation of dinitrogen. Rate-
determining cleavage of just one of the carbon-nitrogen bonds
could be considered only if the resulting alkyl radical and diazenyl radical ($RN_2\cdot$) were stabilized to the extent of 30 kcal/
mole (the average bond energy minus the observed enthalpy of
activation). A tertiary alkyl radical is stabilized to the extent
of only a few kcal/mole; a diazenyl radical usually has been considered to possess no stabilization (6).

With highly unsymmetrical diazenes, and particularly when Ra
is phenyl, decomposition does appear to take place via diazenyl
radicals (eq. 2). Several lines of evidence provide strong support for this view (7), including CIDNP observations of $PhN_2\cdot$,
and studies of the dependence of rate on viscosity and on pressure. Consider the case of phenyltriphenylmethyldiazene (eq. 3).
The activation enthalpy is +27 kcal/mole (1). There will be some

$$Ph-N{=}N-CPh_3 \rightarrow Ph-N{=}N\cdot + \cdot CPh_3 \quad (eq.\ 3)$$

destabilization of the ground state (steric) and considerable
stabilization in the triphenylmethyl radical. Are these two
sources sufficient to account for the observed 50 kcal/mole difference between the average carbon-nitrogen bond energy and the
enthalpy of activation? Probably not; it appears reasonable to
attribute part of this lowering to stabilization in the phenyldiazenyl radical, $PhN_2\cdot$. Although there have been numerous
attempts to observe this species in the esr, they have not been
successful, ascribable in part to the high exothermicity of the
conversion of a diazenyl radical, $RN_2\cdot$, to R$\cdot$ and N_2. However,
a diazenyl radical possesses an aspect that provides considerable
stabilization in other systems, – an odd electron located on an
atom adjacent to an atom containing a lone pair, $\overset{\cdot\cdot}{A}{-}\dot{B}$. The
stabilization provided by this "three-electron" bond is most
easily seen in reference to the "four-electron" counterpart,
$\overset{\cdot\cdot}{A}{-}\overset{\cdot\cdot}{B}$. Interaction between lone pairs results in formation of new
energy levels, one bonding and one antibonding. In the

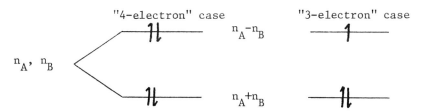

"four-electron" case, the net result is destabilizing (the antibonding combination is somewhat more destabilizing than the bonding combination is stabilizing). In the "3-electron" case, the
net result is stabilizing. The magnitude of the stabilization
obviously depends on the extent of the interaction of the odd

electron with the adjacent lone pair, in turn dependent on the
particular atoms involved, the geometry of the system, and the
state of hybridization at each atom. Some examples are shown in
Chart 1.

Chart 1

$$\ddot{N}-\overset{/}{\underset{\backslash}{C}} \qquad \ddot{N}-\ddot{N}- \qquad \ddot{N}-\ddot{O}:$$

$$-\ddot{O}-\overset{/}{\underset{\backslash}{C}} \qquad -\ddot{O}-\ddot{N}- \qquad -\ddot{O}-\ddot{O}:$$

A qualitative resonance representation of the interaction is
$\ddot{A}-\dot{B} \leftrightarrow \overset{+}{A}-\ddot{B}^{-}$; i.e. one might expect the strength of the interaction to increase with increasing electronegativity of B and
increasing shareability of the lone pair of A. Of those shown in
Chart 1, the most stable "three-electron" species are the
nitroxyls, R_2NO (8), in keeping with the simple generalization
expressed above. Evidence is available on all of the systems
shown in Chart 1 but will not be reviewed here. Suffice it to say
that the adjacency of a lone pair may be a strongly stabilizing
factor in radical stability. The question of stability is a complex one. In some instances the species in question may be subject to direct thermochemical measurement, e.g. some nitroxyls
(9). In some instances, two species in question may be isomeric,
and the question of relative stability may be answered unambiguously, e.g. R_2NO more stable than RNOR. In other cases, apparent
stability may be related to drawbacks to dimerization. For
example, dimerization of nitroxyls would entail the formation of
the weak oxygen-oxygen bond, and the change from "three-electron"
stabilization in nitroxyls to electron destabilization in the
dimer from the four contiguous atoms each holding a lone pair.
Other comparisons of stability in such cases might come from relative reactivity in atom transfer reactions. "Three-electron"
bonding involving carbon radicals deserves further attention.
Much evidence exists to show that hydrogen abstraction by alkoxyl
radicals or halogen atoms is facilitated by the adjacency of
nitrogen or oxygen (10). Much of this facilitation may be due to
ionic contributions to the transition state of the abstraction
process. Few experiments bear directly on the matter of how much
ground state stabilization is present in radicals such as
$R_2\ddot{N}\dot{C}R_2$ (11). Another index of "three-electron" stabilization is
the relative ease of homolytic bond cleavages. In the Stevens
rearrangement (12) a carbon nitrogen bond is broken homolytically
at activation energies well below the carbon-nitrogen bond
strength. The "three-electron" bond in the nitrogen-containing

$$\text{e.g.} \quad RCO\overset{..}{C}H\overset{-}{N}R_2\overset{+}{C}H_2Ph \;\; \overset{\rightarrow}{\underset{\Delta}{}} \;\; RCO\overset{-}{\overset{..}{C}}H\overset{+}{N}R_2 \;\; \leftrightarrow \;\; RCO\overset{..}{C}H\overset{..}{N}R_2$$
$$+ \;\cdot CH_2Ph$$

moiety provides much stabilization thereby lowering the energy needed to effect cleavage. Many related rearrangements are also in this class (12), and provide information on the degree of stabilization in "three-electron" bonds. The scope of "three-electron" stabilization is far broader than the few cases shown in Chart 1. Some additional important charge types are shown below, and include radical anions of olefins and carbonyl compounds (ketyls), both substantially more stable than the corresponding carbon radical or alkoxyl radical.

$$\overset{\backslash}{\underset{/}{\dot{C}}}-\overset{/}{\underset{\backslash}{\ddot{C}}} \qquad \overset{\backslash}{\underset{/}{\dot{C}}}-\ddot{N}- \qquad \overset{\backslash}{\underset{/}{\dot{C}}}-\ddot{O}{:} \qquad {:}\dot{O}-\ddot{O}{:}$$

Mention should be made of "three-electron" bonding involving second row elements, e.g. the effect of adjacent atoms containing lone pairs on phosphorus or sulfur radicals ($\ddot{A}-\dot{B}$ in which B is P or S), and the effect of phosphorus or sulfur adjacent to a radical ($\ddot{A}-\dot{B}$ in which A is R_2P or RS). The great increase in rate of decomposition of symmetrical diazenes containing α-RS substituents (13) is suggestive of stabilization in $RS-\dot{C}R_2$, a conclusion that is also supported by the ease of homolysis of sulfur-carbon bonds in sulfonium ylide rearrangements (12).

Extension to systems in different states of hybridization is also of interest. The odd electron in the iminoxyl radical is

$$\overset{\backslash\ddot{}}{\underset{/}{N}}-\dot{O}{:} \qquad vs. \qquad R_2C{\Large{\diagup}}^{\ddot{N}-\dot{O}{:}}$$

in an orbital interacting with the nitrogen lone pair and is distributed between the nitrogen and oxygen (14), not in a p orbital. Adjacent "three-electron" arrays are also found in systems containing a multiple bond between A and B. The reduced

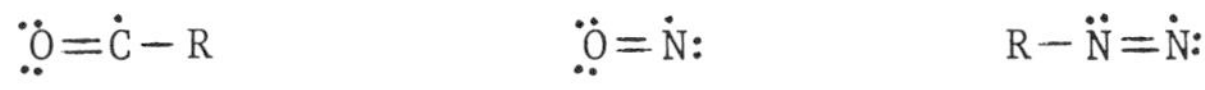

$$\ddot{O}=\dot{C}-R \qquad \ddot{O}=\dot{N}{:} \qquad R-\ddot{N}=\dot{N}{:}$$

strength of the aldehydic carbon-hydrogen bond compared to an alkyl carbon-hydrogen bond (80 vs. 100 kcal/mole) and the ease of abstraction of the aldehydic hydrogen by carbon radicals [e.g. in free radical chain decarbonylation reactions, (15)] are suggestive of greater stabilization in acyl radicals, RCO, than in vinyl or alkyl radicals. "Three-electron" stabilization may also be of consequence in diazenyl radicals, $RN_2\cdot$, the species under discussion at the beginning of this brief survey of "three-electron" stabilization. Some information on diazenyl radicals comes from consideration of the decomposition of diazene N-oxides.

On the Mechanism of Decomposition of Diazene N-Oxides

As indicated above, symmetrical diazenes decompose by a process involving synchronous cleavage of both carbon-nitrogen bonds (eq. 1). Some years ago we were interested in the question of

the possible use of diazene N-oxides as sources of free radicals
(eq. 4). Examination of the literature revealed a compound re-

$$
\begin{array}{c}
R_a \\
\backslash \quad +/O^- \\
N \!=\! N \\
\backslash \\
R_b
\end{array}
\quad ? \quad \rightarrow \quad R_a^{\cdot} \quad N_2O \quad \cdot R_b \quad (eq.\ 4)
$$

ported to be the N-oxide of azobisisobutyronitrile, AIBN-0 (16).
We repeated Aston and Parker's preparation. The compound does
indeed have the structure claimed and is quite stable, requiring
hours at 180° to effect decomposition. We have also prepared dicumyldiazene N-oxide, 1, and compared it with the corresponding
diazene. Decomposition of 1, like AIBN-0, is very slow. The
products of decomposition are ∝-methylstyrene, water, dinitrogen,
and a small amount of cumyl alcohol. Nitrous oxide is not a
product, and decomposition clearly is not proceeding by the process shown in eq. 4. The decomposition is most simply understood
in terms of a cyclic decomposition, eq. 5, related to the Cope
amine oxide decomposition (17).

$$
\begin{array}{c}
PhC(CH_3)_2 - N \\
\quad\quad\quad\quad\quad\quad\searrow N - \overset{CH_3}{\underset{CH_2}{\overset{|}{C}}} - Ph \\
1 \quad\quad\quad O^- \;+\; \\
\quad\quad\quad\quad H
\end{array}
\longrightarrow
\begin{array}{c}
PhC(CH_3)_2 - N \\
\quad\quad\quad\quad\quad\quad\searrow N \\
\quad\quad\quad\quad HO
\end{array}
+ \;
CH_3 \quad Ph
$$

$$
\downarrow
$$

$$
PhC(CH_3)_2OH \quad\quad \longleftarrow \quad\quad PhC(CH_3)_2 \cdot \quad N_2 \quad \cdot OH \quad (eq.\ 5)
$$

$$
PhC(CH_3)CH_2
$$

The azo compound, dicumyldiazene (18), decomposes much faster
than the corresponding N-oxide, kazo/kazozy = 10^5 (est. for 180°).
In summary, diazenes that normally decompose by synchronous two-
bond cleavage (eq. 1) decompose much more rapidly than the corresponding N-oxides. The failure of N-oxides to decompose syn-
chronously to nitrous oxide is apparent from a consideration of
the appropriate energy cycle (Fig. 1). A limiting factor is the
paucity of thermochemical data for diazene N-oxides. The heat of
oxidation of the diazene in Fig. 1 is from the heats of combustion
of diphenyldiazene and the corresponding N-oxide (19). The energy
cycle indicates that the decomposition of a diazene N-oxide into
nitrous oxide and two radicals would be +65 kcal/mole, approximately 40 kcal/mole more endothermic than the decomposition of the
related diazene into the same two radicals and dinitrogen. The
major difference lies in the great stability of dinitrogen; oxidation of dinitrogen is endothermic whereas oxidation of the
diazene is exothermic. Snyder has called attention to the great
difference in stability of cyclic diazenes and diazene N-oxides
in retrocycloaddition reactions (20). For example, 2,3-diazabi-
cyclo [2.2.1]hepta-2,5-diene and the corresponding N-oxide are

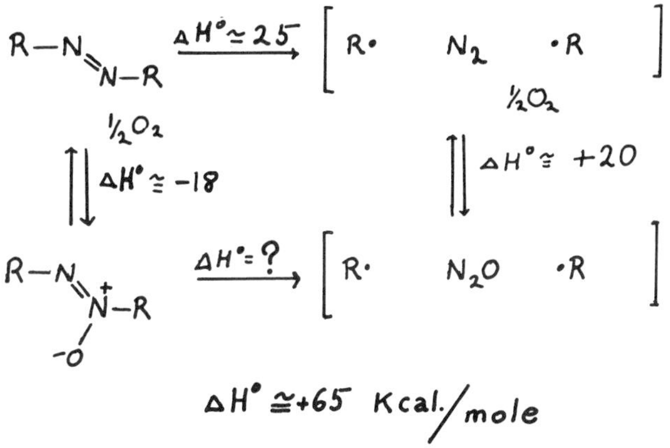

$$R-N \overset{\displaystyle \Delta H^{\circ} \simeq 25}{\underset{N-R}{\longrightarrow}} \left[R\cdot \quad N_2 \quad \cdot R \right]$$

$$\Big\updownarrow \; \Delta H^{\circ} \simeq -18 \qquad \qquad \Big\updownarrow \; \Delta H^{\circ} \simeq +20$$

$$R-N \overset{\displaystyle \Delta H^{\circ} = ?}{\underset{\overset{+}{N}-R}{\longrightarrow}} \left[R\cdot \quad N_2O \quad \cdot R \right]$$
$$\underset{-O}{}$$

$$\Delta H^{\circ} \simeq +65 \; Kcal./mole$$

Figure 1. Estimated heat of reaction for conversion of a diazene N-oxide into radicals and nitrous oxide

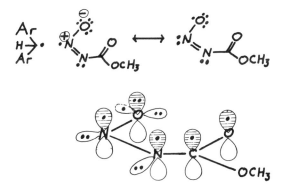

Figure 2. Representation of the transition state for decomposition of diazene N-oxide 7

converted to cyclopentadiene and N_2 and N_2O, respectively; the
diazene reacts seventeen powers of ten faster than the diazene
N-oxide (20). Here, also, it seems likely that the types of differences noted in Fig. 1 are important. However, the synchronous
nature of the decomposition of 2,3-diazabicyclo[2.2.1]hepta-2,5-
diene and the corresponding N-oxide makes it more difficult to
estimate differences in activation energies for these systems
than for the endothermic, homolytic fragmentation reactions of
diazenes and diazene N-oxides of Fig. 1.

We have indicated that diazenes with sufficiently different
groups may undergo one-bond cleavage in the rate-determining step
(eq. 2). How do these systems compare in rate with the corresponding N-oxides? The comparisons in eq. 6 are of interest.

$$R_a—N{\Large\searrow}_{N—R_b} \qquad R_a—\overset{+}{N}{\Large\diagdown}^{O^-}_{N—R_b} \qquad R_a—N{\Large\diagdown}_{{\,\,N—R_b \atop {}_{-O}{\diagup}^+}} \quad (eq.\ 6)$$

$$\Big\downarrow k_1 \qquad\qquad \Big\downarrow k_2 \qquad\qquad \Big\downarrow k_3$$

$$R_a—N_2^{\cdot} \quad {\cdot}R_b \qquad R_a—\overset{+}{N}{\Large\diagdown}^{O^-}_{N\cdot} \quad {\cdot}R_b \qquad R_a—N{\Large\diagdown}_{{\,N\cdot \atop {}_{-O}{\diagup}}} \quad {\cdot}R_b$$

An ideal system in which to make these comparisons would
appear to be phenyltriphenylmethyldiazene, 2, and the corresponding N-oxides. Oxidation of the diazene with peroxytrifluoro-
acetic acid affords a diazene N-oxide, assigned structure 3 on
the basis of spectroscopic data (principally, ultraviolet ~

$$Ph—N{\Large\searrow}_{N—CPh_3} \qquad Ph—\overset{+}{N}{\Large\diagdown}^{\bar O}_{N—CPh_3} \qquad Ph—N{\Large\diagdown}_{{\,N—CPh_3 \atop {}_{-O}{\diagup}^+}}$$

$$\underset{\sim}{2} \qquad\qquad\qquad \underset{\sim}{3} \qquad\qquad\qquad \underset{\sim}{4}$$

comparisons). Compound 3 is a very stable substance decomposing
only slowly at 290°C (half-life ~ 50 min, a minimum value of
43 kcal/mole for $\Delta G^{\mp}$). Here, also, the diazene N-oxide is much
more stable than the corresponding diazene, although the reason
must be different from that operative with the symmetrical, two-
bond-cleaving diazenes described above. The comparisons of
interest are shown in Chart 2. There are many uncertainties, including a lack of knowledge of the mode of decomposition of
diazene N-oxide, 3! One point is clear: for diazenes that decompose by one-bond cleavage in the rate-determining step, placement
of an oxygen on the beta-nitrogen greatly decreases the rate of

Chart 2

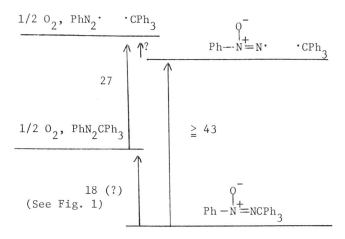

decomposition, i.e., in eq. 6, $k_1 \gg k_2$. Much of this difference may be attributable to the difference between $\underset{\sim}{5}$ and $\underset{\sim}{6}$. The

$$Ph-\ddot{N}\!=\!\dot{N}$$

$$\underset{\sim}{5}$$

$$Ph-\overset{+}{N}\overset{O^-}{\underset{N\cdot}{\diagdown}}$$

$$\underset{\sim}{6}$$

"three-electron" bonding possible in $\underset{\sim}{5}$ is precluded in $\underset{\sim}{6}$ by the presence of the oxygen. The uncertainties in Chart 2 are too great to allow placement of a number on the amount of "three-electron" stabilization in $\underset{\sim}{5}$, but one should note in particular that a localized (non-stabilized) phenyldiazenyl radical would be expected to lie at a considerably higher level than that shown in Chart 2 (on the assumption that the heat of oxidation of <u>locali-</u> <u>zed</u> $PhN_2\cdot$ to $PhN(O)=N\cdot$ would be comparable to the heat of oxida- tion of a diazene to a diazene N-oxide).

The third comparison of interest is that of a one-bond-cleav- ing diazene with the corresponding N-oxide in which the oxygen is on the alpha nitrogen (eq. 6, k_1 vs. k_3). A desirable point of comparison would be triphenylmethylphenyldiazene N-oxide, $\underset{\sim}{4}$. Several approaches to prepare this compound have been unsuccess- ful. An alternative was the methyl triphenylmethyldiazene- carboxylate, $Ph_3CN_2CO_2CH_3$, a diazene that decomposes into radi- cals at a rate close to that of phenyltriphenylmethyldiazene (<u>21</u>). Efforts to obtain the N-oxide of the triphenylmethyldiazene- carboxylate were also unsuccessful. Consequently, we sought to answer the question by examination of a benzhydryl system. Efforts to prepare benzhydrylphenyldiazene N-oxide afforded only the unwanted isomer (oxygen on the nitrogen attached to phenyl). Success was achieved with methyl benzhydryldiazenecarboxylate. Oxidation of this diazene afforded the N-oxide in which the oxygen is on the benzhydryl nitrogen, 7a. Decomposition was carried out at 150° in o-dichlorobenzene, affording the products summarized in eq. 7 (the numbers are moles per mole of reactant). The products

$$7a, \quad Ph_2CHN(O)=NCO_2CH_3 \quad \xrightarrow{\quad 150° \quad}$$

$$Ph_2CHOCO_2CH_3 \qquad Ph_2CHOCH_3 \qquad Ph_2CO \qquad (eq. 7)$$

$$0.30 \qquad\qquad 0.27 \qquad\quad 0.13$$

$$CH_3OH \qquad\qquad Ph_2CHCHPh_2 \qquad Ph_2C=CPh_2 \qquad Ph_2CH_2$$

$$0.12 \qquad\qquad\quad 0.04 \qquad\qquad\quad 0.04 \qquad\quad trace$$

are strongly suggestive that decomposition involves radicals. The decomposition is first order in reactant, and the rate is not increased by the addition of bisbenzhydryldiazene or methyl triphenylmethyldiazenecarboxylate, sources of benzhydryl and carbomethoxy radicals, - two likely candidates for chain carriers in a possible free radical chain decomposition. Consequently, the decomposition of diazene N-oxide 7a appears to proceed by a unimolecular path. Is the rate-determining step simply a homolysis of the benzhydryl - N(O) bond, or are ionic contributions important in the transition state of this reaction? The electron-withdrawing nature of the carbomethoxy group might make this example susceptible to ionic contributions to the transition state of the carbon-nitrogen cleavage. Two related diazene N-oxides were

$$[Ph_2CH\cdot \quad \overset{-O}{\overset{+}{N}}=CO_2CH_3 \quad \longleftrightarrow \quad Ph_2\overset{+}{C}H \quad \overset{-O}{\overset{\cdot}{N}}=NCO_2CH_3]$$

prepared, the p, p´-dimethyl (7b) and the p, p´-dichloro (7c) derivatives. Decomposition of each afforded a product mixture similar to that from 7a (eq. 7). Rates of decomposition are summarized in Table 2. The effect of the para-methyl and the para-chloro substituents is small, both groups exerting a small rate acceleration. [For comparison, the corresponding rel k's for

Table 2

Decomposition of $(p-RC_6H_4)_2CHN(O)=NCO_2CH_3$
in o-Dichlorobenzene at 150°

Cpd	R	rel k	σ
7a	H	1.0	0
7b	CH_3	5.0	-0.17
7c	Cl	1.2	+0.23

$(p-RC_6H_4)_2CHCl$ in ethanol, 25°C are H (1.00), CH_3 (413), Cl (0.15)]

(22). Thus, in decomposition of diazene N-oxides, $\underset{\sim}{7}$, there is little change in the degree of charge at the benzhydryl carbon in going from ground state to transition state. The small rate effects are in accord with simple homolysis of the benzhydryl carbon-nitrogen bond [eg. the rel k's for decomposition of $\text{p-RC}_6\text{H}_4\text{C(CH}_3)_2\text{N=NC(CH}_3)_2\text{C}_6\text{H}_4\text{-p-R}$ at 43° in toluene are H (1.00), CH_3(1.46), Cl (2.67)] ($\underline{23}$). A representation of the transition state for decomposition of diazene N-oxide $\underset{\sim}{7}$ is shown in Fig. 2. Homolytic cleavage of the carbon-nitrogen bond at a cost of only 33 kcal/mole is possible in this case by the combined help of benzhydryl radical stabilization and "three-electron" stabilization in the diazenoxyl radical (Fig. 2).

How does the rate of decomposition of diazene N-oxide 7 compare with the corresponding diazene, $\underset{\sim}{8}$? Direct measurement of the rate of decomposition of methyl benzhydryldiazenecarboxylate, $\underset{\sim}{8}$, was complicated by tautomerization to the corresponding hydrazone, $\text{Ph}_2\text{C=NNHCO}_2\text{CH}_3$. The rate of decomposition of 8 was estimated from the measurement rate of decomposition of benzhydryl-phenyldiazene ($\underline{24}$) and the observation that the rates of decomposition of phenyltriphenylmethyldiazene and methyl triphenyl-methyldiazenecarboxylate are the same ($\underline{21}$). Relative rates of 7, 8, and a related nitrone, 9 ($\underline{25}$) are summarized below.
$\underset{\sim}{}$ $\underset{\sim}{}$

		rel k, 150°	$\Delta G^{\ddagger}$, 150°
$\underset{\sim}{7}$	$\text{Ph}_2\text{CHN(O)=NCOOCH}_3$	1.0	33.3
$\underset{\sim}{8}$	$\text{Ph}_2\text{CHN=NCOOCH}_3$	($\sim$10.)	($\sim$31.5)
$\underset{\sim}{9}$	$\text{Ph}_2\text{CHN(O)=CPh}_2$	$\sim$1	33

Diazene $\underset{\sim}{7}$, diazene N-oxide $\underset{\sim}{8}$ and the nitrone $\underset{\sim}{9}$ all decompose at a comparable rate. The diazene N-oxide and the nitrone should both be aided by "three-electron" stabilization in the diazenoxyl and the iminoxyl moieties, respectively (see Fig. 2). The implication (by reference to the type of diagram shown in Chart 2 but not repeated for the present case) is that considerable "three-electron" stabilization also may be present in the diazenyl radical. The comparison of $\underset{\sim}{7}$ with 8 suffers from some uncertainty in the matter of one-bond vs. two-bond cleavage in the rate-determining step of the model system, benzhydrylphenyldiazene; there may be some weakening of the phenyl-nitrogen bond in the transition state. The greater rate for 8 (estimated) than for $\underset{\sim}{7}$ may be more a consequence of some synchronous character to the decomposition of $\underset{\sim}{8}$ than a consequence of comparable amounts of "three-electron" stabilization in the diazenyl radical, $\text{RN}_2\cdot$ and the diazenoxyl radical, $\text{RN}_2\text{O}\cdot$. In either case, the conclusion from this example is that diazenes which decompose by one-bond cleavage in the rate-determining step (eq. 6) may decompose at a rate comparable to (or perhaps, less than) that of the corresponding N-oxide when the oxygen is on the nitrogen of the carbon-nitrogen bond undergoing homolysis (i.e. in eq. 6, $k_1 \lesssim k_3$).

In summary, the type of stabilization on which attention has

been focussed here, "three-electron" bonding, appears to be an
important factor in radical stabilization, applicable to many
situations, and highly worthy of further investigation.

Acknowledgment. This work was supported, in part, by Public
Health Service Research Grant CA-16592 and Training Grant 5T32
CA-09112 from the National Cancer Institute.

Literature Cited.

(1) Koenig, T. in "Free Radicals", Vol. 1, Ch. 3, J. K. Kochi,
    Ed., Wiley-Interscience, New York, N.Y.,(1973). See also
    "The Chemistry of the Hydrazo, Azo, and Azoxy Groups", Vols.
    1 and 2, S. Patai, Ed., Wiley-Interscience, New York, N.Y.,
    (1975).
(2) Engel, P.S. and Bishop, D.J., J. Am. Chem. Soc., (1975), 97,
    6754.
(3) Hinz, J.,Oberlinner, A., and Ruchardt, C., Tetrahedron Lett.,
    (1973), 1975.
(4) Engel, P.S., Wood, J.L., Sweet, J.A., Margrave, J.L., J. Am.
    Chem. Soc., (1974), 96, 2381.
(5) Martin, J.C. and Timberlake, J.W., J. Am. Chem. Soc., (1970),
    92, 978.
(6) The opposite assumption has also been made i.e. the assumption
    that the gas phase decomposition of azo compounds proceeds by
    initial one-bond cleavage (eq.2), [e.g. see Benson, S.W. and
    O'Neal, H.E., Nat. Stand. Ref. Data Ser., Nat. Bur. Stand.
    (1970), No. 21]; this assumption (not established) would
    require substantial stabilization in diazenyl radicals.
(7) Porter, N.A., Green, J.G., and Dubay, G.R., Tetrahedron Lett.,
    (1975), 3363. Pryor, W.A. and Smith, K., J. Am. Chem. Soc.,
    (1970), 92, 5403. Neuman, R.C., Lockyer, G.D., and Amrick,
    M.J., Tetrahedron Lett., (1972), 1221.
(8) Forrester, A.R., Hay, J.M., and Thomson, R.H., "Organic
    Chemistry of Stable Free Radicals", Academic Press, New York,
    N.Y., (1968), p. 180-246.
(9) Mahoney, L.R., Mendenhall, G.D., and Ingold, K.U., J. Am.
    Chem. Soc., (1973), 95, 8610.
(10) Russell, G.A. in "Free Radicals", Vol. 1, Ch. 7, J. K. Kochi,
     Ed., Wiley-Interscience, New York, N.Y., (1973).
(11) Nelson, S.F. in "Free Radicals", Vol. 2, Ch. 21, J. K. Kochi,
     Ed., Wiley-Interscience, New York, N.Y., (1973).
(12) Lepley, A.R., in "Mechanisms of Molecular Migrations", Vol. 3,
     p. 297, B.S. Thyagarajan, Ed., Wiley-Interscience, New York,
     N.Y., (1971).
(13) Timberlake, J.W. and Hodges, M.L., Tetrahedron Lett., (1970),
     4147.
(14) Gilbert, B.C. and Norman, R.O.C., J. Chem. Soc. (London) B,
     (1966), 86. Also, see ref. 9 (above) and references cited
     therein.

(15) Wilt, J.W. in "Free Radicals", Vol. 1, Ch. 8, J. K. Kochi,
     Ed., Wiley-Interscience, New York, N.Y., (1973).
(16) Aston, J.G. and Parker, G.T., J. Am. Chem. Soc., (1934), 56,
     1387.
(17) Cope, A.C. and Trumbull, E.R., "Organic Reactions", Vol. 11,
     p. 317, Wiley, New York, N.Y., (1960).
(18) Nelson, S.F. and Bartlett, P.D., J. Am. Chem. Soc., (1966),
     88, 137.
(19) Swietoslawski, W. and Popow, M., J. Chim. Phys., (1925), 22,
     397.
(20) Olsen, H. and Snyder, J.P., J. Am. Chem. Soc., (1977), 99,
     1524.
(21) Zabel, D.E. and Trahanovsky, W.S., J. Org. Chem., (1972),
     37, 2413.
(22) Norris, J.F. and Banta, C., J. Am. Chem. Soc., (1928), 50,
     1804.
(23) Shelton, J.R. and Liang, C.K., J. Org. Chem., (1973), 38,
     2301.
(24) Cohen, S.G. and Wang, C.H., J. Am. Chem. Soc., (1955), 77,
     3628.
(25) Grubbs, E.J., Villarreal, J.A., McCullough, J.D., Jr., and
     Vincent, J.S., J. Am. Chem. Soc., (1967), 89, 2234.

RECEIVED December 23, 1977.

9

Cage Effects in Peresters and Hyponitrites

T. W. KOENIG

Department of Chemistry, University of Oregon, Eugene, OR 97403

A. Introduction

The studies of Bartlett and his coworkers ($\underline{1}$) on decompositions of $\underline{t}$-butyl peresters of a wide range of carboxylic acids clearly showed two mechanistic pathways for the production of free radicals. The first mechanism is the simple homolysis of the O-O bond to give an acyloxy $\underline{t}$-butoxy (pair) intermediate which can decarboxylate in a subsequent step. This mechanism (A-1) pertains to the <u>unimolecular</u> free radical pro-

$$\text{A-1} \quad \text{R-CO}_3\underline{t}\text{Bu} \xrightarrow{k_1} [\text{R-CO}_2\cdot\cdot\text{O}\underline{t}\text{Bu}] \longrightarrow \begin{array}{l} \longrightarrow \text{Free Radicals} \\ \quad\quad\quad\quad\quad\uparrow \\ \xrightarrow{k_2} \text{R}\cdot\cdot\text{O}\underline{t}\text{Bu} \end{array}$$

duction for methyl, aryl, cyclopropyl and 1-norbornyl decarboxylation residues (R, equation A-1).

The second mechanism involves concerted cleavage of both the O-O and C-C bonds to give, through a single transition state, alkyl $\underline{t}$-butoxy radical pairs. Activation parameters ($\underline{2}$), isotope effects ($\underline{3}$) and substituent effects ($\underline{4}$) all indicate a change from stepwise (A-1) to concerted (A-2) mode of reaction when the decarboxylation residue has the stability of secondary alkyl or more. Hammet substituent constants

$$\text{A-2} \quad \text{R-CO}_3\underline{t}\text{Bu} \xrightarrow[\substack{\text{Concerted} \\ k_1{}'}]{[\text{R}\cdot\text{CO}_2\cdot\text{O}\underline{t}\text{Bu} \leftrightarrow \text{R}^{\oplus}\text{CO}_2{}^{\ominus}\text{O}\underline{t}\text{Bu}]^{\ddagger}} \begin{array}{l} \text{Free Radicals} \\ \quad\quad\uparrow \\ [\text{R}\cdot(\text{CO}_2)\cdot\text{O}\underline{t}\text{Bu}] \\ \quad\quad\downarrow k_c \\ \text{Ethers} \end{array}$$

indicated a significant dipolar character (4) of the
transition state for the concerted process ($\overline{A}$-2) as
shown.

The major factors determining the rates of free
radical formation, as a function of the structure of
the acyl group in peresters, are certainly well explained by these two mechanisms. Our investigation of
the reactions of nitrosohydroxylamines (2), which rearrange to O-acyl-O'-t-butyl hyponitrites (3), were
initiated with aim of providing a deaminative entry
into the same formal intermediates as given by the
corresponding peresters. The questions we hoped to
answer were (i) to what extent would the deaminative
process give rise to radicals which were atypical in
reactions such as decarboxylation (k_2, A-1) and (ii)
to what extent would the physical presence of the
nitrogen molecule affect the observed yields of combination products (eg. k_c, A-2).

Our initial study (5) involved the low temperature nitrosation of the benzoyl hydroxyamate (1, R=∅)
to give the N-nitroso derivative (2, R=∅) which, in
turn, was expected to rearrange to the O-benzoyl-O'-
t-butyl hyponitrite (3, R=∅) by analogy with nitroso-
amide rearrangements (6). Solutions of the nitroso
hydroxamate were easily obtained and rearrangement to
the hyponitrite was rapid at room temperature. It was
somewhat surprising to find that the hyponitrite rear-

$$
\begin{array}{ccc}
\underset{\text{H}}{\overset{\displaystyle\overset{\text{O}}{\overset{\|}{\text{R-C-NOtBu}}}}{\underset{\displaystyle\text{H}}{\underset{|}{}}}
& \xrightarrow[-20°]{\text{NOCl}}
& \left[\overset{\text{O}}{\overset{\|}{\text{R-C-N-OtBu}}} \atop \underset{\text{O=N}}{\underset{|}{}} \right]
\xrightarrow{+25°}
\overset{\text{O}}{\overset{\|}{\text{R-C}}} \underset{\text{O-N}}{\overset{\text{N-OtBu}}{\underset{\|}{|}}}
\end{array}
$$

$\qquad$ 1 $\qquad\qquad\qquad$ 2 $\qquad\qquad\qquad\qquad$ 3

rangement product was relatively stabile at room temperature in view of the short lifetime of aryl diazo-
acylates towards dissociation to diazonium ions (7).

The decomposition of the benzoyl hyponitrite occurred at a slow rate at room temperature giving products typical of benzoyloxy and t-butoxy free radicals.
A careful comparison (8) of the abstraction/β-scission
competition showed that the radicals formed from the
deamination give the same product ratios as the perbenzoate when extrapolated to a common temperature. Thus,
the deaminative nature of·this reaction does not give
free radicals with any unusual tendency for decarboxylation or β-scission as might be expected if they were
vibrationally excited above the barrier for the fragmentation.

B. Acyloxy t-Butoxy Cage Reactions.

The most interesting product from the decomposition of this hyponitrite was t-butyl perbenzoate. The
fact that free radicals accounted for the balance of
the reaction products made it seem likely that the
perester was formed by cage combination. This was
supported by the observation that the yield of this
product increased with increased solvent viscosity (5,
8). Furthermore, oxygen-18 labelling studies showed
that the perester was formed with essentially complete
randomization of the benzoyloxy oxygen atoms and that
this randomization did not occur in any step preceding
the decomposition of the hyponitrite (8). All of
these observations are consistent with a cage combination process for the formation of the perester product.
      One implication of cage combination of benzoyloxy
t-butoxy pairs in the hyponitrite reaction was that the
O-O bond homolysis of peresters, not proceeding by the
concerted reaction (A-2), must be reversible. Conversely, hyponitrites corresponding to peresters which
did undergo the concerted reaction should not give
perester products. Finally, it was clear that comparison of ether yields, formed by cage combination of
alkyl-t-butoxy pairs from peresters and hyponitrites,
should shed light on the importance of the intervening
small molecules (CO_2 or N_2 and CO_2) initially present.
The phenomenological scheme which characterizes this
kinetic situation is shown below (Scheme I).

Scheme I

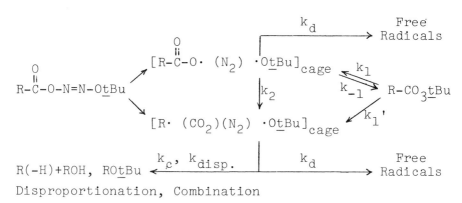

This scheme shows the intervention of one class of
cage radical pair which can decay exclusively by first
order reactions (irreversible diffusive separation to

free radicals (k_d), combination k_{-1} or k_c), disproportionation ($k_{disp.}$) or decarboxylation (k_2)). The observables from the two types of initiators may be related through the cage combination efficiency ratio F. These observables are the isolated yields of per-

$$F \equiv (1/(\text{Fraction cage efficiency})) - 1$$

ester (y_p) from the hyponitrite, the yield of ether (y_e) from the perester or the hyponitrite, the observed rate constant for decomposition of the perester (k_o) and the observed rate constant for scrambling of oxygen-18 from carbonyl labelled perester (k_s). The algebraic connection between these observables and Scheme I are as shown.

$$\text{B-1} \quad \frac{k_o}{k_1 - k_o} = \frac{k_o}{k_s} = \left(\frac{1}{y_p}\right) - 1 = \frac{k_d}{k_{-1}} + \frac{k_2}{k_{-1}} = F_{\text{perester}}$$

$$\text{B-2} \quad k_1 = k_o + k_s$$

$$\text{B-3} \quad \left(\frac{1}{y_e}\right) - 1 \qquad = \frac{k_d}{k_c} + \frac{k_{disp.}}{k_c} = F_{\text{ether}}$$

If it is assumed that k_d (diffusive destruction of the cage pair) is the only solvent (viscosity) sensitive rate constant, then the qualitative implications of the scheme are obvious. The value of F should follow the fluidity (f = 1/viscosity). This means that the values of k_o for a perester undergoing simple O-O bond homolysis should increase with fluidity (viscosity decrease) while k_s should decrease with solvent fluidity. The yield of a perester product, from the corresponding hyponitrite, should increase with fluidity decrease (viscosity increase). The yields of ethers from either perester or hyponitrite should likewise increase with fluidity decrease. All of these qualitative expectation were in fact observed (Table I).

While these qualitative observations substantiate the gross features of the cage scheme, quantitative comparisons are necessary if one is interested in the effects such as those due to the (variable) intervening molecules on cage reactions. The value of k_1, the fundamental rate constant for O-O bond homolysis of a nonconcerted perester, is not directly available from any single experiment in condensed media. According to Scheme I and the assumption that k_d is the only viscos-

Table I

Cage Product Yields[a] from $R-CO_2-N_2-OtBu$, 32°C

R	% Perester (RCO_3tBu)	% O^{18} Scrambling[b]	% Ether (ROtBu)	Ref.
C_6H_5	9.0, 38.0	98.	<.1	5,8
Cyclopropyl[c]	21.0, 46.0	90.	-	9,21
CH_3	2.7, 10.2	93.	8.9, 41.7	10
i-Propyl	<.1	-	10.0, 25.0	11,21
2-Butyl	<.1	-	11.3, 27.8	12,13

a) Low viscosity (hexane, n ∿0.3cp) followed by high
   viscosity (Nujol, n ∿17.cp), percentages based on
   nitrogen evolved.
b) % oxygen-18 scrambling in the isolated and purified
   perester product from reaction of carbonyl-O^{18}
   labelled hyponitrite.
c) Denitrosation complicates the cyclopropyl products;
   T. Koenig, J. A. Hoobler and W. R. Mabey, J. Am.
   Chem. Soc., 94, 2514 (1972).

ity sensitive rate constant, k_1 should be obtained from
extrapolation of values of k_0 to infinite fluidity
(zero viscosity) (14). Figure 1 shows such an extrapolation for t-butyl perbenzoate against the square root
of solvent viscosity (8). The value of k_1, obtained
from this extrapolation, can be used to obtain F_p
values for (re)combination of benzoyloxy t-butoxy pairs
with no intervening molecule using the values of k_0.
Figure 2 shows a comparison of the combination efficiency ratios (F_p's) for the hyponitrite (squares) and
perester (circles). Both plots appear to be linear
with square root of fluidity and both pass through the
origin. The zero intercept in such a plot implies that
no other reaction (like decarboxylation, k_2, Scheme I)
competes with combination except diffusive destruction
of the cage pair and that the yield of cage product at
infinite viscosity would be 100%. The difference in
the two slopes in Figure 2 could be ascribed to the intervening nitrogen molecule (and temperature difference) in the hyponitrite reaction, amounting to a reduction in the k_c of the N_2 (hyponitrite) separated
pair by a factor of one fourth.

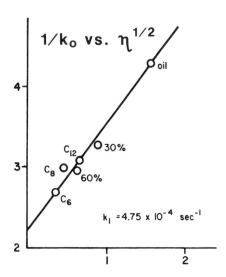

Figure 1. $1/k_o$ vs. square root viscosity. tert-Butyl perbenzoate, $130°C$ (8).

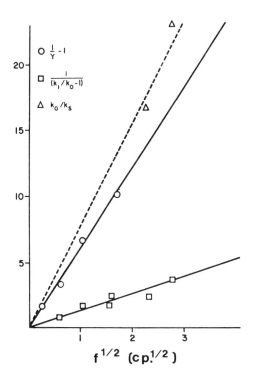

Figure 2. tert-Butyl perbenzoate. F_p from 3 at $32°C$ ($\bigcirc$), K_o ($\square$), K_s ($\triangle$) (8).

A second means of obtaining k_1 is from the sum of
k_O and k_S. If k_d is really the only viscosity sensitive rate constant, then the k_O+k_S sum should be constant in all solvents and equal to the value obtained
by extrapolation. This was not the case for either the
perbenzoate ($\underline{8}$) or the peracetate ($\underline{3}$). Figure 2 also
shows the F_p values from k_O/k_S which takes into account
the variation of both k_O and k_S with solvent (triangles). The descrepancy the two means of estimating k_1
is very large and negates any conclusions concerning
the N_2 effect.

During the course of this work, Neuman and his co-
workers ($\underline{15}$) measured activation volumes for a number
of free radical initiators including $\underline{t}$-butyl perbenzoate. It was clear that our hydrocarbon solvents of
variable viscosity were also of variable internal pressure. The variations in k_O values for $\underline{t}$-butyl perbenzoate were in accord with the +12cc/mole activation
volume which Neuman found by external pressure variation. In order to remove objections concerning methods
of estimating internal solvent pressure for our mixed
solvent series, we used Neuman's +12cc/mole value for
$\underline{t}$-butyl perbenzoate to establish a set of differential
solvent pressures for our mixed solvent system (using
the experimental values of k_O) ($\underline{16}$). The k_O+k_S sums
for the perbenzoate and peracetate were then plotted
(Fig. 3) versus the derived DSP solvent parameters to
give activation volumes for the simple homolyses process (k_1). The activation volumes were within the
range expected from external pressure studies. The
variation in the k_O+k_S sums with varying solvent are
entirely explainable in terms of the internal solvent
pressure effect on k_1 which may be viscosity independent but is not solvent independent. The extrapolative
method for estimating k_1 is therefore incorrect since
there can be no constant value for this rate constant
unless the corresponding activation volume is zero.
Scheme I survives but the assumption that k_d is the
only solvent sensitive rate constant in it, must be
abandoned.

In order to avoid the large extrapolation to common temperature, required for comparison of the thermolysis of both $\underline{t}$-butyl perbenzoate and the corresponding hyponitrite, we decided to study the photolysis
of the perester under the same conditions of solvent
and temperature as convenient for the thermal hyponitrite reaction (27°, hexane) ($\underline{17}$). Product studies
showed hexyl benzoate was formed from a chain induced
reaction. The direct photolysis gave phenyl $\underline{t}$-butyl
ether and high relative yields of carbon dioxide. Both

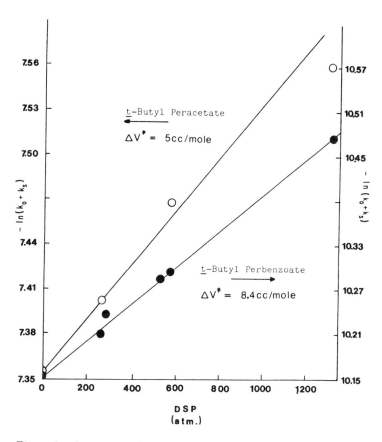

Figure 3. Activation volume from empirical differential solvent pressures (16)

are indications that the reactive singlet excited state
of the perester gives phenyl t-butoxy pairs, bypassing
the benzoyloxy t-butoxy pair. Fortunately, the availability of the carbon dioxide yield data from the
hyponitrite under nearly identical conditions allowed
a correction for the fraction of the total photolysis
which procedes in this manner. The fraction oxygen-18
scrambling of reisolated material after known fractions
of overall perester disappearance and corrected for
induced and concerted reactions, gave a minimum estimate of the fraction combination (0.18) which was
twice the observed yield of perester from the hyponitrite (0.09) under nearly identical conditions.
Acetophenone sensitized photolysis of the perester
(presumably triplet) gave oxygen-18 scrambling corresponding to 8% cage reaction. The intervening nitrogen
molecule and the spin barrier both appear to reduce the
cage combination efficiency of benzoyloxy t-butoxy pair
by a factor of one half.

A series of ring substituted benzoyl hyponitrites
showed the rho value for combination of benzoyloxy t-
butoxy pairs was +0.4 (18). The sign and magnitude of
this reaction constant are both in accord with the O-O
bond dipole repulsion (19) explanation of the rho
value observed in perbenzoate homolysis ($\rho = -0.7$).
The nonzero rho value observed for the benzoyloxy t-
butoxy combination process implies a nonzero activation
energy for combinations through the sign of the activation energy is left ambiguous.

C. Alkyl t-Butoxy Cage Reactions.

Isotope effects indicated that t-butyl isobutyrate
underwent the concerted reaction (A-2) (20). Oxygen-18
scrambling of reisolated material, after partial decomposition showed the nonconcerted (A-1) process contributed only 2% to the overall reaction, exactly the
amount estimated for the relative rate of t-butyl peracetate at the temperature used. The corresponding
hyponitrite gave no detectable perester even at 0° in
paraffin oil solvent (9). The isobutyryl system is
thus one which gives i-propyl t-butoxy cage pairs with
CO_2 or CO_2 and N_2 as intervening molecules. The cage
efficiency ratio (F_e) for ether fromation gave a fairly
linear correlation with square root fluidity ($1/\eta^{1/2}$)
(21). According to Scheme I, the intercepts of such
plot should give the ratio of k_{disp}/k_c. The observed
intercepts (~2.3) was very close to the propylene-ether
ratio found by Kochi and Sheldon in their studies of
the photolysis of the perester in decalin solution. It

should be noted that intercepts of F_e correlations at
zero fluidity give the ratio of k_{disp}/k_c through measurement of only the ether yield and the good numerical
agreement of these intercepts with the value obtained
from the measured ether olefin ratio lends strong support to the methodology associated with Scheme I.

These results on the isobutyryl system convinced
us that both the perester and the hyponitrites with
secondary alkyl side chains underwent concerted homolysis to give secondary alkyl-t-butoxy cage pairs in a
single step. We decided to investigate the chiral 2-
methylbutyrate systems which would allow an examination
of the relative efficiency of the tumbling process of
the 2-butyl radical compared with the combination and
disproportionation. We also wanted to examine the
effects of added scavengers on the ether yields and
optical purities. These additional kinetic channels
are added to the phenomenological cage model and shown
as Scheme II.

Scheme II

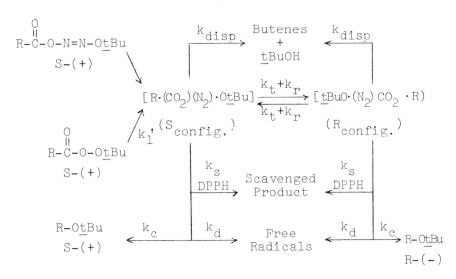

Scheme II generates several new algebraic relationships between the observables of the system and the
interpretative rate constants. The observables are the
yields of 2-butyl t-butyl ether (y_e) and its optical
urity (OP). The interpretative rate constants are that
for combination to give the ether (k_c), that for disproportionation (k_{disp}), that for diffusive destruction of the cage pair (k_d), that for the tumbling of
the 2-butyl radical with respect to the t-butoxy radi-

cal which racemizes the cage pair (k_s), that for internal rotation of the radical center about the ethyl group which racemizes the cage pair (k_r) and that for scavenging of the cage pair by DPPH. The algebraic relationships are shown below.

C-1 $$F_e = \frac{k_{disp.}}{k_c} + \frac{k_d}{k_c} + \frac{k_s[DPPH]}{k_c}$$

C-2 $$[F_e]_{[DPPH]} - [F_e]_o = \frac{k_s[DPPH]}{k_c}$$

C-3 $$\frac{1}{Y_e}(\frac{1-OP}{2(OP)}) = \frac{k_r}{k_c} + \frac{k_t}{k_c}$$

C-4 $$\left[\frac{2(OP)}{1-OP}\right]_{[DPPH]} - \left[\frac{2(OP)}{1-OP}\right]_o = \frac{k_s[DPPH]}{k_t+k_r}$$

Figure 4 shows the results of the combination efficiency ratio (F_e) for the perester at 102° (22) and the hyponitrite at 30° (12) against fluidity to the 0.75 power (eq. C-1, [DPPH]=0). The intercepts again measure the $k_{disp.}/k_c$ ratio. The difference in the intercepts can be explained by a 0.9 kcal/mole difference in the activation energies for disproportionation and combination. The slope of the perester line (1.49) is lower than that for the hyponitrite (1.96), suggesting the additional nitrogen molecule which intervenes between the two radicals of the cage pair reduces k_c to some extent. The higher temperature required for the perester tends to mask this effect.

The optical purity functions are again linear with $f^{0.75}$ (Figure 5, eqn. C-3) (12). The slopes are interpreted as measuring the ratio of 2-butyl radical tumbling to combination (k_t/k_c). The difference in the two provides an independent suggestion that k_c is reduced by the nitrogen molecule formed in the cage from the hyponitrite. The extent of reduction is the same as indicated by the relative slopes in Figure 4. The intercepts are interpreted as the ratio of the rate constants for internal rotation of the radical center to combination k_r/k_c and are again suggestive of a reduced k_c for the deaminative reaction.

Finally, the effect of DPPH on the yields of ether and its optical purity are shown in Figure 6 (12). The important result is that the cage scheme treats the scavenging reaction as a simple pseudo first order process and the linear plots are in good accord with these expectations. The observed ratio k_s/k_c is 3.66

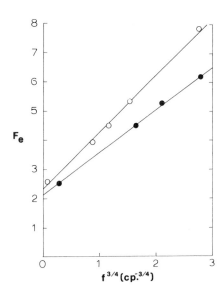

Figure 4. (F_e) vs. $f^{3/4}$ for ether from 2-butyl-tert-butoxy radical pairs generated from perester (−●−) and hyponitrite (−○−) (13, 22)

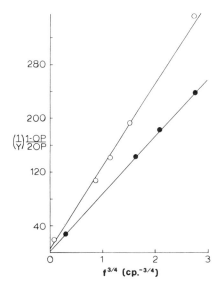

Figure 5. $(1/Y)[(1 − OP)/2OP]$ vs. $f^{3/4}$ for ether product produced from optically active hyponitrite (−○−) and perester (−●−) (22)

ℓ/mole in chloroform (η = 0.56 cp.). A stabile radical scavenger is capable of trapping about half the cage intermediate at concentrations around 1 M. The optical purity also responds in a systematic way to added DPPH. The ratio $k_s/(k_t+k_r)$ is 1.97×10^{-2} ℓ/mole in chloroform.

If the Debye-Stokes relationship and a tumbling radius for the 2-butyl radical of 2.0Å are assumed, the ratios of rate constants found here can be put in terms of absolute rate constants. These values are given in Table II (12) and we believe they correspond fairly well to the actual time scales of the reacting system. The nitrogen molecule effect is shown in the two values of k_c (1.0×10^9 for the hyponitrite at 32° vs. 1.8×10^9 for the perester at 102°). The temperature difference in the two sets of experiments makes this comparison less direct than one would like. The effective diffusion radius for destruction of the cage pair reflects the degree of internal consistency which these data possess.

Table II

Rate Constant Derived[a] from Scheme II

Reaction	Constant	Hyponitrite, 32°	Perester, 102°	E_a
Tumbling	k_t(1 cP)	$(1.1 \times 10^{11} \text{sec}^{-1})$[a]	$(1.5 \times 10^{11} \text{sec}^{-1})$[a]	0.0
Internal rotation	k_r	$5.8 \times 10^{10} \text{sec}^{-1}$	$6.7 \times 10^{10} \text{sec}^{-1}$	0.5 kcal/mole
Combination	k_c	$1.0 \times 10^9 \text{sec}^{-1}$	$1.8 \times 10^9 \text{sec}^{-1}$	0.0
Disproportionation	k_2	$2.4 \times 10^9 \text{sec}^{-1}$	$3.8 \times 10^9 \text{sec}^{-1}$	0.9
Scavenging	k_s(DPPH)	3.6×10^9 1./(mol sec)		
Diffusion	k_d(1 cP)	$2.0 \times 10^9 \text{sec}^{-1}$	$2.7 \times 10^9 \text{sec}^{-1}$	0
$r_{diffusion}$[b]		3.3 Å	3.2 Å	
$\tau_{combination}$[c]		$1.8 \times 10^{-10} \text{sec}$	$1.2 \times 10^{-10} \text{sec}$	
$\tau_{retention}$[d]		$6.0 \times 10^{-12} \text{sec}$	$4.5 \times 10^{-12} \text{sec}$	

a) Assuming 2.0 Å as the effective tumbling radius of the 2-butyl radical and the Debye-Stokes relationship.
b) The effective radius for diffusion using the Stokes-Einstein relationship and the derived k_d.
c) The lifetime for all pairs giving combination.
d) The lifetime of the pairs giving retention.

D.   A Suggested Model for the Cage Effect.

Our general conclusion from all of the above experiments was that a phenomenological scheme, which recognizes a single class of cage radicals undergoing cage (re)combination (k_c,k_{-1}) in competition with first order (irreversible) diffusive destruction to free radicals (k_d) and possibly other first order reactions (such as disproportionation (k_{disp}.), decarboxylation (k_2), molecular tumbling (k_t), internal rotation (k_r) and scavenging (k_s[Scavenger]) is the best way to view these reactions. The intercepts of F-fluidity correlations (at f=0) appear to give reasonable estimates of relative rates of competitive chemical reactions. Intervening molecules have a small but definite effect as evidenced by the slopes and intercepts of the F-fluidity correlations.

One additional observation from these experiments is that the common assumption that the rate constant for diffusive destruction of the cage pair (k_d) is the only one which responds to changes in solvent is untenable. A second is that the power of fluidity which is needed to give linear correlations with F ratios is highly variable. Our results at the low fluidity end gave correlations with powers less than 1 (0.5, Figure 2, 0.75, Figures 4, 5). Other cases (23,24) give powers equal to one out to fluidity of about 5 cp.$^{-1}$ with powers greater than unity applying to data obtained at fluidities greater than 5 cp.$^{-1}$ (more fluid, less viscous than hexane at 102°). First power fluidity correlations are expected on the basis of the model of Noyes (25) for geminate combination. However, a number of features make the Noyes model distinct from the cage idea of Frank and Rabinowitch (26) and it is difficult for it to accommodate the less than first power fluidity dependence which is sometimes observed. We therefore felt it worthwhile to attempt to develope a different mathmatical model for the cage effect which remained as close as possible to the phenomenological scheme discussed above.

The model begins with the postulate that a region of space in a solution may be defined as the cage. A pair of radicals inside this region are "caged" and those outside this region are "free". The radius of the cage is called ρ. For simplicity, we consider the cage to be spherical. We therefore begin with the equation (D-1) for a spherical source of particles diffusing into an infinite sink. This equation requires

$$D-1 \quad P_{(r,t)} = \frac{A}{2}\left[\text{erf}\left(\frac{r_N-r}{\sqrt{4Dt}}\right)+\text{erf}\left(\frac{r_N+r}{\sqrt{4Dt}}\right)+\sqrt{\frac{4Dt}{\pi r^2}}\left\{\exp\left(\frac{-(r+r_N)^2}{4Dt}\right)-\exp\left(\frac{-(r_N-r)^2}{4Dt}\right)\right\}\right]$$

a second radius which contains all of the pair at t=0. This we call the normalization radius (r_N, A=P(r_N, t=0)).

The probability that a cage pair is destroyed by diffusive separation is the integral of the source equation (D-1) from the outside boundary of the cage (ρ) to infinity (equation D-2).

D-2
$$P_D(t) = \int_\rho^\infty 4\Pi r^2 P(r,t)dr$$

The probability the cage pair exists at time t ($\Theta(t)$, D-3) is taken as the product of the probability ($1-P_{D(t)}$) that it has not yet diffused times the probability that it has not yet reacted by any chemical reaction (k_i) ($e^{-\Sigma k_i t}$ were k_i pertains to any of the first order chemical reactions mentioned above; combination, disporportionation, scavenging, etc.)

D-3
$$\Theta(t) = e^{-\Sigma k_i t}(1-P_D(t))$$

The probability of forming combination product is simply obtained by integration as in any first order reaction. The final results of these integrations can

D-4
$$Y_c = \int_0^\infty k_c e^{-\Sigma k_i t}(1-P_D(t))dt$$

be written in fairly simple form using the parameters P (which is the ratio ρ/r_N) and B (which is $\sqrt{D/k_c r_N^2}$). Two solutions (D-5 and D-6) are required dependent on P(P>1 or P<1). The value of a cage product yield (Y_c)

$$P = \frac{\rho}{r_N} \qquad\qquad B = \sqrt{\frac{D}{\Sigma k_1 r_N^2}}$$

D-5 $P > 1$

$$Y_c = \frac{k_c}{\Sigma k_1}\left\{1 - \frac{3B}{2}(P+B)\left[(1+B)\exp\left(-\frac{(P+1)}{B}\right) + (1-B)\exp\left(-\frac{(P-1)}{B}\right)\right]\right\},$$

D-6 $P < 1$

$$Y_c = \frac{k_c}{\Sigma k_1}\left\{P^3 - \frac{3B}{2}(1+B)\left[(P+B)\exp\left(-\frac{(P+1)}{B}\right) + (B-P)\exp\left(-\frac{(1-P)}{B}\right)\right]\right\}$$

depends on four variables, the effective diffusion coefficient for separation of the pair (D), the first order rate constants for chemical reactions ($k_i = k_c$, k_2, $k_{disp.}$, k_S[DPPH], etc.), the radius of the cage (ρ) and normalization radius require to enclose the pair as of t=0 when random motion begins (r_N). It is clear from D-5 that F $((1/Y)-1)$ values give $\Sigma k_i/k_c$ ($i \neq c$) when D=0 (fluidity of zero or infinite viscosity) so long as $\rho > r_N$ (P>1). This consideration applies to the (f=0) intercepts shown in Figure 4 where k_i is disproportionation (k_{disp}.). If all chemical reactions are negligibly slow compared to combination ($\Sigma k_i = k_c$ or k_i ($i \neq c$) = 0) and P>1, then the intercept of the F-fluidity correlation (f=0) should be zero as indicated for t-butyl perbenzoate (Figure 2).

On the other hand, with Σk_i($i \neq c$) = 0, a nonzero intercept (f=0, D=0) is predicted if p<1 in which case it should be interpreted as the ratio of the cage volume to the normalization volume at zero fluidity. This condition could apply to radicals studied by matrix isolation which experimentally require an intervening molecule in the generating reaction. In rigid media the effective radius of the cage could be smaller than the radius given by the generating reaction and a fraction of the pair could be infinitely stabile as predicted by D-6 (Figure 7). For a case with P<1 and $\Sigma k_i \neq k_c$, the value of F at D=0 would not be simple to interpret since it would be the product $P^3(\Sigma k_i)/k_c$.

It should be emphasized that D-5 and D-6 are general in effective diffusion coefficient (D), cage radius (ρ) and normalization radius (r_N). Analysis of the D=0 (f=0) limits was examined first since values for the yield, with no translational motion permitted, are independent of the connection one chooses between the effective diffusion coefficient of the pair and the macroscopic viscosity of the medium. We have used the Stokes-Einstein equation (D-7) for this purpose where f is fluidity (1/η, cp.) and b is the effective radius for diffusive separation of the pair.

$$D-7 \qquad D = \frac{kTf}{6\pi b} = \frac{kT}{6\pi\eta b}$$

This allows the basic parameters of the model to be expressed as three distinct radii, that for the cage (ρ), that for the normalization (r_N) and that for diffusive separation (b). One might hope that a single value of ρ and r_N would suffice for a given cage pair source in every solvent. However, consideration of the Noyes' model indicates this cannot be the case. The physical basis here is the distance which a pair separates during the short time while the relative velocity vectors of the cage partners are oriented towards separation. This relative velocity orientation is intrinsic to the initiation event. Variations in combination efficiency with solvent fluidity, in Noyes' model, are ascribed entirely to these small initial

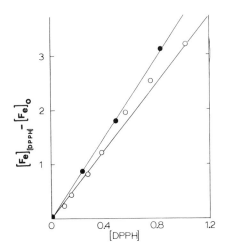

Figure 6. $[F_e]_{[DPPH]} - [F_e]_o$ *vs.* $[DPPH]$
from decomposition of optically active
hyponitrite $(-\bullet-)$ *(22) and ketenimine*
$(-\bigcirc-)$ *(30)*

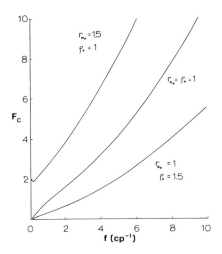

Figure 7. Calculated F_c vs. fluidity for
a hypothetical initiator, $b = 1\text{Å}$, mass =
10, $k_b{}^{\circ} = 2 \times 10^{11}$ sec^{-1} various values
of Δr_m and $\Delta \rho_m$

separations of the pair which vary with the frictional drag (viscosity) of the solvent. Translated into terms of the present model this means that the normalization radius will vary with solvent and we have taken this variation (Δr_N) to be that which Noyes derived (D-8).

D-8
$$\Delta r_N = \sqrt{\frac{3mD^2}{2kT}}$$

A second factor in the normalization radius is the effect of the intervening molecules which, when bonded to the radical partners in the starting material, will impart an initial separation (Δr_m). For the same cage pair partners, the normalization radius will thus vary with both the structure of the initiator and the solvent (D-9).

D-9
$$r_N = b + \Delta r_m + \sqrt{\frac{3mD^2}{2kT}}$$

The radius of the cage (ρ) should likewise be affected by solvent and intervening small molecules. We have approximated the small molecule(s) effect by an increment ($\Delta \rho_m$) equal to the radius of a sphere of the volume occupied by any small molecule eliminated by the initiation event. The solvent dependent factor ($\Delta \rho_n$, D-10) is taken from the mean increment in separation of Brownian particles during the time τ.

D-10
$$\Delta \rho_\eta = \sqrt{6D\tau}$$

The inclusion of the latter factor (D-10) is suggested by the studies of Walling and Lepley (27) on random jumps by a pair of particles in a cubic lattice. These workers showed that the pair has no chance of reencounter after 20 such jumps. During the timescale of the first 20 jumps, reencounters are possible. Alternatively, Noyes defined secondary recombination as those combination reactions involving a pair which had been separated and returned to collision. Either point of view leads to a sensitivity of the cage size (or radius of escape) to the effective diffusion coefficient like that expressed as D-10. The value of τ would roughly correspond to the time of 20 lattice jumps in the Walling-Lepley work and must be fixed empirically in the present model. The final equation for the cage radius (D-11) takes Noyes' secondary combination into account but, in the present model, it cannot be dis-

D-11 $\rho = b + \Delta\rho_m + \sqrt{6D\tau}$

tinguished from the "primary" events.
The other parameters of the model are the first
order rate constants for chemical reaction of the cage
pair. Each has an associated activation volume and is
therefore subject to a small solvent effect over and
above solvation energy considerations. We have ignored such effects except those for k_c, k_{-1} and k_{disp}.,
all of which depend on collision between the cage partners. Analysis of the limiting behavior of equations
D-5 and D-6 indicates that inclusion of the Noyes increment in normalization radius (D-8) <u>demands</u> that the
rate constant for combination vary to an inverse power
of fluidity greater than one. Equation D-12, which
sets the variation in k_c in proportion to the normalization volume, satisfies this requirement. Here $k_b{}^o$

D-12 $k_c = k_b{}^o (b/r_N)^3$

refers to the combination rate constant for the pair in
a sphere of radius b at zero fluidity.
While the parameterization of equation D-5 and D-6
is lengthy, it is also straightforward. The effective
diffusion coefficient is approximated by the Stokes-
Einstein relationship. This equation (D-7) requires
the diffusion radius (b) of the pair which is simply
estimated from Van der Walls radii. The normalization
radius is simply the same value of b plus the increment
of separation of the radical fragments given by the
structure of the covalent initiator (Δr_m) and the viscosity term derived by Noyes (Δr_n). The latter term
ultimately depends again of the radical size (b). The
cage size is simply the same diffusion radius (b) plus
an increment due to the excluded volume of any small
molecules eliminated ($\Delta\rho_m$) plus a diffusion dependent
term ($\Delta\rho_n$) requiring the time τ. Finally, the first
order rate constants for chemical reaction of the pair
in a sphere the size of the particle (b) are required.
These rate constants also vary by the dilution volume
of the intervening molecules and Δr_n.
Figure 7 demonstrates the sensitivity of the model
to the increments in ρ and r_N for a hypothetical pair
of mass 10 and diffusion radius 1 Å, reacting exclusively by combination and diffusive separation ($\Sigma k_i = k_c$
or $\Sigma k_i = 0$) with $k_b{}^o$ equal to $1 \times 10^{12} sec^{-1}$ at 25°C.
  $i \neq c$
These give a purely theoretical look at the possible
effects of differing effective size of the cage and
normalization volumes.

To test the model against experimental data, only two parameters were varied: $k_b{}^o$ (D-12) and τ (D-10). The values of b, Δr_m and $\Delta \rho_m$ were fixed from geometry considerations (Table III). We initially thought that τ could be obtained by a fit to one set of data and would be a constant for every pair. We chose the data of Szwarc and his coworkers (23) on perfluoroazomethane photolysis since these are the most extensive yield-fluidity results available for a single system. Figure 8 shows the fit of calculated (solid line) and experimental (eliptical points) values for F_c vs. fluidity at 65°C. The greater than first power fluidity dependence of the high fluidity region is easily reproduced. The upward curvature of the F_c correlation in the present model due to the downward curvature of P (ρ/r_N) and reduced values of k_c at high fluidity. The present model confirms the intuitive conclusion of Bartlett (27) and Szwarc (23) that highly fluid solvents act to separate the cage pair to a greater extent than simple viscosity considerations would suggest.

Figure 9 shows the fit of the data (28) for the same pair in a single solvent (2-methylbutane) from +90°C to -190°C. The values of all parameters were the same as those in Figure 8 except τ which varied inversely with the temperature (with the value at 65° being set by the fit of Figure 8). The very strong upward curvature (greater than first power fluidity dependence) in both Figure 8 and Figure 9 is directly attributable to the Noyes term in r_N (Δr_η) which causes decreasing values of P (ρ/r_N) and k_c at high fluidity.

Figure 10 shows the fit obtained to our data (22) on the ether formation from the 2-methylbutyryl perester. In this case, the calculated line shows downward curvature when plotted vs. fluidity to the first power. This downward curvature is related to the increase in the value of P at fluidities between 0 and 1 cp^{-1}. This feature, in the present model, is directly related to the inclusion of the fluidity dependent term in the cage radius (Δr_η) which is responsible for the initial rise in P (ρ/r_N) values. The physical interpretation is that, at low fluidities there is a relatively important increase in the chance for slight separation and return of the pair while at high fluidities the separation achieved while passing the transition state for homolysis (Δr_η) makes the start of random motion nearer the boundary of the cage limit.

Figure 11 shows the fit obtained for the hyponitrite source (12) of the 2-butyl-t-butoxy pair using the same value of $k_b{}^o$ as obtained in Figure 10. That

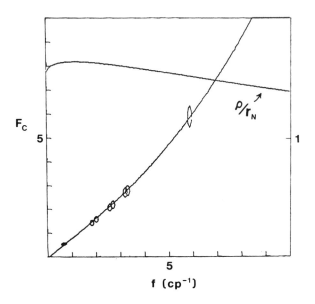

Figure 8. F_c *vs. fluidity for* $[CF_3 \cdot (N_2)CF_3 \cdot]$ *(23)*

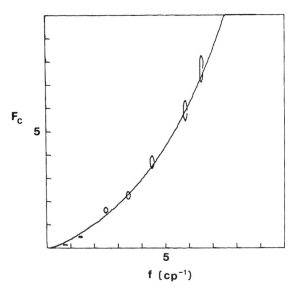

Figure 9. F_c *vs. fluidity for* $[CF_3 \cdot (N_2) \cdot CF_3]$ *in 2-methylbutane (24)*

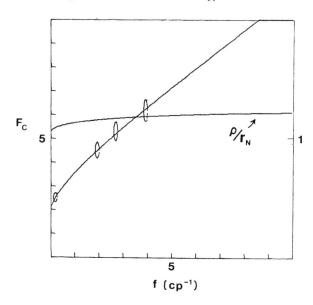

Figure 10. F_{ether} *vs. fluidity for* tert-*butyl 2-methylbutyl ether, perester 102°C (12).*

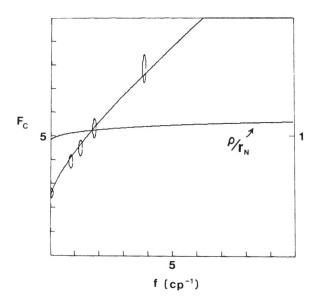

Figure 11. F_{ether} *vs. fluidity for* tert-*butyl 2-methylbutyl ether formation, hyponitrite (22) 32°C*

is, the same value of k_c for a pair in a volume of
radius b was used in both data sets (Figures 10 and 11).
The actual k_c for the hyponitrite is lowered at all
viscosities, due to the larger r_N resulting from the
extra nitrogen molecule in the hyponitrite initiator.
The intervening molecule effect deduced in the discussion of Figure 4 and Table II is thus reproduced here.

Figure 12 shows a fit to our data (5, 8) on the
O-benzoyl-O-t-butyl hyponitrite. These data require
an anomalously low value for $\Delta\rho_m$ (Table III) but, with
that adjustment, a square root dependence of F_c on
fluidity is observed in the theoretical curve over the
viscosity range which was investigated.

Figure 13 shows a fit which was obtained for the
data pertaining to di-tert-butyl peroxide formation
from the corresponding hyponitrite (29). These data
show the most variable apparent fluidity dependence of
any available. The high viscosity (low fluidity) region is approximately linier with square root fluidity
while the high fluidity (low viscosity) region varies
with a power of fluidity greater than one. The fit of
the calculated line, shown in Figure 15, required a
fluidity sensitive factor in r_N (r_η, eqn. D-8) 10 times
larger than the Noyes relationship. This relationship
(D-8) gives a minimum value for the pair separation
during the homolysis event since it assumes no excess
translational energy (above kT) is available to the
separating fragments. The deaminative nature of the
hyponitrite reaction suggests that as much as 30 kcal/
mole could be available for translational separation of
the fragments (occurring in a time less than 10^{-13}sec).
A magnification of the Noyes factor is not entirely
unreasonable in such cases but additional experimental
verification is needed before this factor can be considered a certainty.

Table III contains a summary of the parameters
entering these calculations and shows the range of k_c^0
and τ values. The cage combination of benzoyloxy-t-
butoxy and t-butoxy-t-butoxy pairs have significantly
lower rate constants for combination than that for
perfluoromethyl pairs. An additional observation,
which was interesting to us, is that the product of
k_c^0 and τ is nearly constant. In other words, the τ
required in D-10 is approximately a constant divided
by Σk_i^0 for the pair undergoing reaction. A pair
undergoing a slow combination reaction should thus
have a larger effective cage radius. There is an intuitive basis for this to be a real feature of cage
reactions.

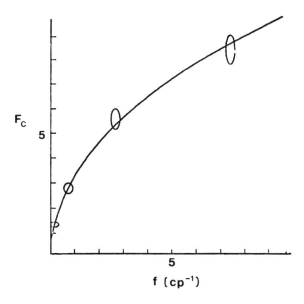

Figure 12. $F_{perester}$ vs. fluidity for tert-butyl *perbenzoate from the hyponitrite (8)*

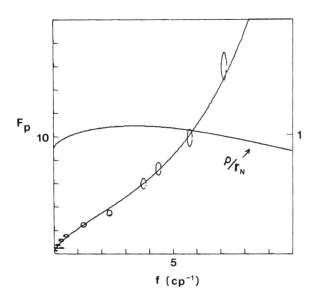

*Figure 13. F_c vs. fluidity for di-*tert-butyl *peroxide from the hyponitrite (9)*

Table III

Cage Model Summary

Fig.	Pair	T°c	k_b^o x 10^{-11}sec.[f]	$k_c^{2cp^{-1}}$ 10^{-11}sec.[a]	τ psec.	b	Δr_m	$\Delta \rho_m$	Σk_1^o·τ[f]
8	[CF3· N2 ·CF3]	65	2.70	0.59	0.14	1.35Å	0.67Å	1.74Å(N2)	0.038[b]
9	[CF3· N2 ·CF3]	Vari-able	2.70	0.59	48/T°K	1.35Å	0.67Å	1.74Å(N2)	
10	[2-Bu· CO2 ·OtBu]	102	0.39	0.11	0.39	3.20Å	1.60Å	1.84Å(CO2) [/]	0.040[c]
11	[2-Bu· CO2 N2 ·OtBu]	32	0.39	0.06	0.48	3.20Å	2.26Å	2.26Å(CO2,N2)	0.057[d]
12	[tBuO· N2 ·OtBu]	45	0.20	0.03	1.5	2.70Å	1.60Å	1.74Å(N2)	0.030[b,e]
13	[ØCO2· N2 ·OtBu]	32	0.007	0.002	6.3	2.70Å	1.60Å	0.70Å(N2)	0.043[b]

a) k_o^{2cp}· is the rate constant for combination at a fluidity of 2 cp.$^{-1}$.
b) $\Sigma k_1^o = k_b^o$.   c) $\Sigma k_1^o = k_b^o + k_{disp}^o$. = 3.3 k_b^o.   d) $\Sigma k_1^o = k_b^o + k_{disp}^o$. = 3.1 k_b^o.
e) The value for Δr_n was x 10 that given by D-8.   f) Super zero refers to f=0.

We believe the present model is distinct and superior to the version of Noyes. It uses structures and rate constants in familiar terms. It appears to deal with intervening molecule effects in a manner which is in keeping with experimental observations. It provides a rationalization for the variable power of fluidity which is needed to correlate F_c values. It supports the notion that the intercepts (at f=0) in F_c - fluidity correlation are generally due to chemical reaction competing with combination. Overall it provides a new framework for designing experimental tests of the effects of the several variables which may be controlled in chemical reactions.

On the negative side, it should be emphasized that the fits between experimental and calculated results shown in Figures 8-13 are not unique. A different criterion (within limits) for estimating b, Δr_m and/or $\Delta \rho_m$ would give equally acceptable fits using different values for k_b^o and τ. Application of picosecond laser methods or adaptation of the present model to CIDNP theory could remove some of this ambiguity and would serve as a more stringent test of the model than existing data allow.

Acknowledgement: Appreciation is due to Prof. R. M. Mazo for his assistance in the integrations involved in the development of the present model and to Drs. J. M. Owens, J. G. Huntington, W. R. Mabey, J. A. Hoobler and Max Deinzer whose doctoral theses contain much of the experimental work of this paper. Appreciation is also expressed to the National Science Foundation for financial support.

Literature Cited

1. Bartlett, P. D. and Hiatt, R. R., J. Am. Chem.
   Soc., (1958), 80, 1398; Bartlett, P. D. and Tidwell, T. T., ibid, (1968), 90, 3294.
2. For reviews see: Swern, D., "Organic Peroxides
   Vol. II", Wiley, New York (1971); Kochi, J., "Free
   Radicals Vol. I", Wiley-Interscience, New York,
   1973.
3. Koenig, T., and Wolf, R., J. Am. Chem. Soc.,
   (1969), 91, 2569; Koenig, T., Huntington, J. G.
   and Cruthoff, R., J. Am. Chem. Soc., (1970), 92,
   5413.
4. Bartlett, P. D. and Rüchardt, Ch., J. Am. Chem.
   Soc., (1960), 82, 1756.
5. Koenig, T. and Deinzer, M., J. Am. Chem. Soc.,
   (1966), 88, 4518.
6. Huisgen, R. and Reinertshofer, Annalen, (1952),
   575, 174, 197; Huisgen, R. and Krause, T., ibid.,
   (1951), 574, 157; Hey, D. H., Webb, J. and Williams, J., J. Chem. Soc., (1952), 4657.
7. Rüchardt, Ch. and Tan, C., Chem. Ber., (1970), 103,
   1774.
8. Koenig, T., Hoobler, J. A. and Deinzer, M., J. Am.
   Chem. Soc., (1971), 93, 938.
9. Mabey, W. R., Ph.D. Thesis, University of Oregon,
   1972.
10. Koenig, T. and Deinzer, M., J. Am. Chem. Soc.,
    (1968), 90, 7014.
11. Koenig, T., Huntington, J. G. and Mabey, W. R.,
    Tetrahedron Letters, (1973), No.36, 3487.
12. Koenig, T., and Owens, J., J. Am. Chem. Soc.,
    (1974), 96, 4052.
13. Owens, J. M., Ph.D. Thesis, University of Oregon,
    1976.
14. Pryor, W. A. and Smith, K., J. Am. Chem. Soc.,
    (1970), 92, 5403.
15. Neuman, R. C., Jr., Accounts Chem. Res., (1972),
    5, 381.
16. Owens, J. and Koenig, T., J. Org. Chem., (1974),
    39, 3153.
17. Koenig, T. and Hoobler, J. A., Tetrahedron Letters,
    (1972), No.18, 1803.
18. Koenig, T., Tetrahedron Letters, (1973), No.36,
    3487.
19. Bloomquist, A. T. and Bernstein, I., (1961), J.
    Am. Chem. Soc., 73, 5546.
20. Koenig, T. and Huntington, J. G., J. Am. Chem. Soc.,
    (1974), 96, 592.

21. Huntington, J. G., Ph.D. Thesis, University of Oregon, 1974.
22. Koenig, T. and Owens, J., J. Am. Chem. Soc., (1973), 95, 8484.
23. Dobis, O., Pearson, J. M. and Szwarc, M., J. Am. Chem. Soc., (1968), 90, 278; Chakravorty, K., Pearson, J. and Szwarc, M., ibid., (1968), 90, 283.
24. Herkes, F., Friedman, J. and Bartlett, P. D., Int. J. Chem. Kinetics, (1969), 1, 193.
25. Noyes, R. M., Prog. Reaction Kinetics, (1963), 1, 129.
26. Frank, J. and Rabinowitch, E., Trans. Faraday Soc., (1934), 30, 120.
27. Walling, C. and Lepley, A., Int. J. Chem. Kinetics, (1971), 3, 97.
28. Chakrovorty, K., Pearson, J. M. and Szwarc, M., Int. J. Chem. Kinetics, (1969), 1, 357.
29. Kiefer, H., and Traylor, T., J. Am. Chem. Soc., (1969), 89, 6667.
30. Waits, H. P. and Hammond, G. S., J. Am. Chem. Soc., (1964), 86, 1911.

RECEIVED December 23, 1977.

Gamma Radiation-Induced Free Radical Chain Reactions

ABRAHAM HOROWITZ

Soreq Nuclear Research Center, Yavne, Israel

The use of gamma radiation for the initiation of free radical chain reactions in liquids offers several advantages over the more conventional methods of initiation. With this technique radical reactions that are not readily accessible by other methods can be studied. Because of the wide temperature range within which radiolytic initiation can be applied this method allows an accurate determination of Arrhenius parameters and therefore can bring better understanding and deeper insight to the factors that control the rates and mechanism of free radical reactions.

At Soreq gamma radiolysis of liquid systems has been used in studies of various free radical chain reactions. The purpose of this work is to illustrate and demonstrate, through a survey of those studies, the advantages and limitations of the radiolytic method.

Radiolytic Initiation

General aspects. Most of the chain reactions that will be discussed in this work were initiated in alkane solutions by ^{60}Co gamma radiation. Therefore, the following rather brief and somewhat simplified discussion of the primary radiolytic processes will be limited mainly to alkanes and alkane solutions. An understanding of these processes is necessary in order to explain how radicals are generated by ionizing radiation and to understand the causes for the advantages and shortcomings of this technique. A detailed discussion of the primary steps that take place in a liquid exposed to ionizing radiation is beyond the scope of this work and can be found in several excellent texts (1-4).

In a pure alkane (RH) the radiolytic generation of radicals can be described by reaction 1. Surprisingly, this rather schematic formulation of the initiation

$$2RH \rightarrow 2R + H_2 + products \qquad (1)$$

process applies also to alkane solutions of various compounds
provided that a long chain reaction is initiated by the absorption
of gamma radiation.

How are those radicals formed? In a liquid that is exposed
to ionizing radiation the formation of radicals is preceded by a
number of rather complex steps. Basically, in the primary steps
that follow the absorption of ionizing radiation the molecules of
the absorbent become ionized or electronically excited. In nonpolar liquids such as alkanes, the charge neutralization processes
are fast and therefore the lifetime of the charged species,
molecular cations (RH^+) and electrons, is very short. The directly formed excited molecules (RH^*) or those that are created in
the neutralization processes (RH^{**}) can lose their energy in
processes such as radiative and nonradiative conversion to the
ground state, collisional deactivation, etc. These processes do
not result in a chemical change in the system. Alternatively,
these excited molecules can dissociate into molecular products
and free radicals. This latter chain of events, that leads to
the formation of radicals, is summarized by reactions 2-11.

$$RH \rightarrow RH^+ + e \qquad\qquad \text{(ionization)} \qquad\qquad (2)$$

$$RH \rightarrow RH^* \qquad\qquad \text{(excitation)} \qquad\qquad (3)$$

$$RH^+ + e \rightarrow RH^{**} \qquad\qquad \text{(neutralization)} \qquad\qquad (4)$$

$$RH^*, RH^{**} \rightarrow R + H \qquad\qquad\qquad\qquad\qquad (5)$$
$$\text{(radical formation)}$$
$$RH^*, RH^{**} \rightarrow R' + R'' \qquad\qquad\qquad\qquad\qquad (6)$$

$$RH^*, RH^{**} \rightarrow H_2 + \text{olefin} \qquad\qquad\qquad\qquad\qquad (7)$$
$$\text{(molecular products)}$$
$$RH^*, RH^{**} \rightarrow P_1 + P_2 \qquad\qquad\qquad\qquad\qquad (8)$$

$$H + RH \rightarrow H_2 + R \qquad\qquad\qquad\qquad\qquad (9)$$

$$R' + RH \rightarrow R'H + R \qquad\qquad \text{(hydrogen transfer)} \qquad (10)$$

$$R'' + RH \rightarrow R''H + R \qquad\qquad\qquad\qquad\qquad (11)$$

It can be seen that the overall change caused by reactions
2-11 will be given by reaction 1, provided that the excited
molecules RH^* and RH^{**} decompose primarily by the rupture of
C-H bonds as in reaction 5. A priori, such selective bond
cleavage seems to be rather unexpected since the gamma photons of
^{60}Co have a mean energy of 1.25 million electron volts (MeV),
which by far exceeds the energy of any chemical bond. In fact,
however, selective C-H bond rupture is indeed observed. For
example, cyclohexene and bicyclohexyl comprise more than 95% of
the carbon-containing products that are formed in the gamma

radiolysis of cyclohexane. A similar product distribution is observed in the radiolysis of other alkanes.

Chain reactions are usually studied in mixtures, where the situation is more complicated. In mixtures, processes such as electron capture (reaction 12), charge transfer (reaction 13) and energy transfer (reaction 14) might result in the preferential decomposition of the solute (SX) instead of the alkane solvent. However, even in that case, the transfer reactions 9-11 are replaced by reactions 19-21 so that the net effect is reduction in the yield of hydrogen and the regeneration of alkyl radicals. If reactions 19 and 20 are faster

$$e + SX \rightarrow S + X^- \text{ or } SX^- \qquad \text{(electron capture)} \qquad (12)$$

$$RH^+ + SX \rightarrow RH + SX^- \qquad \text{(charge transfer)} \qquad (13)$$

$$RH^*, RH^{**} + SX \rightarrow RH + SX^* \qquad \text{(energy transfer)} \qquad (14)$$

$$X^- + RH^+ \rightarrow RH + X \qquad (15)$$

$$SX^- + RH^+ \rightarrow RH + SX^* \qquad (16)$$

$$SX^+ + e \rightarrow SX^* \qquad (17)$$

$$SX^* \rightarrow S + X \qquad (18)$$

$$S + RH \rightarrow HS + R \qquad (19)$$

$$X + RH \rightarrow HX + R \qquad (20)$$

$$X^- + RH^+ \rightarrow HX + R \qquad (21)$$

than the termination reactions of S and X radicals and if R radicals participate in a long chain reaction, then the schematic description of the initiation step by reaction 1 can still be used.

Let us now consider some quantitative aspects of radiolytic initiation. The radiolytic yield, G value, is defined as the number of molecules that 'react' per 100 eV of absorbed energy. In alkanes and alkane solutions, G(radicals) does not exceed 10. This value is practically independent of temperature ($\underline{5}$). Therefore, product formation with a G value that exceeds 10 indicates that this product is formed, at least in part, in a chain reaction.

The amount of γ radiation energy absorbed per unit volume of an absorbent is determined by the number of electrons in that volume. This number is known as electron density (E_d) and is given by expression I, where Z is the apparent atomic number of a molecule,

(I) $E_d = dZN/M$

i.e. the sum of the atomic numbers of the elements in the molecule
[$Z(CCl_4)$ = 6 + 4x17 = 74], N is Avogadro's number and d is the
density of the absorbent. Since many of the organic compounds
have nearly the same electron density, the energy absorbed per
unit volume of pure solvent and the energy absorbed by the same
volume of a concentrated solution are almost equal. In other
words, the rate of radiolytic initiation is practically independent
of the composition of the irradiated solution.

The advantages and limitations of radiolytic initiation. Radiolytic initiation is particularly useful in kinetic studies of
liquids phase free radical chain reactions. The advantages of
this technique can best be seen by comparing it with other commonly used methods of initiation, photolysis and thermal decomposition of initiators.

The determination of the time dependence of the rate of a
chain reaction and the study of the effects of reactants concentrations and of incident light intensity are the most commonly
used techniques for the elucidation of the kinetics and mechanism
of chain reactions. The best way to establish the effect of these
three parameters is to change them separately. However, only radiolytic initiation allows such 'one at a time' variation of
parameters. In this case the rate of initiation is independent of
the composition of the solution and it is constant in time and
in space (neglecting 'spur reactions'). In comparison, photochemical initiation is possible only when one of the components of the
solution absorbs ultra violet or visible light. The concentration
of this absorbent is constantly depleted. In solutions with low
absorbance this change in concentration is accompanied by a
decrease in the rate of initiation. This is a particularly
serious problem when the light absorbent is also one of the reactants in the chain reaction. While it is true that the concentration change in solutions that absorb a large fraction of the
incident light has a small effect on the overall rate of initiation, the rate of initiation in regions that are removed from the
light source is considerably lower than in regions that are close
to it. In other words, the rate of initiation is not uniform in
space. Thermal initiation is uniform in space but not in time.
As in the case of photochemical initiation, thermal decomposition
of the initiator and its consumption in a subsequent chain reaction present the most complicated situation.

Because of its high penetrating power, gamma radiation induced
reactions can be carried out in opaque vessels and metal autoclaves. Consequently, the temperature range within which a chain
reaction can be studied in a given solution, extends from its
freezing point to its critical temperature. Thermally induced
reactions can also be studied in autoclaves. However, the tempe-

rature range in which thermal initiation can be employed is usually quite narrow as it is determined by the thermal stability of the initiator. Furthermore, even within this narrow region, the rate of initiation is temperature dependent. In contrast, the rate of radiolytic initiation is independent of temperature. Photochemical initiation on certain cases is also independent of temperature. However, the need for transparent cells limits the pressure and therefore the temperature at which photochemical studies can be carried out.

Finally, side reactions of the absorbent or the initiator may compete with and inhibit the chain reaction under investigation. The products of these side reactions interfere with the determination of the main reaction products. When radiolytic initiation is used the solvent is the main source of radicals and consequently these complications are minimized.

Radiolytic initiation is not free from limitations and shortcomings. We have previously seen that ions are formed in the initial steps that follow the absorption of gamma radiation. While in pure alkanes the lifetime of the ions is very short, this is not necessarily the situation in the presence of solutes. In some cases, solute anions and cations formed by electron capture and charge transfer, respectively, may have considerably long lifetimes. If this occurs, it is possible that ionic reactions take place and therefore the free radical mechanism of the chain reaction has to be established. This can be done by various methods such as the addition of electron, charge and radical scavengers, and the determination of the effect of these additives on the chain reaction. Another method is to show that the same reaction can also be initiated thermally or photochemically.

The second complication encountered in radiolytic studies in which a radical derived from the solvent participates in the reactions, arises from the fact that the termination reaction of this radical and the initiation can result in the formation of the same product. Further clarification of this problem seems to be in order and it brings us back to the details of the initiation step. It has been found that formation of some of the radiolytic products cannot be suppressed by the addition of radical scavengers. These products are known as 'molecular products', which is misleading since not all these products are formed by a true molecular mechanism. There are various reasons for the inability of radical scavengers to suppress their formation, but they will not be discussed here. For the purpose of this work, the important consequence of the formation of 'molecular products' lies in the fact that if a certain product is formed both by a molecular mechanism and in a radical termination reaction, then the yield of this product cannot, as such, be used for the estimation of the steady state concentration of radicals. For example, in cyclohexane a significant portion of the radiolytic yield of cyclohexene and bycyclohexyl is 'molecular'. These 'molecular yields have to be subtracted from the total yields of those products and only the corrected values represent the amounts of cyclohexene and bicyclohexyl that are formed in the termination reactions of the cyclohexyl radicals. It should be noted that the complications arising from the formation of 'molecular products' are similar to those encountered in photochemically and thermally induced reactions in which the 'cage' effect is important.

Experimental

Thoroughly degassed solutions in glass ampules were irradiated in a ^{60}Co gamma source. Irradiation at elevated temperatures was made possible by the use of metal autoclaves in which the ampules were immersed in solution of the same composition as the sample. This procedure ensures that an increase in the irradiation temperature brings about an equal change in pressure on both sides of the ampule. This irradiation apparatus was devised by Katz (6) and is shown in Figure 1. Gas chromatography was the main method used for product analysis.

Chlorovinylation of Alkanes

Gamma radiolysis of alkane (RH) solutions of chloroethylenes (ECl$_2$) i.e., C_2Cl_4 (7-10), C_2Cl_3H (9-11), cis- and trans- $C_2Cl_2H_2$ (9,10,12) and C_2Cl_3F (13), results in the formation of hydrogen chloride and chlorovinyl substituted alkanes. These products are formed in almost equal yields with G values that exceed 1000 at 200°C (9,13), clearly indicating that a chain mechanism is operative. The free radical character of this mechanism was established in experiments with added inhibitors (7) in which it was shown that the radical scavengers, iodine and pentene, strongly inhibit the chain reaction, while SF$_6$ and N$_2$O, known to be efficient electron scavengers, do not. The propagation step of the radical chain mechanism that is consistent with these findings is given by reactions 22 through 25 and the termination step by reactions 26 through 31. This reaction sequence is identical to the one suggested for peroxide (14, 15) and photo (16) initiated chlorovinylation.

$$R + ECl_2 \rightarrow RECl_2 \qquad\qquad (22)$$

$$RECl_2 \rightarrow RECl + Cl \qquad\qquad (23)$$

$$RECl_2 + RH \rightarrow RECl_2H + R \qquad\qquad (24)$$

$$Cl + RH \rightarrow HCl + R \qquad\qquad (25)$$

$$2R \rightarrow R_2 \qquad\qquad (26)$$

$$2R \rightarrow cyclohexene + RH \qquad\qquad (27)$$

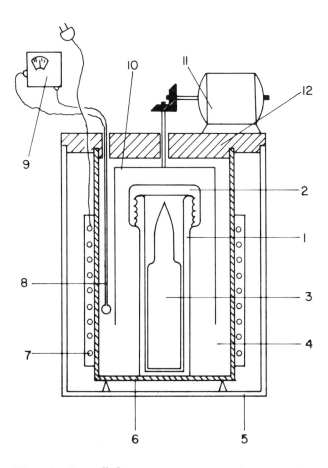

Figure 1.   Controlled temperature apparatus for gamma irradiation. 1. autoclave, 2. cover, 3. ampule, 4. silicon–oil bath, 5. glass beaker 6. aluminum beaker, 7. heater, 8. thermocouple, 9. potentiometer, 10. stirrer, 11. motor, 12. cover.

$$R + RECl_2 \rightarrow RCl + RECl \tag{28}$$

$$R + RECl_2 \rightarrow \text{cyclohexene} + RE \tag{29}$$

$$2RECl \rightarrow (RECl)_2 \tag{30}$$

$$2RECl \rightarrow RE + RECl_2 \tag{31}$$

It should be noticed that all possible reactions do not always take place. Thus chlorine elimination (reaction 23) from $c\text{-}C_6H_{11}CCl_2CClF$ and $c\text{-}C_6H_{11}C_2Cl_3$ radicals is so fast that reaction 24 cannot compete with it. Also, cyclohexyl radicals add only to the less substituted end in trichloroethylene. When reaction 24 does occur, it is of minor importance, as is indicated by the fact that $G(RECl) \gg G(RECl_2H)$.

We will now demonstrate how the radiolytic method can be used for the determination of the rate constants for the addition of cyclohexyl radicals to chloroethylenes in terms of the above outlined reaction scheme. Using the steady state assumption under conditions of a long chain reaction it can be shown that the rate of formation of $RECl_2$ is given by expression II. According to this expression k_{22} can be determined

$$(II) \qquad [R_{RECl}] = k_{22}[R][ECl_2] = k_{23} \, RECl_2$$

provided that the steady state of the cyclohexyl radicals can be estimated from the rates of formation of the termination products cyclohexene and bicyclohexyl. Unfortunately, both products are also formed by 'molecular processes' and, therefore, only the corrected values, obtained by the subtraction of the molecular yields from the experimental values, can be used for the estimation of the steady state concentration of cyclohexyl radicals. This is a rather inaccurate method, particularly when the difference between the experimental and the molecular yields is small, and should therefore be used only when there is no other alternative.

The alternative method for the derivation of the steady state concentration of cyclohexyl radicals utilizes the fact that the rate of radiolytic generation of cyclohexyl radicals is not affected by the presence of chloroethylenes. Assuming that the sum of the steady state concentrations of the R and $RECl_2$ radicals is constant and that it is equal to the steady state concentration of R radicals in pure alkane, it can be shown that $G(RECl_2)$ should be given by the following expression:

$$(III) \qquad \frac{1}{G(RECl_2)} = \left[\frac{2\alpha(k_{26}+k_{27})}{G(R_o)}\right]^{\frac{1}{2}} \left[\frac{1}{k_{23}} + \frac{1}{k_{22}[ECl_2]}\right]$$

where $G(R_0)$ is the yield of cyclohexyl radicals in pure cyclohexane, which is equal to 5.7 (_17_)and α is a constant that converts G values into rates of formation in units of mole ℓ^{-1} sec^{-1}. The applicability of this relation was verified for $ECl_2 = C_2Cl_4$, C_2Cl_3H and C_2Cl_3F.

Since $G(R_0)$ is known, the ratio $k_{22}/(k_{26}+k_{27})^{\frac{1}{2}}$ can be determined from the slope of the plot of $1/G(RECl_2)$ versus the reciprocal of the concentration of chloroethylene. The advantage of this method is that it does not require the determination of the termination products, which can be difficult for a long chain reaction. When such a reaction occurs the chloroethylene may be consumed long before a detectable amount of termination products is formed. Furthermore, at sufficiently low ECl_2 concentrations, when $k_{23} \gg k_{22}[ECl_2]$, expression III reduces to the linear form given in expression IV. Reversing this argument, it can be seen that if $G(RECl_2)$

$$\text{(IV)} \qquad G(RECl_2) = k_{22}\left[\frac{G(R_0)}{2\alpha(k_{26}+k_{27})}\right]^{\frac{1}{2}}[ECl_2]$$

is proportional to $[ECl_2]$, then $k_{23} \gg k_{22}[ECl_2]$ and therefore $[R] \gg [RECl_2]$. In other words, when expression IV is obeyed the steady state concentration of R radicals is constant. This approach was used in the derivation of the rate constants and Arrhenius parameters for the addition of cyclohexyl radicals to C_2Cl_4. The Arrhenius expression for the rate constants for the addition of cyclohexyl radicals to the other chloroethylenes were obtained from competitive studies. These results are summarized in Table I.

Inspection of the results given in Table I reveals that the reactivity trends in the addition reaction are determined by steric and inductive effects. Thus, trichloroethylene is more reactive than the dichloroethylenes because of the electron withdrawing effect of the additional chlorine atom. However, in tetrachloroethylene this effect is offset by the steric effect and therefore this compound is less reactive than trichloroethylene. Particularly interesting is the comparison between C_2Cl_4 and C_2Cl_3F. Fluorine has a strong inductive effect. Unlike chlorine it does not sterically hinder the reaction and therefore the addition of $c-C_6H_{11}$ radicals to C_2Cl_3F requires an activation energy that is lower by 1.2 kcal/mole than the activation energy needed for the addition of those radicals to C_2Cl_4. On the other hand, the same inductive effect lowers the reactivity of the CCl_2 site in C_2Cl_3F to the extent that it requires an energy of activation that is higher by 1.2 kcal/mole than the activation energy for the addition to C_2Cl_4.

Table I. Arrhenius Parameters and Relative Rate Constants for the
Addition of $c-C_6H_{11}$ Radicals to Chloroethylenes (ECl_2)

ECl_2 (a)	$\log k_{22}/k_{22}(C_2Cl_4)$ at $150°$	$\log A_{22}$ (b) $M^{-1}sec^{-1}$	E_{22} (b) kcal/mole	Ref.
$cis-C_2Cl_2H_2$	4.92	8.72	7.30	10
$trans-C_2Cl_2H_2$	5.35	9.13	7.25	10
$\underline{HCCl}=CCl_2$	5.58	9.10	6.75	10.
C_2Cl_4	4.88	8.68	7.30	10
$\underline{HCF}=CCl_2$	5.32	8.64	6.38	13
$HCF=\underline{CCl}_2$	4.30	8.72	8.49	13

(a) the site of addition is underlined.
(b) derived by taking $2(k_{26}+k_{27}) = 2x10^9$ $M^{-1}sec^{-1}$ (16) and
$E_{dif} = 4.56$ kcal (19)

Chain Reactions of Halomethanes and Haloethanes

General. The free radical chain decomposition of halogen (mainly chlorine) substituted methanes (XM) and haloethanes (XEtY), was studied in cyclohexane. The propagation step of the chain reaction of halomethanes is given by reactions 32 and 33 while reactions 34 through 37 describe the chain decomposition of haloethanes. These reactions can be

$$R + XM \rightarrow RX + M \tag{32}$$

$$M + RH \rightarrow MH + R \tag{33}$$

$$R + XEtY \rightarrow RX + EtY \tag{34}$$

$$EtY \rightarrow Et + Y \tag{35}$$

$$EtY + RH \rightarrow EtYH + R \tag{36}$$

$$Y + RH \rightarrow HY + R \tag{37}$$

divided into three groups: halogen transfer, hydrogen transfer and halogen elimination. In addition, in some cases, hydrogen transfer from XM and XEtY to the R radical also becomes important.

Chlorine transfer reactions. The rate constants and Arrhenius parameters of chlorine transfer reactions, given in Table II, were determined by competition studies in solutions that contained another compound, SX, in addition to the chloro-methane (ClM) or chloroethane (EtCl$_2$) under investigation. The reaction of SX with cyclohexyl radicals (reaction 38) served as a reference.

$$R + SX \rightarrow P \tag{38}$$

In these systems, the rate constant for chlorine transfer, k_{Cl} , is given by expression V. At low conversions, expression V reduces to expression VI. The indices i,f and av indicate initial, final and average concentrations, respectively. The same equations

$$\text{(V)} \qquad \frac{k_{Cl}}{k_{38}} = \frac{\log[ClM]_i/[ClM]_f}{\log[SX]_i/[SX]_f}$$

$$\text{(VI)} \qquad \frac{k_{Cl}}{k_{38}} = \frac{G(RCl)[SX]_{av}}{G(P)[ClM]_{av}} = \frac{G(MH)[SX]_{av}}{G(P)[ClM]_{av}}$$

are used for solutions containing chloroethane, by substituting
$[EtCl_2]$ for $[ClM]$.

The chlorine transfer data given in Table II can be treated
in various ways and grouped accordingly. Thus the compounds
$ClCCl_3$, $HCCl_3$, CF_3CCl_3, CH_3CCl_3, CCl_3CCl_3, CH_2ClCCl_3 and CCl_3CN
are included in a series having the general form $XCCl_3$, while
$ClCCl_2H$, $HCCl_2H$, $HCCl_2CH_2Cl$, CH_3CHCl_2 and $CHCl_2CN$ are in the
$XCHCl_2$ series. As can be seen from Figure 2, within each series,
the chlorine transfer rate constants correlate well with the Taft
polar substituent constants σ_x^*.

Another way to treat the results of Table II is to compare
reactivity trends in pairs such as $CCl_4 - CCl_3CN$, $CHCl_3 - CHCl_3CN$
and $CH_2Cl_2 - CH_2ClCN$. This approach shows that substitution of
a chlorine atom by the cyano group lowers the activation energy
E_{Cl} by 3.6 kcal/mole. This effect reflects the lower C-Cl bond
strength in chloroacetonitriles, and the stabilization of the
radicals formed by chlorine transfer reactions of those compounds.

Hydrogen transfer reactions. Hydrogen transfer reactions
of CCl_3 (25,26), CCl_2CN (24) and CH_2CN (27) radicals, generated
by chlorine and bromine atom transfer from CCl_4, CCl_3CN and
$BrCH_2CN$, respectively, were studied in alkanes. In these systems
combination is the only termination reaction. Therefore, $G(MH)$
and $G(RX)$ are given by expression VII, where k_t

$$M + M \rightarrow M_2 \qquad\qquad\qquad (39)$$

$$(VII) \qquad G(RX) = G(MH) = \frac{k_H}{k_t^{\frac{1}{2}}} \alpha^{\frac{1}{2}} G(M_2)^{\frac{1}{2}} [RH]$$

$= k_{39}$ and $k_H = k_{33}$ and α is a known coefficient that converts G
values into rates of formation.

It should be noticed that the products M_2 are derived from
the solute and therefore are not formed by 'molecular processes'.
Consequently the experimentally determined values of $G(M_2)$ can
be substituted directly into expression VII. Since [RH] and α
are known, this expression can then be used to obtain the rate
constant ratio $k_H/k_t^{\frac{1}{2}}$ and the respective relative Arrhenius parameters $E_H - 0.5E_t$ and $A_H/A_t^{\frac{1}{2}}$. This method was used in the derivation of rate constants and Arrhenius parameters of the H
transfer reactions of trichloromethyl radicals in cyclohexane and
n-hexane.

Another situation exists in $BrCH_2CN$ and CCl_3CN solutions,
where at very low solute concentrations, $G(CH_3CN)$ and $G(CHCl_2CN)$
reach constant values. Since the alkane concentration is
constant this can happen only if $[M]>>[R]$. Under these conditions
expression VIII can be used for the derivation of $k_H/k_t^{\frac{1}{2}}$. The
relative Arrhenius parameters of the H transfer reactions of

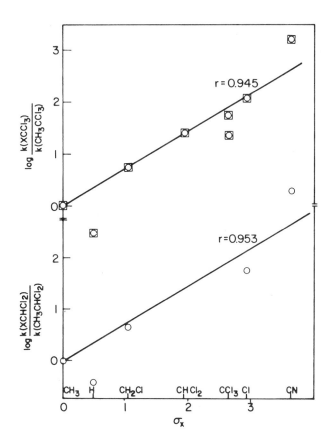

Figure 2. Plot of relative values of log k_{Cl} at 80° vs. Taft σ*
parameters, r = linear correlation coefficient.

Table II. Rate Constants and Arrhenius Parameters of Cl Atom
          Transfer Reactions of Cyclohexyl Radicals [a]

Compound	$\log A_{Cl}$ $M^{-1}sec^{-1}$	E_{Cl} kcal/mole	Temp. ^oK	$\log k_{Cl}$ [c] $M^{-1}sec^{-1}$	Ref.
CCl_4	9.40	5.88	333–453	5.76	20
CCl_3H	9.45	10.16	392–492	3.16	20
CCl_2H_2	9.24	13.67	463–523	0.98	20
CF_3CCl_3	8.97	6.33	333–473	5.05	21
CCl_3CCl_3	9.56	6.17	295–463	5.74	21
$CHCl_2\underline{CCl}_3$	8.69	5.80	333–473	5.10	21
$CH_2Cl\underline{CCl}_3$	9.46	8.18	373–493	4.40	21
CH_3CCl_3	9.46	9.32	373–498	3.69	21
$\underline{CHCl}_2CH_2Cl$	8.98	11.17	423–523	2.01	22
$\underline{CHCl}_2CH_3$	8.81	11.93	403–523	1.42	22
CH_2ClCH_2Cl	9.18	14.87	423–523	−0.62	22
CH_2ClCN	8.58	9.99	413–483	2.40	23
$CHCl_2CN$	8.80	6.55	423–513	4.75	23
CCl_3CN	8.20	2.09	373–523	6.91	24

(a) The absolute values are based on $\log k_{22}(M^{-1}sec^{-1}) =$
8.68−7300/2.303RT.
(b) The site of attack by the cyclohexyl radical is underlined.
(c) at 353^oK.

CCl_2CN radicals in cyclohexane, given in Table III, were

$$(VIII) \qquad G(RX) = G(MH) = \frac{k_H}{(2k_t)^{\frac{1}{2}}} \alpha^{\frac{1}{2}} G(R_o)^{\frac{1}{2}} [RH]$$

determined using both expressions VII and VIII. The good agreement between the two results supports the validity of the assumptions that were made in the derivation of the two expressions.

Table III.   Relative Arrhenius Parameters of H Atom Transfer Reactions

Radical	Alkane	$\log A_H - \frac{1}{2}\log A_t$ $M^{-\frac{1}{2}} \sec^{-\frac{1}{2}}$	$E_H - \frac{1}{2}E_t$ kcal/mole	Ref.
CCl_3	cyclohexane	3.28	8.81	25
CCl_3	n-hexane	3.69	9.62	26
CCl_2CN (a)	cyclohexane	3.68	13.24	24
CCl_2CN (b)	cyclohexane	3.95	13.70	24
CH_2CN	cyclohexane	4.22	11.96	27

(a) Determined using expression VII.
(b) Determined using expression VIII.

Now, in order to obtain E_H and A_H from the relative values of Table III, the absolute values of E_t and A_t have to be known. Unfortunately, in liquids, rate constants of radical termination reactions have, at best, been determined at only one temperature. If k_t is known, as it is for CCl_3 radicals (28), then it is possible to calculate A_t, provided that E_t can be estimated. According to Patrick (29), radical metathesis reactions should have the same activation energies in the gas and liquid phase. This theoretical prediction is satisfied when the self-reaction of trichloromethyl radicals is assumed to be diffusion controlled, that is, when E_t is equated to the activation energy of diffusion in cyclohexane of 4.54 kcal. The Arrhenius expression for k_H obtained from the results of Table III, assuming that $E_t = 4.54$ kcal/mole is:

$$\log k_H(CCl_3) = 8.8 - 11,100/2.303RT \quad (M^{-1}sec^{-1})$$

The assumption that $E_t = E_{diff}$ is further supported by the fact that the difference between $E_H - 0.5E_t$ values in cyclohexane and n-hexane is 0.8kcal/mole. This is very close to half the difference between the diffusion energies in these two alkanes. The termination rate constants for the reactions of CCl_2CN and CH_2CN radicals are not known. However, for $(CH_3)_2CCN$ radicals $k_t = 4 \times 10^9 M^{-1}sec^{-1}$ at 25° (30). Therefore, it is safe to assume that the termination reactions of CH_2CN and CCl_2CN radicals are diffusion controlled. In these cases the results of Table III give E_H values of 14.23 and 15.97 kcal/mole, respectively. Comparison of these values with the E_H value for CCl_3 shows that the cyano substituted radicals require a considerably higher activation energy for hydrogen abstraction. This effect can be ascribed to the stabilization of the abstraxting radical by the cyano group. The stability of cyano-substituted radicals thus manifests itself in two ways — they react slowly but can be generated easily in reactions such as halogen transfer.

Chlorine elimination reactions. Chloroethyl radicals (ClEt) formed by Cl or Br transfer reactions in cyclohexane can subsequently eliminate a Cl atom by reaction 34 or abstract a H atom from the solvent by reaction 35. The rate constant ratio $k_{34}/k_{35} = k_{el}/k_H$ and the related Arrhenius parameters can be obtained with the use of expression IX. These data

$$(IX) \qquad \frac{k_{el}}{k_H} = \frac{G(ClEH)}{G(Et)[RH]}$$

are summarized in Table IV. The absolute Arrhenius parameters of Cl elimination, given in the table, were derived by assuming that the Arrhenius parameters of H atom transfer are the same for all the chloroethyl radicals and equal to the gas phase Arrhenius parameters of H atom transfer from cyclohexane to the CCl_3 radical. Small differences between the reactivity of various chloroethyl radicals are expected. However, gas phase data shows that these differences in E_H are small when compared to the differences in E_{el}.

In the gas phase addition of Cl atoms to chloroethylenes, reaction-35, requires

$$Cl + Et \rightarrow ClET \qquad\qquad (-35)$$

activation energies that are very close to zero and therefore $E_{el} \sim D(Et-Cl)$. In cyclohexane one would expect E_{-35} to be slightly higher than in the gas phase. Hence, the cyclohexane

Table IV. Arrhenius Parameters for Cl Atom Elimination Reactions of Chloroethyl Radicals

Radical	$\log A_{el}/A_H$ [a] M	$E_{el}-E_H$ [a] kcal/mole	$\log A_{el}$ sec^{-1}	E_{el} kcal/mole	Ref.	$D(C-Cl)$ [b] kcal/mole
CCl_3CCl_2	5.52	6.45	14.3	17.17	31	17.77
$CHCl_2CCl_2$	5.28	7.60	14.1	18.27	32	21.38
$CHCl_2CHCl$	5.53	6.26	14.3	16.98	32	19.66
CH_2ClCCl_2	5.00	8.69	13.8	19.37	33	21.85
$CH_2ClCHCl$	5.74	8.43	14.5	19.10	34	22.57

(a) Data from references (31)-(33) corrected for the change in the density of cyclohexane.

(b) Gas phase data.

E_{el} values are the upper limit of the C-Cl bond dissociation
energies in chloroalkyl radicals. Inspection of the results of
Table IV shows that E_{el} in cyclohexane is consistently lower than
$D(C-Cl)$. Solvation of the Cl atoms in cyclohexane can be envisa-
ged as the possible cause of the lower $D(C-Cl)$ bond dissociation
energies in cyclohexane.

Radiolytic Decomposition of Alkanesulfonyl Chlorides

The free radical mechanism of the gamma radiation induced
decomposition of alkanesulfonyl chlorides (RSO_2Cl) in cyclohexane
was studied (35-37) to obtain kinetic and thermochemical information on the reactions of alkylsulfonyl radicals (RSO_2). Previously discussed systems are relatively simple and their kinetic
analysis is almost straightforward. The radiolysis of alkanesulfonyl chlorides presents a more complex situation. The problems
encountered in the radiolysis of alkanesulfonyl chlorides will be
illustrated in the $c-C_6H_{12}$ - $MeSO_2Cl$ system (36,37).
The main products of the radiolysis of methanesulfonyl
chloride solutions in cyclohexane are methane, SO_2 and chloro-
cyclohexane. In addition, small amounts of methylchloride are
also formed. The yields of these products satisfy the material
balance relations XI and XII.

(XI) $G(MeCl) + G(RCl) = G(SO_2) = G(-MeSO_2Cl)$

(XII) $G(RCl) = G(MeH)$

Two mechanims are consistent with these findings. The first
mechanism is given by reactions 40-47 while the second mechanism
includes, in addition to these reactions, reactions 48 through 51,
as well. However, since the formation of MeCl by a chain

$Me + MeSO_2Cl \rightarrow MeCl + MeSO_2$ (40)

$R + MeSO_2Cl \rightarrow RCl + MeSO_2$ (41)

$MeSO_2 \rightarrow Me + SO_2$ (42)

$Me + RH \rightarrow MeH + R$ (43)

$R + SO_2 \rightarrow RSO_2$ (44)

$2MeSO_2 \rightarrow products$ (45)

$2RSO_2 \rightarrow products$ (46)

$MeSO_2 + RSO_2 \rightarrow products$ (47)

$$MeSO_2 + RH \rightarrow [MeSO_2H] + R \tag{48}$$

$$[MeSO_2H] \xrightarrow{\text{fast}} MeH + SO_2 \tag{49}$$

$$RSO_2 + RH \rightarrow RSO_2H + R \tag{50}$$

$$[RSO_2H] \xrightarrow{\text{fast}} RH + SO_2 \tag{51}$$

reaction was observed, reactions 48 and 49 cannot completely replace reactions 42 and 43 as a source of methane. Furthermore, the ratio [MeCl]/[MeH] is independent of $MeSO_2Cl$ and SO_2 concentrations. Kinetic analysis shows that this can occur only if methane formation through the intermediacy of methanesulfinic acid is negligible. Incidentally, the [MeCl]/[MeH] ratios observed in the thermal decomposition of $MeSO_2Cl$ in cyclohexane at 150° are the same as in the radiolytic experiments at this temperature (36), although the rate of the thermal reaction is markedly lower than the rate of the gamma radiation induced reaction. This observation further supports the radical mechanism of the radiolytic reaction.

In the RH–$MeSO_2Cl$ system the rates of MeH, RCl and SO_2 formation sharply decrease with irradiation time, even at very low conversions of $MeSO_2Cl$. This effect can be explained with the help of expression XIII that applies when MeH>>MeCl, as observed. According to

$$\text{(XIII)} \quad -R_{MeSO_2Cl} = R_{MeH} = \frac{k_{-42}\,k_{43}[RH][MeSO_2]}{k_{42}[SO_2] + k_{43}[RH] + k_{41}[MeSO_2Cl]}$$

this expression R_{MeH} decreases because $k_{42}[SO_2]$>>$k_{43}[RH]$ + $k_{41}[MeSO_2Cl]$. In other words, even when very small amounts of SO_2 are formed R_{MeH} is given by expression XIV. At these

$$\text{(XIV)} \quad R_{MeH} = \frac{k_{-42}k_{43}[RH]}{k_{41}\,SO_2}[MeSO_2]$$

SO_2 concentrations the equilibrium of reaction 42 is far to the left, i.e. [$MeSO_2$]>>[Me] and therefore R_{MeH} should be proportional to the reciprocal of SO_2 concentration. This behavior was verified in experiments in which small amounts of SO_2 were added to the RH–$MeSO_2Cl$ system. With sufficiently large amounts of added SO_2, $MeSO_2Cl$ decomposes at a constant rate. Expression XIV shows, that under those conditions, the ratio $k_{-42}k_{43}/k_{42}$ can be determined provided that $MeSO_2$ is known.

Obviously, MeSO$_2$ concentration can be estimated from the rate of formation of its termination products. Unfortunately, these products are known to be unstable and therefore their determination is difficult. The advantage of radiolytic systems of a constant and known rate of initiation can again be utilized to derive the concentration of MeSO$_2$ radicals. This estimation is based on the assumption that [MeSO$_2$]$\gg$[Me] and [RSO$_2$]$\gg$[R] so that

$$[R]_o = [RSO_2] + [MeSO_2]$$

where [R]$_o$ is the steady state concentration of cyclohexyl radicals in cyclohexane.

When the necessary kinetic analysis, based on this assumption and expression XIV is carried out (37), the temperature dependence of K$_{-42}$= k$_{-42}$/k$_{42}$ is found to be given by the following expression:

$$\log K_{-40}(M^{-1}) = 4.99 \pm 0.52 - (14.94 \pm 0.92)/2.303RT$$

Accordingly, the bond dissociation energy D(Me-SO$_2$) in cyclohexane is 14.94 kcal/mole. This value is lower by 6 kcal/mole than the average of the gas phase estimations (38-40). It is worth noting that the difference of 6 kcal/mole between the gas and liquid phase bond dissociation energies is almost equal to the heat of vaporization of SO$_2$.

Radiolytic Studies of the Reactions of Cl$_3$Si and Et$_3$Si Radicals

The use of gamma radiation for the initiation of free radical chain reactions is not limited to alkane solutions. Reactions of other radicals can be studied by this method. For example, silyl radicals are generated in silane solutions and their subsequent reactions with added solutes can be investigated.

In trichlorosilane the chain decomposition of chloromethanes, ClM, proceeds by reactions 52 and 53. In the Cl$_3$SiH-ClM systems the relative Arrhenius parameters of the Cl atom transfer, reaction 52, were determined in competitive studies (41). The same method was

$$Cl_3Si + ClM \rightarrow Cl_4Si + M \qquad\qquad (52)$$

$$M + Cl_4Si \rightarrow MH + Cl_3Si \qquad\qquad (53)$$

used for the determination of relative Arrhenius parameters of the analogous Cl transfer reactions of Et$_3$Si radicals with chloromethanes and chloroethanes (42).

In general, chlorine transfer reactions of silyl radicals were found to be much faster and less selective than the analogous reactions of cyclohexyl radicals. The higher reactivity of silyl radicals can be ascribed to the fact that Si-Cl bond is stronger than the C-Cl bond while the Si-H bond is considerably weaker than the C-H bond. Consequently, the Cl atom transfer reactions of silyl radicals are more exothermic than the reactions of alkyl radicals and have lower activation energies. However, the trends in reactivity cannot be explained entirely by changes in activation energies since in both the Cl_3SiH-ClM and Et_3SiH-ClM systems large differences between the preexponential Arrhenius coefficients were observed.

Particularly interesting is the chain mechanism of the decomposition of chloroethylenes in triethylsilane (43). For C_2Cl_3H solutions in Et_3SiH the propagation step of the decomposition reaction is given by reactions 54 through 58.

$$Et_3Si + HCCl=CCl_2 \rightarrow Et_3SiCl + HCCl=CCl \qquad (54a)$$

$$Et_3Si + HCCl=CCl_2 \rightarrow Et_3SiCl + HC=CCl_2 \qquad (54b)$$

$$Et_3Si + HCCl=CCl_2 \rightarrow Et_3SiCHClCCl_2 \qquad (55)$$

$$Et_3SiCHClCCl_2 \rightarrow Et_3SiCH=CCl_2 + Cl \qquad (56)$$

$$HCCl=CCl + Et_3SiH \rightarrow (cis, trans)-C_2Cl_2H_2 + Et_3Si \qquad (57a)$$

$$HC=CCl_2 + Et_3SiH \rightarrow CH_2=CCl_2 + Et_3Si \qquad (57b)$$

$$Cl + Et_3Si \rightarrow HCl + Et_3Si \qquad (58)$$

At $80°$ $k_{55}/k_{54a} = 1.09 \pm 0.11$ and $k_{54a}/k_{54b} = 37 \pm 14$.

The ability of triethyl silyl radicals to remove the strongly bound vinylic chlorine from trichloroethylene (reaction 54) is a characteristic feature of the reactions of triethylsilyl radicals with other chloroethylenes. Radical addition to acetylenes is the method used most frequently to generate vinyl radicals. The formation of vinyl radicals by Cl atom transfer from chloroethylenes to silyl radicals thus offers an interesting alternative to this method.

Literature Cited

(1) Spinks J.W.T. and Woods R.J., "An Introduction to Radiation Chemistry", Wiley, New York, 1964.
(2) Holroyd R.A., in: "Fundamental Processes in Radiation Chemistry" P. Ausloos, Ed., p.413, Interscience, New York, 1968.
(3) Swallow A.J., "Radiation Chemistry", Longman, London, 1973.
(4) Gaumann T. and Hoigne J., "Aspects of Hydrocarbon Radiolysis", Academic Press, London, 1968.
(5) Burns W.G. and Reed C.R.V., Trans. Faraday Soc. (1970) 66 , 2159.

(6) Katz M.G., M.Sc. Thesis, Hebrew University, Jerusalem, 1970.
(7) Rajbenbach L.A. and Horowitz A., Chem. Commun. (1966) 769.
(8) Horowitz A. and Rajbenbach L.A., J. Amer. Chem. Soc. (1969) 91, 4626.
(9) Horowitz A. and Rajbenbach L.A., 'Proceedings of Symposium on Large Radiation Sources', I.A.E.A., Vienna, p.21, 1969.
(10) Horowitz A. and Rajbenbach L.A., J. Amer. Chem. Soc. (1973) 95, 6308.
(11) Horowitz A. and Rajbenbach L.A., J. Amer. Chem. Soc. (1969) 91, 4631.
(12) Horowitz A. and Rajbenbach L.A., J. Amer. Chem. Soc. (1970) 92, 1634.
(13) Horowitz A.,Mey-Marom A. and Rajbenbach L.A., Int. J. Chem. Kinet. (1974) 6, 265.
(14) Schmerling L. and West J.P., J. Amer. Chem. Soc. (1949) 71, 2015.
(15) Tanner D.D., Lewis C.S. and Wada N., J. Amer. Chem. Soc. (1972) 94, 7034.
(16) Horowitz A. and Rajbenbach L.A., J. Amer. Chem. Soc. (1968) 90, 4105.
(17) Schuler R.H. and Bansal K.M., J. Phys. Chem. (1970) 74,3924.
(18) Sauer M.C. Jr. and Mani M., J. Phys. Chem. (1968) 72, 3856.
(19) McCall D.W., Douglass D.C. and Anderson, E.W., Ber. Busenges Phys. Chem. (1963) 67, 336.
(20) Katz M.G., Horowitz A. and Rajbenbach L.A., Int. J. Chem. Kinetics (1975) 7, 183.
(21) Katz M.G. and Rajbenbach L.A., Int. J. Chem. Kinet. (1975) 7, 785.
(22) Katz M.G., Baruch G. and Rajbenbach L.A., Int. J. Chem. Kinet. (1976) 9, 55.
(23) Gonen (Geliebter) Y., Horowitz A. and Rajbenbach L.A., J.C.S. Faraday Trans. (1977) 73, 866.
(24) Gonen (Geliebter) Y., Horowitz A. and Rajbenbach L.A., J.C.S. Faraday Trans. (1976) I 72, 901.
(25) Katz M.G., Baruch G. and Rajbenbach L.A., Int. J. Chem. Kinet. (1976) 8, 131.
(26) Katz M.G., Baruch G. and Rajbenbach L.A., Int. J. Chem. Kinet. (1976) 8, 599.
(27) Gonen (Geliebter) Y., Rajbenbach L.A. and Horowitz A., Int. J. Chem. Kinet. (1977) 9, 362.
(28) Melville H.W., Robb J.C. and Tutton R.C., Discuss. Faraday Soc. (1953) 14, 154.
(29) Patrick G.R., Int. J. Chem. Kinet. (1973) 5, 769.
(30) Weiner S.A. and Hammond G.S., J. Amer. Chem. Soc. (1969) 91, 986.
(31) Horowitz A. and Rajbenbach L.A., J. Phys. Chem. (1970) 74, 678.
(32) Katz M.G., Horowitz A. and Rajbenbach L.A., Trans. Faraday Soc. (1971) 67, 2354.

(33) Aloni R., Katz M.G. and Rajbenbach L.A., Int. J. Chem. Kinet. (1975) 7, 669.
(34) Katz M.G., Baruch G. and Rajbenbach L.A., J.C.S. Faraday Trans (1976) I, 72, 1903.
(35) Horowitz A. and Rajbenbach L.A., J. Amer. Chem. Soc. (1975) 97, 10.
(36) Horowitz A., Int. J. Chem. Kinet. (1975) 7, 927.
(37) Horowitz A., Int. J. Chem. Kinet. (1976) 8, 709.
(38) Busfield, W.K., Ivin K.J., Mackle H. and O'Hare P.A.G., Trans. Faraday Soc. (1971) 57, 1064.
(39) Good J.A. and Thynne J.C., Trans. Faraday Soc. (1967) 63, 2708.
(40) Good J.A. and Thynne J.C., Trans. Faraday Soc. (1967) 63, 2720.
(41) Aloni R., Horowitz A. and Rajbenbach L.A., Int. J. Chem. Kinet. (1976) 8, 673.
(42) Aloni R., Rajbenbach L.A. and Horowitz A. (to be published).
(43) Aloni R., Rajbenbach L.A. and Horowitz A. (to be published).

RECEIVED December 23, 1977.

Structures and Rearrangements of Free Radicals

Some Applications of Free Radical Rearrangements

K. U. INGOLD

Division of Chemistry, National Research Council, Ottawa, Canada

I have chosen to discuss some applications of free-radical rearrangements in solution for a number of reasons. In the first place, this Symposium is intended to honour Paul Bartlett and Cheves Walling for their many contributions to free-radical chemistry. Early work on free-radical rearrangements by both Bartlett (1) and Walling (2) played vital roles in the subsequent development of the subject (3-5). Secondly, this Symposium is being held in the city in which the first free-radical rearrangement, the neophyl rearrangement, was discovered by Urry and Kharasch in 1944 (6). Third and finally, our own wish to put free-radical chemistry on a quantitative footing has led us to measure the rate constants and Arrhenius parameters for a number of well-known rearrangements using the technique of kinetic electron spin resonance spectroscopy (7-11).

Rate constants for radical rearrangements (9) [and for radical scission reactions (12)] can be measured directly by E.S.R. However, it is usually simpler to obtain such rate constants in two separate experiments (7, 8, 10-16). It is first necessary to measure the concentrations of the unrearranged radical, U, and the rearranged radical, R, formed from some source of U under steady-state conditions. Provided the radicals react according to the following scheme,

$$U \xrightarrow{k_i} R$$

$$
\left.
\begin{array}{l}
U + U \xrightarrow{k_t^{U}} \\[4pt]
U + R \xrightarrow{k_t^{UR}} \\[4pt]
U + R \xrightarrow{k_t^{R}} \\[4pt]
R + R \xrightarrow{\phantom{k_t^{R}}}
\end{array}
\right\} \quad \text{nonradical products}
$$

the usual steady-state treatment (12-14) yields equation I.

$$\frac{1}{[R]} = \frac{2k_t{}^R{}_{\sim}[R]}{k_i[U]} + \frac{2k_t{}^{UR}{}_{\sim\sim}}{k_i} \qquad (I)$$

The relative and absolute concentrations of U and R can be varied
by changing the rate of radical production. A plot of $1/[R]$
against $[R]/[U]$ will yield a straight line of slope $2k_t{}^R/k_i$. The
rearrangement rate constant is then obtained following direct
measurement of $2k_t{}^R{}_{\sim}$ by the usual kinetic E.S.R. method ($\underline{17}$, $\underline{18}$)
under similar experimental conditions [solvent, temperature, etc.].
In certain cases, it can be safely assumed that $k_t{}^{UR}{}_{\sim\sim} = k_t{}^R{}_{\sim}$, in
which case equation I can be simplified to

$$k_i/2k_t{}^R{}_{\sim} = ([R]^2/[U]) + [R] \qquad (II)$$

and only one measurement of the U and R concentrations is
necessary ($\underline{7}$).

The Neophyl Rearrangement
        Urry and Kharasch's pioneering discovery of the neophyl
rearrangement arose from an investigation into the factors
influencing the course and mechanism of Grignard reactions ($\underline{6}$).
It was found that although neophyl chloride did not react with
pure phenyl magnesium bromide, it reacted vigorously in the
presence of cobaltous chloride to give tert-butylbenzene (27%),
isobutylbenzene (15%), 2-methyl-3-phenyl-propene (9%), β,β-
dimethylstyrene (4%), and a mixture of dimers and biphenyls - all
of these products having to be identified by boiling point,
refractive index, etc. It was concluded that the neophyl free
radical, $\underline{1U}$, had rearranged to the β,β-dimethylphenethyl radical,
$\underline{1R}$.

$$C_6H_5C(CH_3)_2\overset{\bullet}{C}H_2 \xrightarrow{\quad o \quad} C_6H_5CH_2\overset{\bullet}{C}(CH_3)_2$$

$$\underset{\sim\sim}{1\ U} \qquad\qquad\qquad \underset{\sim\sim}{1\ R}$$

The occurrence of this rearrangement can, nowadays, be very simply
verified by uv photolysis of a di-tert-butyl peroxide solution of
tert-butylbenzene in the cavity of an E.S.R. spectrometer. At
room temperature and below the spectrum due to $\underline{1\ U}$ is observed
($\underline{8}$, $\underline{19}$).

$$(CH_3)_3COOC(CH_3)_3 \xrightarrow{\quad h\nu \quad} 2(CH_3)_3CO^{\bullet}$$

$$(CH_3)_3CO^{\bullet} + C_6H_5C(CH_3)_3 \xrightarrow{\qquad\qquad} (CH_3)_3COH + \underset{\sim\sim}{1\ U}$$

At higher temperatures the spectrum due to $\underline{1\ R}$ is also present and so the rate constant for the rearrangement can be measured by the method described above. Although this was the first free-radical rearrangement to be discovered it requires quite a large activation energy (see Table I) and so it is relatively slow $[k_i^{\underset{\sim}{1}} = 59 \text{ sec}^{-1}$ at 25°C ($\underline{8}$)].

The neophyl rearrangement must proceed through a spiro [2.5] octadienyl type of intermediate or transition state, viz.,

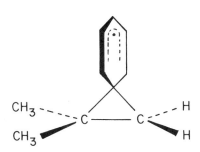

It is clear that the rearrangement will be accelerated if this structure is stabilized by better delocalization of the unpaired electron into the aromatic ring. For this reason, the analogous rearrangements involving the β-naphthyl system,

$$\underset{2U}{\begin{array}{c} \end{array}} \text{—} C(CH_3)_2\dot{C}H_2 \xrightarrow{\quad\circ\quad} \text{—} CH_2\dot{C}(CH_3)_2 \quad (2)$$

$$k_i^{\underset{\sim}{2}} = 2.9 \times 10^3 \text{ sec}^{-1} \text{ at } 25°C \text{ (8)}$$

and 4-pyridyl ring,

$$\underset{3U}{N} \text{—} C(CH_3)_2\dot{C}H_2 \xrightarrow{\quad\circ\quad} N \text{—} CH_2\dot{C}(CH_3)_2 \quad (3)$$

$$k_i^{\underset{\sim}{3}} = 1.4 \times 10^3 \text{ sec}^{-1} \text{ at } 25°C \text{ (8)}$$

are appreciably faster than the neophyl rearrangement (see also Table I). It is, I think, worth noting that the first report of a substituent effect on the migratory abilities of aromatic rings in free-radical rearrangements is due to Bartlett and Cotman ($\underline{1}$).

The Benzoylmethyl Rearrangement

In 1948 Kharasch, McBay, and Urry ($\underline{20}$) found that the reaction of aliphatic ketones with acetyl peroxide, at temperatures where the latter undergoes thermal decomposition, provided a convenient synthetic route to 1,4-diketones.

$$CH_3 - C \underset{O-O}{\overset{O\ \ \ O}{\diagdown\diagup}} C - CH_3 \xrightarrow{125°C} 2CH_3^{\bullet} + 2CO_2$$

$$CH_3^{\bullet} + R - \overset{O}{\overset{\|}{C}} - CH_3 \longrightarrow CH_4 + R - \overset{O}{\overset{\|}{C}} - \overset{\bullet}{C}H_2$$

$$2R - \overset{O}{\overset{\|}{C}} - \overset{\bullet}{C}H_2 \longrightarrow R - \overset{O}{\overset{\|}{C}} - CH_2 - CH_2 - \overset{O}{\overset{\|}{C}} - R$$

However, all attempts to synthesize 1,2-dibenzoylethane and related diketones by dehydrodimerization of the appropriate alkyl phenyl ketones using the same procedure were unsuccessful, only resinous polymeric material being obtained. This was subsequently attributed ($\underline{21}$) to a "neophyl-like" rearrangement of the benzoylmethyl radical, $\underline{4U}$, via an intermediate 1-keto spiro [2.5] octadienyl radical, $\underline{4I}$, to form the phenylacetyl radical, $\underline{4R}$.

$$C_6H_5C\overset{\bullet}{O}CH_2 \longrightarrow \boxed{} \longrightarrow C_6H_5CH_2\overset{\bullet}{C}O \qquad (4)$$

$$\underset{\sim\sim}{\underline{4U}} \qquad\qquad \underset{\sim\sim}{\underline{4I}} \qquad\qquad \underset{\sim\sim}{\underline{4R}}$$

The phenylacetyl radical then undergoes an α-scission to yield the benzyl radical, 5, and carbon monoxide.

$$C_6H_5CH_2\overset{\bullet}{C}O \xrightarrow{k_5} C_6H_5\overset{\bullet}{C}H_2 + CO \qquad (5)$$

$$\underset{\sim\sim}{\underline{4R}} \qquad\qquad\qquad \underset{\sim}{5}$$

We have recently reinvestigated this sequence of reactions by E.S.R. spectroscopy ($\underline{11}$). Our kinetic data show that the formation of resinous products rather than dimers in the reaction of alkyl phenyl ketones with acetyl peroxide is unrelated to the rearrangement that benzoylalkyl radicals can undergo.

The benzoylmethyl radical was generated in the E.S.R. spectrometer by two different photolytic methods.

(a) $(\underline{n}\text{-Bu}_3\text{Sn})_2 \xrightarrow{h\nu} 2\ \underline{n}\text{-Bu}_3\text{Sn}^{\bullet}$

$\underline{n}\text{-Bu}_3\text{Sn}^{\bullet} + C_6H_5COCH_2Br \longrightarrow \underline{n}\text{-Bu}_3\text{SnBr} + C_6H_5CO\overset{\bullet}{C}H_2$

(b) $(\underline{t}\text{-BuO})_2 \xrightarrow{h\nu} 2\ \underline{t}\text{-BuO}^{\bullet}$

$\underline{t}\text{-BuO}^{\bullet} + C_6H_5COCH_3 \longrightarrow \underline{t}\text{-BuOH} + C_6H_5CO\overset{\bullet}{C}H_2$

At temperatures of $-50°C$ and lower the benzoylmethyl radicals decayed with second order kinetics at a rate that was close to the diffusion-controlled limit ($2k_t^{\underset{\sim}{4U}} = 5 \times 10^8\ \underline{M}^{-1}\ sec^{-1}$ at $-70°C$). The product was 1,2-dibenzoylethane.

$$2C_6H_5CO\overset{\bullet}{C}H_2 \xrightarrow{2k_t^{\underset{\sim}{4U}}} C_6H_5COCH_2CH_2COC_6H_5$$

At temperatures above about $80°C$, $\underset{\sim}{4U}$ decayed with first order kinetics ($k = 1.7 \times 10^3\ sec^{-1}$ at $100°C$). At $120°C$ the spectrum due to $\underset{\sim}{4U}$ was replaced by that of the benzyl radical, $\underset{\sim}{5}$. At no temperature between 100 and $200°C$ was it possible to detect the spiro-octadienyl radical, 4I, (which may be only a transition state in the rearrangement rather than a discrete intermediate) or the phenylacetyl radical, $\underset{\sim}{4R}$. However, $\underset{\sim}{4R}$ can be detected when it is generated at low temperatures (22). Photolysis of a cyclopropane solution of di-$\underline{tert}$-butyl peroxide and phenyl-acetaldehyde at $-116°C$ gave the E.S.R. spectra of $\underset{\sim}{4R}$ and $\underset{\sim}{5}$ simultaneously (11). At this temperature $k_5 \approx 9 \times 10^2\ sec^{-1}$, while at $125°C$, the temperature at which the original attempt was made to prepare 1,2-dibenzoylethane from acetophenone (20), we can estimate that k_5 would be ca. $10^9\ sec^{-1}$ (11). It is clear that at $125°C$, no bimolecular reaction of $\underset{\sim}{4R}$ could be fast enough to compete with its decarbonylation. Hence, radical $\underset{\sim}{4R}$ cannot itself be directly involved in any of the reactions that lead to resinous ·materials in the acetophenone-acetyl peroxide reaction.

The rearrangement of the 2-benzoyl-2-propyl radical, $\underset{\sim}{6}$, is very much slower than the rearrangement of the benzoylmethyl radical. In fact, $\underset{\sim}{6}$ decays with second order kinetics even at temperatures as high as $130°C$.

$$2C_6H_5CO\overset{\bullet}{C}(CH_3)_2 \xrightarrow{2k_t^{\underset{\sim}{6}}} \text{Products} \qquad (6)$$

$$2k_t^{\underset{\sim}{6}} = 2.4 \times 10^9\ \underline{M}^{-1}\ sec^{-1} \text{ at } 130°C(11).$$

Since the reaction of acetyl peroxide with phenyl isopropyl ketone
at 115°C did give resinous material (20) we must conclude that
isomerization products of 6 could not have been involved in the
formation of this material.

Since the rearrangement of benzoylalkyl radicals is unrelated
to the production of resinous materials (instead of the
anticipated dimers) in the acetyl peroxide - alkyl phenyl ketone
reactions, we must look elsewhere for an explanation. Once we
come to this conclusion the fact that acetophenone is isoelectronic with α-methylstyrene immediately becomes significant. That
is, radical additions to the carbonyl oxygen of phenyl ketones
will be facilitated by resonance stabilization of the adduct, just
as is the case with α-methylstyrene or styrene, i.e.

The overall route to the formation of resinous materials can, we
believe, therefore be represented as:

A mixture of dimers and of disproportionation products.

The formation of resinous material at the temperatures employed by Kharasch, McBay, and Urry (20) in their attempted dehydrodimerizations of alkyl phenyl ketones is, after the event, perhaps not too surprising.

The 5-Hexenyl Cyclization.[23]

In the foregoing we have seen how kinetic E.S.R. spectroscopy has been used to solve a long-standing problem in reaction mechanism. I now want to turn to a different radical rearrangement reaction, the cyclization of the 5-hexenyl radical, 7U, to the cyclopentylmethyl radical, 7R.

$$
\begin{array}{ccc}
\begin{array}{c}
CH_2 \\
\parallel \\
CH \\
\diagup \quad \\
CH_2 \quad \dot{C}H_2 \\
| \qquad | \\
CH_2 \text{——} CH_2
\end{array}
&
\xrightarrow{O}
&
\begin{array}{c}
\dot{C}H_2 \\
| \\
CH \\
\diagup \quad \diagdown \\
CH_2 \qquad CH_2 \\
| \qquad\quad | \\
CH_2 \text{——} CH_2
\end{array}
\end{array}
\qquad (7)
$$

$$\underset{\sim\sim}{7U} \qquad\qquad\qquad \underset{\sim\sim}{7R}$$

Our initial kinetic E.S.R. study of this reaction (7) has now led us into a quantitative investigation of the "spin-trapping" (24) of 7U and other primary alkyl radicals.

Free-radical cyclizations have a long history (2) but of all known cyclizations that of the 5-hexenyl radical has proved most useful as a mechanistic probe. This is because the anion corresponding to 7U does not cyclize (25) while, as Bartlett has shown (26), the corresponding cation cyclizes to yield products having a six-membered ring. Therefore, in reactions that involve the 5-hexene moiety, the formation of products containing the five-membered ring is diagnostic of a homolytic process and any potential heterolytic mechanism can be ruled out. This criterion has been used extensively by Walling (27-29) and by Lamb and Garst (25,30,31) and their coworkers to investigate reactions which might, in principle, proceed by homolysis or by heterolytic or concerted pathways. A case in point, is the autoxidation of Grignard reagents (29,30). Attempts to inhibit such autoxidations by conventional free-radical inhibitors have not been successful, probably because the inhibitors that have been tried are rapidly removed by reaction with the Grignard reagents. However, the autoxidation of 5-hexenylmagnesium bromide with an excess of oxygen at low temperatures yields 5-hexenylperoxymagnesium bromide whereas, if the oxygen supply is restricted, or if the temperature is raised, a substantial amount of cyclopentylmethylperoxymagnesium bromide is formed. The overall reaction can therefore be represented as:

The 5-hexenyl cyclization is clearly under kinetic rather than under thermodynamic control since the thermodynamically favoured product would be the cyclohexyl radical. It has been suggested that the 5- rather than the 6-membered ring is produced because bond formation requires the approach of the radical within the plane of the π-orbital and along an axis extending above one of the terminal atoms of the double bond either approximately vertically (32, 33), or at an angle of about 109° to the double bond (34). In the latter case, the angle subtended between the three interacting atoms is maintained during the reaction pathway and becomes the angle between these atoms in the product. On the basis of a series of elegant experiments on the cyclization of related radicals, Beckwith and Gara (35) have, however, concluded that intramolecular radical cyclizations are kinetically favoured if the newly formed bond and the semi-occupied orbital at the new radical center can become completely coplanar. The 5-hexenyl radical and analogous species yield the five-membered ring because coplanarity is readily achieved by the free rotation of the exocyclic C-C bond. Complete coplanarity cannot be achieved in forming the six-membered ring and so cyclization to the six-membered ring is kinetically disfavoured ($\underline{35}$).

Our own interest in reaction 7 began when we were able to estimate its rate constant at ambient temperatures ($\underline{36}$). This first estimate of k_i^7 was obtained by combining some product studies of Walling ($\underline{28}$) with a rotating-sector kinetic study of our own on the trialkyltin hydride – alkyl halide reaction. We have subsequently investigated reaction 7 by kinetic E.S.R. spectroscopy ($\underline{7}$) and have confirmed our earlier rate constant (see also, Table I):

$$k_i^7 = 1 \times 10^5 \text{ sec}^{-1} \text{ at } 25°C \quad (\underline{7}).$$

Since the rate constants for this, and for related ($\underline{36}$), cyclizations are known these reactions can be used to determine the rates of reaction of primary alkyls with such diverse species as cupric ion ($\underline{32}$, $\underline{37}$) and tert-butyl hypochlorite ($\underline{38}$). However, there can be no doubt that some of the most interesting radical-molecule reactions are those involved in "spin-trapping". This is a technique which has been used qualitatively for several years to detect and identify transient free-radicals ($\underline{24}$). Its quantitative use in mechanistic studies has been hampered by the paucity of data available for the rate constants, k_a, for the addition of radicals to the spin traps, $\underset{\sim}{T}$.

$$R^{\cdot} + \underset{\sim}{T} \xrightarrow{\quad k_a \quad} R\underset{\sim}{T}^{\cdot} \qquad (a)$$

What little rate data is available rests on competitive experiments with reactions having "known" rate-constants but, unfortunately, even these "known" rate constants are somewhat uncertain. In view of the great potential of spin-trapping we have begun a program to determine accurate rate constants for the trapping of some commonly encountered radicals, starting with primary alkyls ($\underline{39}$).

Since the 5-hexenyl and cyclopentylmethyl radicals are both primary alkyls the spin adducts that they form with a trap will have similar kinetic and thermodynamic stabilities, which is an advantage, and similar E.S.R. spectra, which is a handicap. However, a nice distinction between the spectra of the two spin adducts, $\underset{\sim}{7}U\underset{\sim}{T}^{\cdot}$ and $\underset{\sim}{7}R\underset{\sim}{T}^{\cdot}$, can be obtained by labelling the 5-hexenyl radical with carbon-13 ($I = \frac{1}{2}$) in the 1-position. In most cases, hyperfine splitting by this C-13 should be detectable in the E.S.R. spectrum of $\underset{\sim}{7}U\underset{\sim}{T}^{\cdot}$ because of the proximity of the C-13 to the orbital containing the unpaired electron in this adduct. However, in $\underset{\sim}{7}R\underset{\sim}{T}^{\cdot}$ the C-13 will be too remote from the unpaired electron to produce any appreciable hyperfine splitting. The rate constant for the addition of $\underset{\sim}{7}U$ to the trap can be calculated from the trap concentration and the measured ratio of the concentrations of the two spin-adducts in the early stages of the reaction, i.e.

$$k_a^7 = k_i^7 \, [T] \, \frac{[7\,UT\cdot]}{[7\,RT\cdot]}$$

The spin-adduct ratio must be extrapolated to zero time if it shows any variation and the total concentration of adducts must not be allowed to approach its steady-state level (see below).

Our source of C-13 labelled $7U$ is di(2-^{13}C-6-heptenoyl) peroxide (90 atom % ^{13}C) which can be decomposed thermally or photochemically in an inert solvent, such as benzene, containing a known concentration of a spin trap (39).

$$[CH_2=CH(CH_2)_3{}^{13}CH_2C(O)O]_2 \xrightarrow[h\nu]{\Delta} 2CH_2=CH(CH_2)_3{}^{13}CH_2^{\cdot} + 2CO_2$$

We have investigated a number of the usual nitroso traps,

$$R^{\cdot} + O=N-R' \longrightarrow R-\overset{\overset{\displaystyle O^{\cdot}}{|}}{N}-R'$$

and nitrone traps,

$$R^{\cdot} + R''CH=\overset{\overset{\displaystyle O}{\uparrow}}{N}-R' \longrightarrow \overset{\displaystyle R''}{\underset{\displaystyle R}{\diagdown}}CH-\overset{\overset{\displaystyle O^{\cdot}}{|}}{N}-R'$$

as well as a few less commonly employed spin traps such as di-tert-butyl thioketone,

$$R^{\cdot} + S=C[C(CH_3)_3]_2 \longrightarrow R-S-\overset{\cdot}{C}[C(CH_3)_3]_2$$

Values of k_a^7 at 40°C for some of the spin-traps studied are listed in Table II. Some of the data in this Table has been obtained with the n-hexyl radical, 8, (from n-heptanoyl peroxide) by competitive experiments in the presence of two spin traps that give readily distinguishable E.S.R. spectra.

$$CH_3(CH_2)_5{}^{\cdot} + T_1 \longrightarrow CH_3(CH_2)_5T_1^{\cdot} \qquad (8)$$
$$\underset{8}{}$$

$$CH_3(CH_2)_5^{\cdot} + T_2 \longrightarrow CH_3(CH_2)_5T_2^{\cdot}$$

In all cases where confirmation has been sought, it has been found that the rate constant ratios $(k_a^7)_{T_1}/(k_a^7)_{T_2}$ and $(k_a^8)_{T_1}/(k_a^8)_{T_2}$ are equal, as we would expect since $7U$ and 8 are both primary alkyls.

By monitoring the initial rate of formation of the spin-adduct it is possible to measure the rate, V_i, at which the

Table I. Rate Constants and Arrhenius Parameters for the Rearrangement of Some Primary Alkyl Radicals.

Reaction	Radical U	k_i, sec^{-1} at 25°C	log A_i, sec^{-1}	E_i, kcal/mol	Ref.
1	⟨O⟩—C(CH$_3$)$_2$ĊH$_2$	59	11.7$_5$	13.6	8
2	⟨OO⟩—C(CH$_3$)$_2$ĊH$_2$	2.9×10^3	11.7$_5$	11.3	8
3	N⟨O⟩—C(CH$_3$)$_2$ĊH$_2$	1.4×10^3	11.8	11.8	8
4	⟨O⟩—ĊOCH$_2$	10^a	11.8^b	14.7^a	11
7	CH$_2$=CH(CH$_2$)$_4^{\bullet}$	1.0×10^5	10.7	7.8	7
9	▷—ĊH$_2$	1.3×10^8	12.5	5.9	10

^{a}Calculated from the measured rate constant of 1.7×10^3 sec^{-1} at 102°C and the assumed pre-exponential factor. ^{b}Assumed.

Table II. Some Rate Constants for the Spin-Trapping of Primary Alkyl Radicals in Benzene at 40°C

Spin-Trap	$k_a \times 10^{-5}$, $\underline{M}^{-1}$ sec^{-1}
(CH$_3$)$_3$CNO	90
⟨O⟩—NO (2,4,6-trimethyl)	394
CH$_2$=N(O)C(CH$_3$)$_3$	31
C$_6$H$_5$CH=N(O)C(CH$_3$)$_3$	1.3
[(CH$_3$)$_3$C]$_2$CS	0.45

peroxide is decomposing to form free-radicals ($\underset{\sim}{40}$). We have found ($\underset{\sim}{39}$) that if the reaction is allowed to continue until the spin-adduct concentration reaches a steady-state then, by measuring this concentration, it is possible to obtain the rate constants for the two processes that normally will lead to its consumption, viz.,

$$R^{\bullet} + R\underset{\sim}{T}^{\bullet} \xrightarrow{k_b} \text{Products} \qquad (b)$$

$$R\underset{\sim}{T}^{\bullet} + R\underset{\sim}{T}^{\bullet} \xrightarrow{2k_c} \text{Products} \qquad (c)$$

Radical formation at a rate V_i, followed by reactions a, b, and c yields the relation ($\underset{\sim}{32}$):

$$\left\{ \frac{k_a[\underset{\sim}{T}]}{[R\underset{\sim}{T}^{\bullet}]_{ss}} - k_b \right\} \Bigg/ \left\{ \frac{k_a[\underset{\sim}{T}]}{[R\underset{\sim}{T}^{\bullet}]_{ss}} + k_b \right\} = \frac{2k_c}{V_i}[R\underset{\sim}{T}^{\bullet}]^2_{ss} \qquad (III)$$

This application of the kinetic E.S.R. technique can be illustrated for $R^{\bullet}$ = $\underline{n}$-hexyl and $\underset{\sim}{T}$ = 2-methyl-2-nitrosopropane in benzene at $40°C$. Measurement of the initial rate of $R\underset{\sim}{T}^{\bullet}$ formation gave V_i = 2×10^{-8} $\underline{M}$ sec^{-1}. Taking k_a = 9.0×10^6 $\underline{M}^{-1}$ sec^{-1} (see Table II), the steady-state concentrations of $R\underset{\sim}{T}^{\bullet}$ at various trap concentrations could best be correlated by taking k_b = 3.5×10^8 $\underline{M}^{-1}$ sec^{-1} and $2k_c \approx 125$ $\underline{M}^{-1}$ sec^{-1}. The value found for k_b is in range generally observed for the addition of other alkyl radicals to sterically hindered, persistent, nitroxides such as di-$\underline{tert}$-butyl nitroxide ($\underline{41}$, $\underline{42}$).

$$CH_3(CH_2)_5^{\bullet} + CH_3(CH_2)_5\overset{O^{\bullet}}{\underset{|}{N}}C(CH_3)_3 \xrightarrow{3.5 \times 10^8} CH_3(CH_2)_5\overset{CH_3(CH_2)_5O}{\underset{|}{N}}C(CH_3)_3$$

This technique would, therefore, appear to provide a simple method for studying reactions between alkyl radicals and transient nitroxides and also, possibly, for studying other "cross" radical-radical reactions. The value obtained for $2k_c$ was found to be in reasonable agreement with that obtained by a direct measurement of the bimolecular decay of $\underline{n}$-hexyl-$\underline{tert}$-butyl nitroxide, viz., $\leqslant 250$ $\underline{M}^{-1}$ sec^{-1} ($\underline{39}$):

$$2CH_3(CH_2)_5\overset{O^{\bullet}}{\underset{|}{N}}C(CH_3)_3 \xrightarrow{\leqslant 250} CH_3(CH_2)_5\overset{OH}{\underset{|}{N}}C(CH_3)_3 + CH_3(CH_2)_4CH=\overset{O}{\overset{\uparrow}{N}}C(CH_3)_3$$

The Cyclopropylmethyl Ring-Opening Reaction

Another example of a mechanistically useful rearrangement of a primary alkyl radical is provided by the extremely facile ring-opening of cyclopropylmethyl, $\underset{\sim}{9U}$, to yield the 3-butenyl radical, $\underset{\sim}{9R}$.

$$
\begin{array}{c}
CH_2 \\
| \quad \diagdown \\
\quad \quad CH-\overset{\bullet}{C}H_2 \\
| \quad \diagup \\
CH_2
\end{array}
\longrightarrow_{\Delta}
\begin{array}{c}
\overset{\bullet}{C}H_2 \\
| \\
\quad \quad CH=CH_2 \\
| \quad \diagup \\
CH_2
\end{array}
\qquad (9)
$$

$$9U \qquad\qquad\qquad 9R$$

The effect that ring substituents have on the direction of ring opening, which has been unravelled by Davies and Pereyre and their co-workers (43-49), provides a fascintaing picture of the subtle stereoelectronic factors that control such processes. At low temperatures, cis-2-methylcyclopropylalkyl radicals, 10U, undergo ring opening to give the thermodynamically preferred secondary alkyl radical, 10R.

$$ (10) $$

10U   10R

$R^1=H$; $R^2=H$, OH, $OSn[(CH_2)_3CH_3]_3$, $OSi(CH_3)_3$, etc.

However, trans-2-methylcyclopropylalkyl radicals, 11U, undergo ring opening at low temperatures to give the thermodynamically disfavoured primary alkyl, 11R. [At high temperatures and low concentrations of radical trapping agents, 11U, gives an increasing proportion of the thermodynamically preferred products from 10R because the ring opening is reversible (48)].

$$ (11) $$

11U   11R

The remarkably selective formation of the primary alkyl **radical**
from 11U was initially thought to be due to polar contributions
to the transition state which conferred some carbanionic character
on the incipient alkyl (43, 44), e.g.,

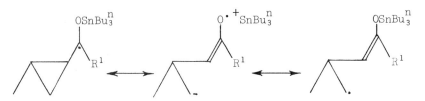

However, further work (49) showed that the regioselectivity of the
ring openings did not depend on the presence of a polar $OSnBu_3^n$ or
OH substituent at the initial radical center. It was therefore
suggested that the methyl substituent interacts with the
polarizable electrons of the cyclopropyl ring to render the C^1-C^3
bond electronically non-equivalent with the C^1-C^2 (and C^2-C^3)
bond. Interaction between the semi-occupied orbital and the C^1-C^3
bonding orbital could cause conformer 11Ua to be more highly
populated than 11Ub, which would lead to preferential cleavage of
the C^1-C^3 bond, and hence to the production of 11R.

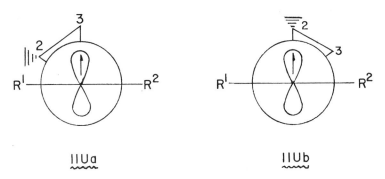

IIUa   IIUb

However, in the case of the cis radical, 10U, electronic factors
could be overwhelmed by steric effects so that conformer 10Ub
(or b')- which is the precursor of the secondary alkyl radical,
10R - would be preferred to conform 10Ua (or a').

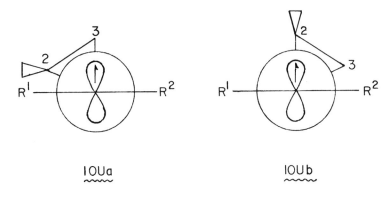

IOUa   IOUb

R^1 and R^2 reversed for conformers 10Ua' and 10Ub'.

Our own work on the ring opening of cyclopropylalkyl radicals has been confined to measuring the rate for the cyclopropylmethyl radical, i.e. reaction 9, by kinetic E.S.R. spectroscopy at low temperatures. Radical 9U was generated by photolysis of tert-butyl cyclopropylperacetate in methylcyclopropane as solvent.

$$\overline{CH_2 CH_2 \overset{\bullet}{C}HCH_2} CO_2 C(CH_3)_3 \xrightarrow{h\nu} \overline{CH_2 CH_2 CH\overset{\bullet}{C}H_2} + CO_2 + (CH_3)_3 CO^{\bullet}$$

$$(CH_3)_3 CO^{\bullet} + \overline{CH_2 CH_2 \overset{\bullet}{C}HCH_3} \longrightarrow (CH_3)_3 COH + \overline{CH_2 CH_2 CH\overset{\bullet}{C}H_2}$$

Both 9U and 9R could be detected at temperatures in the range -120 to -145°C. Extrapolation of the measured rate constants for the isomerization to room temperature (see Table I) yielded,

$$k_i^{9} = 1.3 \times 10^8 \ sec^{-1} \ at \ 25°C \qquad (10)$$

The Tri-tert-butylphenyl Isomerization

The majority of intramolecular hydrogen atom transfers proceed *via* 6-center cyclic transition states and so involve a 1,5 migration of hydrogen. The 2,4,6-tri-tert-butylphenyl radical, 12U, however, undergoes an intramolecular H-transfer *via* a 5-center cyclic transition state (i.e., a 1,4 H-migration) to produce the 3,5-di-tert-butylneophyl radical, 12R, (9).

$(CH_3)_3C$ ⟶ ∘ ⟶ $C(CH_3)_3$ $(CH_3)_3C$ ⟶ $C(CH_3)_2\overset{.}{C}H_2$

(12)

$C(CH_3)_3$ $C(CH_3)_3$

12U   12R

The unusual feature about this reaction is that it provides the first authentic example of quantum-mechanical tunneling in an intramolecular H-transfer, that is, the H atom is transferred through a potential barrier, rather than over the top of the barrier as would be required by classical mechanics. Both 12U and 12R can be observed by ESR spectroscopy and k_i^{12} has been measured by both the direct and indirect methods for hydrogen and for deuterium atom transfer (9). The occurrence of tunneling is indicated four kinetic phenomena. Firstly, the kinetic deuterium isotope effect, $\left(k_i^{12}\right)_H\Big/\left(k_i^{12}\right)_D$, is much greater at all temperatures than the classically calculated "maximum" value, which is based on the assumption that the maximum isotope effect arises when all zero-point energy is lost in the transition state. Thus, the "maximum" isotope effects at -30 and -150°C are calculated to be 17 and 260, respectively, while the measured values are 80 and 13,000. Secondly, the Arrhenius plots are strongly curved, the temperature dependence of the reaction becoming less as the temperature is lowered. In fact, the hydrogen transfer occurs at an appreciable rate at temperatures not far above absolute zero (38). The third and fourth pieces of evidence for tunneling are the values of the Arrhenius pre-exponential factors and the activation energies. That is, although the usual Arrhenius equation cannot realistically be applied to reaction (12), over any limited range of temperature it is possible to draw straight lines through the data in the normal manner. The A-factors and activation energies derived in this way are enormously different from those which would be expected if the reaction obeyed classical mechanics. They can only be explained satisfactorily by invoking the wave nature of hydrogen and deuterium atoms.

The curved Arrhenius plot for H-transfer can be described by the modified Arrhenius equation:

$$k_i^{12}(T) = A\Gamma(T)\ e^{-V_0/RT}$$

in which V_O is the height of the potential barrier through which tunneling takes place (i.e., V_O = the classical activation energy), A has essentially the same significance as in the usual equation, and $\Gamma(T)$ is a temperature dependent function which gives the ratio of quantum-mechanical to classical transmission rates from one side of the barrier to the other. The data for H-transfer can be fitted extremely well by a barrier having the shape proposed by Eckart ($\underline{50}$) with a height of 14.5 kcal/mol and with an A factor of $10^{11} sec^{-1}$, while for the D-transfer zero-point energy effects cause V_O to increase to 14.9 kcal/mol ($\underline{9}$).

The analogous isomerizations of several other aryl radicals that are structurally related to $\underline{12U}$ also occur slowly and by quantum-mechanical tunneling ($\underline{9},\underline{38}$). In contrast, the 2,4,6-tri-neopentylphenyl radical, $\underline{13U}$, isomerizes by a 1,5 hydrogen transfer which is so fast that only the rearranged radical, $\underline{13R}$, could be detected by ESR spectroscopy even at $-160°C$ ($\underline{9}$).

$$(13)$$

$\underline{13U}$ $\underline{13R}$

Conclusion.

A glance at the k_i values found for the radicals listed in Table I will show that we now have available an extremely versatile STABLE ($\underline{51}$, $\underline{52}$) of primary alkyl radical rearrangements. Within this stable we can find radicals whose rearrangement rates will enable them to compete in a wide variety of radical-molecule races. That is, amongst these radicals we can find entrants suitable for competition against reactions whose speed varies from the gallop of,

$$R^• + O_2 \longrightarrow ROO^•$$

to the walk of,

$$R^• + RH \longrightarrow RH + R^•$$

We would very much like to expand this stable and obtain accurate rates for the rearrangement of additional primary alkyl

radicals. This would enable us both to "plug the gaps" in the
range of k_i values that is available and to provide a wider choice
of radicals which could be used in any particular competition. A
few possible new candidates for this stable are listed below

$$C_6H_5CH_2{}^*CH_2^• \xrightarrow{} C_6H_5{}^*CH_2CH_2^•$$

$$H_2C=CHC(CH_3)_2CH_2^• \xrightarrow{} H_2C=CHCH_2\overset{•}{C}(CH_3)_2$$

$$(C_6H_5)_3CCH_2^• \xrightarrow{} C_6H_5CH_2\overset{•}{C}(C_6H_5)_2$$

$$CCl_3CH_2^• \xrightarrow{} ClCH_2\overset{•}{C}Cl_2$$

Rate constants for one or two of these rearrangements have been
estimated ($\underline{5}$, $\underline{36}$) but for none of these reactions are reliable
Arrhenius parameters available.

The ordered growth and future development of homolytic
chemistry will, to quite a large extent, be dependent on the
building of additional free-radical stables, set up to cater to
the rearrangement or fragmentation of other breeds of free
radicals. Limited data is already available on the α-scission of
certain acyl radicals ($\underline{11},\underline{13},\underline{15},\underline{53}$), on the β-scissions of some
tetra-alkoxyphosphoranyl radicals ($\underline{12},\underline{14}$), some alkoxyalkyls ($\underline{54}$),
and some dialkoxyalkyls ($\underline{16}$, $\underline{54}$), and on the rearrangement of some
β-acyloxyalkyl radicals ($\underline{55},\underline{56}$). An extension of such data to
cover a wider variety of reactive free radicals is an important
task for the future.

Abstract.
Measurement of the rate constants and Arrhenius parameters
for the rearrangement of a variety of primary alkyl radicals by
E.S.R. spectroscopy is briefly described. At 25°C these rate
constants vary from a low of 10 sec^{-1} for the reaction:
$C_6H_5COCH_2^• \xrightarrow{} [C_6H_5CH_2\overset{•}{C}O] \longrightarrow C_6H_5CH_2^• + CO$; to a high of
1.3×10^8 sec^{-1} for the reaction $\overset{\overline{}}{CH_2CH_2CHCH_2} \xrightarrow{} \overset{•}{C}H_2CH_2CH=CH_2$.
These isomerizations provide useful probes in mechanistic studies
and can be used to determine the rates with which primary alkyls
react with a variety of molecules. The latter use is illustrated
by the measurement of the rates of spin trapping of the 5-hexenyl
radical by some common traps. Aryl radical isomerizations that
proceed by quantum-mechanical tunneling are reviewed briefly.

Literature Cited.

($\underline{1}$) Bartlett, P.D., and Cotman, Jr., J.D., J. Am. Chem. Soc.,
      (1950), $\underline{72}$, 3095.
($\underline{2}$) Walling, C., J. Am. Chem. Soc., (1945), $\underline{67}$, 441.
($\underline{3}$) Walling, C., in "Molecular Rearrangements" Vol. 1, P. de Mayo,
      Ed. Interscience, New York, 1963, Chap. 7.

(4)   Freidlina, R.Kh., in "Advances in Free Radical Chemistry",
      Vol. 1, G.H. Williams, Ed., Academic Press, New York, 1965,
      Chap. 6.
(5)   Wilt, J.W., in "Free Radical" Vol. 1., J.K. Kochi, Ed.,
      Wiley, New York, 1973, Chap. 8.
(6)   Urry, W.H., and Kharasch, M.S., J. Am. Chem. Soc., (1944),
      66, 1438.
(7)   Lal, D., Griller, D., Husband, S., and Ingold, K.U., J. Am.
      Chem. Soc., (1974), 96, 6355.
(8)   Maillard, B., and Ingold, K.U., J. Am. Chem. Soc., (1976),
      98, 1224.
(9)   Brunton, G., Griller, D., Barclay, L.R.C., and Ingold, K.U.,
      J. Am. Chem. Soc., (1976), 98, 6803.
(10) Maillard, B., Forrest, D., and Ingold, K.U., J. Am. Chem.
      Soc., (1976), 98, 7024.
(11) Brunton, G., McBay, H.C., and Ingold, K.U., J. Am. Chem.
      Soc., (1977), 99, 4447.
(12) Watts, G.B., Griller, D., and Ingold, K.U., J. Am. Chem.
      Soc., (1972), 94, 8784.
(13) Griller, D., and Roberts, B.P., J. Chem. Soc., Perkin Trans.
      2, (1972), 747.
(14) Davies, A.G., Griller, D., and Roberts, B.P., J. Chem. Soc.,
      Perkin Trans. 2, (1972), 993.
(15) Schuh, H., Hamilton, Jr., E.J., Paul, H., and Fischer, H.,
      Helv. Chim. Acta, (1974), 57, 2011.
(16) Perkins, M.J., and Roberts, B.P., J. Chem. Soc., Perkin
      Trans. 2, (1975) 77.
(17) Adamic, K., Bowman, D.F., Gillan, T., and Ingold, K.U.,
      J. Am. Chem. Soc., (1971), 93, 902.
(18) Watts, G.B., and Ingold, K.U., J. Am. Chem. Soc., (1972), 94,
      491.
(19) Hamilton, Jr., E.J., and Fischer, H., Helv. Chim. Acta,
      (1973), 56, 795.
(20) Kharasch, M.S., McBay, H.C., and Urry, W.H., J. Am. Chem.
      Soc., (1948), 70, 1269.
(21) McBay, H.C., J. Org. Chem., (1975), 40, 1883.
(22) Paul, H., and Fischer, H., Helv. Chim. Acta, (1973), 56,
      1575.
(23) For recent reviews by the two major investigators of this and
      related cyclization reactions see: A.L.J. Beckwith in
      "Essays on Free Radical Chemistry", Chemical Society Special
      Publications, London, (1970), No. 24, p.239; M. Julia
      Accounts Chem. Research (1971), 4, 386; Pure Appl. Chem.,
      (1974), 40, 553.
(24) For recent reviews of spin-trapping see: Perkins, M.J. in
      "Essays on Free Radical Chemistry", Chemical Society Special
      Publications, London, (1970), No.24, p.97; Janzen, E.G.,
      Accounts Chem. Res., (1971), 4, 31.
(25) Garst, J.F., Ayers, P.W., and Lamb, R.C., J. Am. Chem. Soc.,
      (1966), 88, 4260.

(26) Bartlett, P.D., Closson, W.D., and Cogdell, T.J., J. Am.
     Chem. Soc., (1965), 87, 1308.
(27) Walling, C., and Pearson, M.S., J. Am. Chem. Soc., (1964),
     86, 2262.
(28) Walling, C., Cooley, J.H., Ponaras, A.A., and Racah, E.J.,
     J. Am. Chem. Soc., (1966), 88, 5361.
(29) Walling, C., and Ciofarri, A., J. Am. Chem. Soc., (1970), 92,
     6609.
(30) Lamb, R.C., Ayers, P.W., Toney, M.K., and Garst, J.F., J. Am.
     Chem. Soc., (1966), 88, 4261.
(31) Garst, J.F., and Barton, II, F.E., Tetrahedron Letters,
     (1969), 587.
(32) Beckwith, A.L.J., Gream, G.E., and Struble, D.L., Aust. J.
     Chem., (1972), 25, 1081.
(33) Beckwith, A.L.J., Blair, I., and Phillipou, G., J. Am. Chem.
     Soc., (1974), 96, 1613.
(34) Baldwin, J.E., Chem. Commun., (1976), 734.
(35) Beckwith, A.L.J., and Gara, W.B., J. Chem. Soc., Perkin
     Trans. 2, (1975), 593, 795.
(36) Carlsson, D.J., and Ingold, K.U., J. Am. Chem. Soc., (1968),
     90, 7047.
(37) Jenkins, C.L., and Kochi, J.K., J. Am. Chem. Soc., (1972), 94,
     843.
(38) Unpublished results from this laboratory.
(39) Schmid, P., and Ingold, K.U., J. Am. Chem. Soc., (1977), 99,
     0000, and unpublished results.
(40) Janzen, E.G., Evans, C.A., and Nishi, Y., J. Am. Chem. Soc.,
     (1972), 94, 8236.
(41) Asmus, K.D., Nigam, S., and Willson, R.L., Int. J. Radiat.
     Biol., (1976), 29, 211.
(42) Nigam, S., Asmus, K.D., and Willson, R.L., J. Chem. Soc.,
     Faraday Trans. 1, (1976), 72, 2324.
(43) Godet, J.-Y., and Pereyre, M., J. Organometallic Chem.,
     (1972), 40, C23.
(44) Godet, J.-Y., and Pereyre, M., Compt. rend., (1971), 273C,
     1183; (1973), 277C, 211.
(45) Godet, J.-Y., Pereyre, M., Pommier, J.-C., and Chevolleau, D.,
     J. Organometallic Chem., (1973), 55, C15.
(46) Davies, A.G., and Muggleton, B., J. Chem. Soc., Perkin Trans.
     2, (1976), 502.
(47) Davies, A.G., Muggleton, B., Godet, J.-Y., Pereyre, M., and
     Pommier, J.-C., J. Chem. Soc., Perkin Trans. 2, (1976), 1719.
(48) Davies, A.G., Godet, J.-Y., Muggleton, B., and Pereyre, M.,
     Chem. Commun. (1976), 813.
(49) Blum, P., Davies, A.G., Pereyre, M., and Ratier, M., Chem.
     Commun., (1976), 814.
(50) Eckart, C., Phys. Rev., (1930), 35, 1303.
(51) Noun. Building set apart and adapted for lodging and feeding
     horses. "Concise Oxford Dictionary" 5th Ed., Clarendon,
     Oxford (1964).

(52) For comments on the use of "stable" as an adjective in free-radical chemistry see: Griller, D., and Ingold, K.U., Accounts Chem. Research, (1976), 9, 13.
(53) Perkins, M.J. and Roberts, B.P., J. Chem. Soc., Perkin Trans. 2, (1974), 297.
(54) Steenken, S., Schuchmann, H.-P., and von Sonntag, C., J. Phys. Chem., (1975), 79, 763.
(55) Beckwith, A.L.J., and Tindal, P.L., Aust. J. Chem., (1971), 24, 2099.
(56) See also: Beckwith, A.L.J., and Thomas, C.B., J. Chem. Soc., Perkin Trans. 2, (1973), 861.

RECEIVED December 23, 1977.

12

EPR Studies of Radical Pairs. The Benzoyloxy Dilemma.

J. MICHAEL McBRIDE, MICHAEL W. VARY, and BONNIE L. WHITSEL

Department of Chemistry, Yale University, New Haven, CT 06520

For the past ten years we have been using free-radical systems to study the chemistry of organic solids. This work originated with studies of the photolysis of solid azoalkanes in P. D. Bartlett's laboratory at Harvard.(1,2) Much of our effort has been devoted to epr spectroscopy of radical pairs generated in low concentration by photolyzing single crystals of such standard radical initiators as azoalkanes and diacyl peroxides. In several cases this tool has allowed us to determine where molecules in a solid move during chemical reaction and how easily they move.(3,4)

Since single-crystal epr spectra contain information about electron spin distribution, it is necessary to know how the electron spin is distributed about the nuclear framework of a radical in order to infer how the radical is oriented. When the spin distributions of isolated radicals are known from independent sources, epr gives detailed information about the arrangement of the radicals in pairs. Naturally knowing this arrangement can be uniquely helpful in determining the factors which influence reactions of organic solids.

Sometimes radical structure and spin distribution are not known reliably. Epr results can then be useful in determining radical structure as well as in studying the arrangement of pairs. This paper focuses on what solid systems can tell about radical structure rather than on what radical systems can tell about the chemistry of solids. Its first purpose is to show what information about radical structure and pair arrangement is contained in single crystal EPR spectra. Its second is to show, using pairs including benzoyloxy radicals as an example, how the spectra sometimes confuse us by providing more information than appears to be consistent with simple explanations.

EPR of Radical Pairs

The radicals we study are immobilized in pairs separated by 3.5 to 8 Å. At these distances the bonding between them is weak enough that for temperatures above about 10 K (20 cal/mole) the

triplet state predominates over the singlet, which is presumably
the ground state, by the statistical factor of three. The
residual bonding remains strong enough, however, that for most
magnetic purposes (perhaps 0.01 to 0.001 cal/mole) the pair must
be considered as a triplet state with electron spins exchanging
rapidly between the radicals, rather than as a pair of independent
doublet states each with its own spin. (3) In addition to the
normal electron Zeeman splitting there are three types of magnetic
interactions which have a strong influence on the epr spectra of
triplet radical pairs. These are:
    1) interaction between electron spins and nuclear spins, which
gives rise to hyperfine splitting (hfs) and is described by an A
tensor for each magnetic nucleus;
    2) indirect interaction between the electron spins and the
applied field mediated by orbital angular momentum, which
determines how much the field at the center of the pattern differs
from that for a free electron, and is described through the g
tensor; and
    3) interaction between the two electron spins, which gives
rise to spectral fine structure (fs) and is described by the D
tensor.
    What information do the A, g, and D tensors contain? The
six independent elements of each tensor (D, being traceless, has
only five) may be transformed to yield three sets of values
consisting of one, two, and three members. These sets contain
different types of information and may be determined under
different conditions. The set of one member (which D lacks) is
the isotropic or average value of the interaction and may be
determined from samples in any phase. The set with two members
describes the magnitude of the interaction's anisotropy and
contains symmetry information. Its determination usually requires
a rigid sample. The set of three members describes the
orientation of the anisotropy, giving for example the spherical
polar coordinates of a molecular symmetry axis and the phase of
rotation about it. The orientation can only be determined with an
oriented sample, usually a single crystal.

    Hyperfine Splitting. The isotropic value of the hfs depends
on electron spin density at the nucleus in question. Since p
orbitals have a node at the nucleus, it shows how much spin
density resides in the atom's s orbitals. For the 2s orbital of
carbon the proportionality constant is 1110 gauss/electron; for
the 1s orbital of hydrogen, 507. (5) The magnitude of the hfs
anisotropy depends on how much the spin density in the
neighborhood of the nucleus departs from spherical, and how much
from axial symmetry. Since the through-space interaction is
attenuated by the cube of the electron-nuclear distance, it
samples primarily the immediate vicinity of the nucleus. For C-13
the anisotropic hfs commonly has an axis of symmetry and may be
interpreted in terms of spin density in the 2p orbitals of the

atom. The splitting by a C-13 nucleus of an electron localized in
its p orbital would vary over a range of 97 gauss.(5) For
hydrogen the interpretation is in terms of spin density on its
neighbor. The orientational information can then tell which 2p
orbital of carbon bears the electron spin.

For radicals in a triplet pair the observed hfs is the
average of what it would be for the individual radicals. Since
the spin of a nucleus on one radical has negligible direct
interaction with the electron spin on the other, the hfs is half
of the value for the isolated radical. The C-13 hfs of pairs of
bridgehead triptycyl radicals provides an example. Pairs which
persist at -100°C in a photolyzed single crystal of the related
diacyl peroxide show signals with satellite doublets from
splitting by bridgehead C-13 in natural abundance.(6) The
isotropic value of this splitting is 49.9 G and the anisotropy
shows approximate axial symmetry with a range of about 38 G. The
isolated radical thus has (2 x 49.9 / 1110) = 9% spin density in
the 2s orbital of the bridgehead carbon and (2 x 38 / 97) = 77%
density in its 2p orbital. Thus some 86% of the spin density is
localized on the bridgehead carbon in an sp^9 hybrid. The
direction of the symmetry axis shows how the orbital is oriented
in the solid, and the existence of a single pair of satellites
shows that the two radicals in a pair are close enough to
antiparallel that their bridgehead carbons give equivalent
splitting.

Spin-Orbit-Field Coupling. Even a nondegenerate radical can
acquire orbital angular momentum when an applied magnetic field
mixes electronically excited states with the ground state by the
$\underline{H} \cdot \underline{L}$ interaction.(7) In an LCAO-MO picture this mixing is the
resultant of contributions from each atomic p orbital in the
singly occupied MO, which after rotation by 90° about the applied
field direction would overlap a p orbital in another MO of similar
energy. For example, if the singly occupied MO had a large
coefficient from p_x of a particular atom, and a vacant MO had a
large coefficient for p_y of the same atom, a field in the z
direction would tend to mix them through the L_z operator, which
"rotates" p_x into $i\hbar$ p_y (at the same time rotating p_y into $-i\hbar$ p_x,
and destroying p_z). The magnitude of the mixing depends directly
on the extent of rotation and the amount of overlap it generates
(that is on the field strength and the product of the coefficients
of coupled AOs in the two MOs) and inversely on the energy gap
between the MOs. The sign of the mixing and the sense of the
orbital angular momentum depend on the sign of the energy gap
(that is on whether mixing is with a lower-energy doubly occupied
orbital or with a higher vacant one). The influence of the
resulting orbital angular momentum on the electron spin through
$\underline{L} \cdot \underline{S}$ depends on the strength of the spin-orbit coupling constant of
those atoms whose p orbitals make a significant contribution to
the mixing. The following expression for the effective g-value

summarizes these effects:

$$g = 2.0023 - 2 \sum_n \sum_k \left\langle \psi_o \mid \zeta_k L_k \mid c_n \psi_n \right\rangle$$

where 2.0023 is the g value for the free electron; ψ_o is the singly occupied MO; ψ_n is another MO; ζ_k is the spin-orbit coupling constant of atom k; L_k is the orbital angular momentum operator which rotates the p orbitals of atom k about the applied field direction; and c_n is the amount of ψ_n mixed into ψ_o.

$$c_n = \left\langle \psi_n \mid L_k \mid \psi_o \right\rangle / (E_n - E_o).$$

The isotropic value of the shift in g from the free electron value shows the magnitude of spin-orbit-field coupling averaged over all field directions. It gives qualitative information about whether there is significant spin density on atoms of high spin-orbit coupling and whether there exist low-energy vacant or high-energy doubly occupied orbitals. The anisotropy of g shows what the extremes in spin-orbit-field coupling are and gives information about molecular symmetry. The orientation of the g tensor is especially valuable, since by showing the field directions which give strong and weak coupling it helps assign the ground state and low-lying excited states of the radical. Such an application is exemplified by benzoyloxy radical below.

For radicals in a triplet pair the observed g tensor is the average of the appropriately oriented tensors of the constituent free radicals. Thus if the g tensor of one member of the pair is known, that of the other may be found by subtracting the known tensor from twice the observed tensor.

Electron-Electron Splitting. The fine structure splitting in a radical pair, unlike hfs and g, is not an average of values for the separate radicals. Since it arises from interaction of the radicals, it depends more on their separation than on their individual orientations. To first order the fs line separation is 3D:

$$3D = \left\langle (1-3\cos^2\theta)/r^3 \right\rangle \cdot 27,853 \text{ gauss } Å^3,$$

where r is the distance between electron spins and θ is the angle between the spin-spin vector and the applied field.(8) The average is taken over all pairs of sites between the two radicals and weighted according to the product of the spin densities of the sites. If only one site is important in each radical the "point dipole" approximation holds, and the largest splitting is observed when the field lies along the radical-radical vector so that the average in the 3D expression is $-2/r^3$. Spin delocalization in the radicals will affect 3D in obvious ways, and contributions from pairs of sites with opposite spin density may partially cancel contributions from pairs with the same sign. The

point-dipole approximation overestimates the "average" spin-spin
distance when the spins are delocalized in directions
perpendicular to their average separation. If the spin
distribution and orientation of the radicals is known, 3D is
readily calculable for any field direction and interradical
vector.

Since $(1-3\cos^2\theta)$ averages to zero over a spherically
symmetric set of spin-spin vectors within a species, D has no
isotropic value and is unobservable by epr for a rapidly tumbling
sample. The anisotropy of D tells by how much the relative spin
distribution departs from spherical symmetry. If a spherical
distribution is distorted by stretching along one direction and
compressing along the others to preserve the average spin-spin
separation, the splitting measured with the field in the
stretching direction becomes negative (the average $\cos^2\theta$
increases). Splittings measured with fields in the plane
perpendicular to the stretching become positive but, of course,
remain equal to one another. If the cylindrical symmetry is
broken by additional stretching along some direction in the plane,
3D in that direction becomes less positive. The magnitude of the
anisotropy of D (the familiar "D" and "E" values) depends on the
symmetry and interradical distance of the pair, while the
orientation of the tensor assigns directions to the distortions.

The fs of radicals separated by 6 to 10 Å is less sensitive
than hfs to details of their individual internal spin
distributions and, unlike g, is insensitive to the energy gaps
between ground states and various excited states. Given a
reasonably accurate model for the spin distributions and
orientations of two radicals, the fs gives precise information
about the interradical vector. For example the pair of triptycyl
radicals mentioned above has an axially symmetric fs with a
maximum (negative) splitting of 319 G. For point dipoles this
would correspond to a separation of 5.58 Å. The splittings
measured with an accuracy of 2.5 G fix the length of this vector
within 0.02 Å and its direction within 1°. The direction of
maximum splitting lies within 5° of the symmetry axis of the C-13
hfs, so the interaction should be calculated for electrons in
antiparallel sp^9 hybrid orbitals rather than for point dipoles
centered on the nuclei. Using Slater orbitals we calculate a
larger separation of 6.22 Å between the bridgehead nuclei. So the
fs provides highly precise geometrical information, but to
interpret it properly we must know the spin distribution within
the radicals.

Σ or Π ? The Benzoyloxy Dilemma

Benzoyloxy is one radical for which the spin distribution is
questionable. Despite its chemical importance neither the
electronic ground state nor the equilibrium geometry is certain.
Various semi-empirical and *ab initio* SCF-MO methods, with and

without spin and spacial restrictions, have been used to study acyloxy radicals.($\underline{9},\underline{10}$) They agree that most of the spin density (in all cases greater than 75%) resides on oxygen, but how is it distributed between the oxygens? If the radical is symmetrical (C_{2v}), the oxygens should of course share the spin density equally. If, however, the radical possesses only a plane of symmetry (C_s), as would a carboxylic acid with the hydroxyl hydrogen removed, the spin would be unevenly distributed. Figure 1 illustrates the question by showing the form of the molecular orbitals which are singly occupied in various states of the formyloxy radical.($\underline{3}$) For the C_{2v} geometry the candidates are a π orbital (a_2, the out-of-phase combination of the oxygen p π orbitals), and two σ orbitals (a_1, the in-phase, and b_2, the out-of-phase combinations of in-plane oxygen p orbitals). For the C_s geometry the π possibility is a", a distorted version of a_2, and the σ possibility is a', a distorted sum of a_1 and b_2. In either geometry the orbitals which are not singly occupied are filled.

Since distortion from C_{2v} to C_s mixes a_1 with b_2, it is not surprising that all molecular orbital calculations have predicted that the lowest-energy Σ radical is the $^2A'$ state.($\underline{9},\underline{10}$) The energy of the Π radical is less sensitive to distortion, and SCF-MO methods differ in assigning its equilibrium geometry. Newton has found that even with an STO 4-31G basis set, the C_{2v} formyloxy radical shows doublet instability, meaning that the calculated energy is minimized by an unsymmetrical electron distribution.($\underline{10}$) This is physically unreasonable and is but one of many symptoms of the inadequacy of single-determinant methods as applied to this system. The 5-15 kcal/mole differences they predict among the lowest states and geometries vary in sign as well as magnitude and are certainly untrustworthy. At least until CI calculations become available,($\underline{10}$) we can depend on theory only to give us the form of the orbitals in the various states. Only experimental data would allow a confident choice between a Σ state with most of the spin on one oxygen and a Π state with equal spin on both.

To understand such chemical properties of the benzoyloxy radical as the rates of decarboxylation and of oxygen scrambling one should know its electronic structure. We had hoped by studying the epr spectra of pairs including this radical to determine its ground state experimentally.

Acetyl Benzoyl Peroxide

Methyl Motion and the D Tensor. Photolysis of crystalline acetyl benzoyl peroxide with O-18 in the peroxy positions gives methyl benzoate with most of the O-18 bound to methyl rather than scrambled between the carbonyl and ether positions.($\underline{11}$) In attempting to understand this discrimination we studied the epr spectrum of the methyl-benzoyloxy radical pair (M-B) trapped in a photolyzed crystal of the peroxide at low temperature.($\underline{3}$)

The first row of Table I expresses in several ways the two items of non-orientational information from the electron-electron fs of M-B. The first two entries, Z and X, are the extreme values of splitting (in gauss) and correspond to the maximum and minimum components, respectively, of the average spin-spin distance. The third entry, Y, is the splitting in the third orthogonal direction, and its difference from X shows by how much the distribution departs from axial symmetry. The next two entries express this same information in wave numbers as the D and E parameters. The last entry is the separation (in Å) between point dipoles which would give the same Z. As discussed above this provides an upper limit to the distance between delocalized spins.

Table I. Electron-Electron Splittings of Radical Pairs.

Pair	Z	X	Y	D	E	Dist.
M-B	-375	199	176	0.0175	0.0003	5.30
M-P	-259	135	125	0.0121	0.0001	5.99
P-B	-329	173	156	0.0154	0.0003	5.53
P-P	-197	98	92	0.0092	0.0001	6.56
B-B	-1256	943	313	0.0587	0.0098	3.54
B-B'	-1320	1024	296	0.0617	0.0113	3.48

More of the fs information is presented in Figure 2D, where a heavy arrow indicating the direction of maximum spin separation (Z) is shown relative to the framework of the peroxide precursor in two orthogonal projections. The length of this arrow is 5.30 Å, corresponding to the point dipole limit for the observed splitting. The forked line indicates the direction of minimum spin-spin separation (X). Since the site of spin in methyl is obvious, we could use the arrow to locate the radical relative to benzoyloxy, if we knew the spin distribution and orientation of the latter.

On the assumption that benzoyloxy is C_{2v} and that it is immobile in the crystal, we suggested earlier that the methyl radical moves from its position in starting peroxide by 2.4 Å. The motion can be visualized approximately in Figure 2 by translating the heavy arrow so that its origin is centered between the oxygens of benzoyloxy. Benzoyloxy delocalization would shorten the point dipole arrow by about 0.1 Å.

While this motion seems to explain the preference for methyl attack at the original peroxidic oxygen, it is surprising in light of the fs of the methyl-phenyl radical pair (M-P), which can be generated from M-B by photolysis at 800 nm. This fs is summarized in the second entry of Table I, and the point dipole vector it predicts is illustrated by the light arrow in Figure 2D. The origin of this arrow is more certain than that of M-B, since the

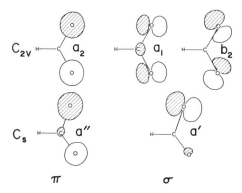

Figure 1. Orbitals that might be singly occupied, low energy states of an acyloxy radical in the C_{2v} and C_s geometries. The plane of the drawing is the nodal plane of the π orbitals. The most likely candidates for single occupancy in the ground state are a_2 (π) and a' (σ).

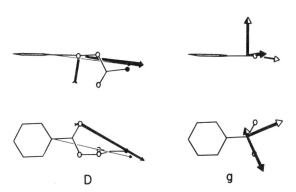

Figure 2. M–B and M–P tensors relative to acetyl benzoyl peroxide.

The two orthogonal projections (above and below) show the orientation of the D tensors of M–B and M–P (left) and of the g tensor of benzoyloxy (right). D. Open circles denote oxygen atoms; a filled circle, the methyl group. The heavy line with a forked tail shows the direction of minimum spin–spin separation of M–B. The heavy arrow represents the point dipole approximation to the spin–spin vector of M–B in length and direction. The light arrow presents the same information for M–P. g. The principal directions for the g shift of benzoyloxy are denoted by heavy vectors 2–Å long. The filled and open triangles denote the most and least positive g shifts, respectively (see Table II).

phenyl radical is mostly localized. (13)   The arrow's head falls
very near the original methyl position.   If methyl moves by 2.4 Å
in forming M-B, it must move back again when benzoyloxy decarboxy-
lates.   This coincidence is conceivable, but suspicious.

The geometry of M-P would have been less surprising if we
had assigned benzoyloxy a C_s Σ electronic state with most of its
spin density on the oxygen which was originally carbonyl.   The
origin of the heavy arrow would then have been as shown in Figure
2D, and methyl would undergo only modest displacements from
peroxide to M-B to M-P.   If the C_s Σ benzoyloxy radical had its
spin on the oxygen which was originally peroxy, the arrow would
originate on that oxygen, and the excursion of the methyl radical
would be even lengthier than for the Π.   If benzoyloxy radical had
a Σ ground state, one might expect the two geometries to coexist
and give a doubling of the peaks in the epr spectrum, which we
have not observed.   But the lattice forces which impose a 5-6
kcal/mole barrier on radical recombination could easily bias an
equilibrium toward one of the Σ geometries.

If we wished to be parsimonious about the amount of methyl
motion in the crystal, we would choose the Σ benzoyloxy radical
with spin on the original carbonyl oxygen, although the direction
of minimum spin-spin distance is more consistent with the Π than
with this Σ radical.

The g Tensor of Benzoyloxy.   By subtracting the isotropic g
value of methyl radical from twice the g tensor of M-B, we
determined an approximate g tensor for benzoyloxy.   Shifts of its
principal values from 2.0023 are presented in the first row of
Table II.   Since the shifts are positive, filled orbitals must
predominate over vacant ones in mixing with the singly occupied
orbital.

Table II.   Benzoyloxy g Shifts.[a]

Partner	iso	X	Y	Z
$CH_3 \cdot$	94	205	59	18
$Ph \cdot$	99	221	61	15
$PhCO_2 \cdot$	86 (89)	$\left\{ \begin{array}{l} 207\,(239) \\ 212\,(209) \end{array} \right.$	113 (83)	$-64\,(-58)$
			37 (36)	9 (20)

a)   g shift = (g - 2.0023) x 10^4, where g is calculated
    from twice the g tensor observed for the radical pair
    of benzoyloxy with a partner minus the g tensor of the
    partner.   The parenthetical values in the $PhCO_2 \cdot$
    entries are for B-B', the others for B-B.   The last
    two rows correspond to different tensor assignments,
    one of which is incorrect, as explained under The B-B
    Pairs in the text.

How the g shifts help choose the ground state can be seen by examining Figure 1. The C_{2v} Π state ($^2A_2$) would be strongly mixed with $^2B_2$ by a field in the oxygen-oxygen direction, since the oxygen orbitals of a_2 rotated about that axis give high overlap with those of b_2 (they have the wrong phase to overlap with a_1). A field along the molecular symmetry axis would in principle mix a_1 with a_2, but in fact the shape of the a_1 orbital is such that very little overlap would result, and only a small g shift would be expected. A field perpendicular to the molecular plane would not mix the Π state with any other state, and would give no g shift. Thus a Π radical should display a large g shift along the O-O direction, a small g shift along the symmetry axis, and no g shift perpendicular to the nuclear plane.

The C_s Σ state ($^2A'$) would be strongly mixed with $^2A''$ by a field along the bond connecting the radical oxygen to the carboxyl carbon. A field perpendicular to the bond and in the nuclear plane would not mix the $^2A'$ state, since it would lie along the axis of the spin-bearing p orbital of oxygen. A field perpendicular to the plane would mix $^2A'$ with higher-energy Σ states. Thus the Σ radical would show a strong g shift along the C-O single bond, a weak shift perpendicular to the plane, and no shift in the third orthogonal direction.

The orientation of the g tensor shown in Figure 2g is just what would be expected for the $^2A'$ benzoyloxy radical with spin on the original peroxy oxygen. Kim, Kikuchi, and Wood have used this agreement to argue for a Σ benzoyloxy radical with spin on this oxygen.(14) Their interpretation may well be correct, but it is disturbing that this assignment is the one for which the D tensor requires maximum motion of the methyl radical. As we suggested earlier, the Σ radical with spin on the original carbonyl oxygen, which gives the least-motion interpretation of the D tensor, gives the worst agreement with the g tensor.(3)

Thus the D and g tensors give conflicting indications of the geometry of M-B and of the ground state for benzoyloxy. To decide on a radical pair geometry which will help us understand the solid state reaction we must resolve this conflict. Unfortunately the additional data we have been able to collect have served more to sharpen the conflict than to resolve it.

Dibenzoyl Peroxide.

The P-B Pair. Partly to assess the reliability of the experimental tensors for M-B, we determined D and g tensors for the analogous phenyl-benzoyloxy pair (P-B) in photolyzed dibenzoyl peroxide. This species, like M-P, was first reported by Lebedev.(15) We used perdeuterated crystals to remove phenyl hfs from the highly overlapped spectra of four symmetry-related versions of each radical pair (there would have been eight, each corresponding to loss of one of the CO_2 groups in the $P2_12_12_1$ unit cell, if it were not for an approximate, non-crystallographic

two-fold symmetry axis, which almost converts a screw axis to a
translation). Box has determined the D tensor for the phenyl-
phenyl pair (P-P) in dibenzoyl peroxide. (16) Figure 3 presents
our D and g tensors for P-B, and Box's D tensor for P-P in the
same format as Figure 2. Comparison of Figures 2 and 3 and of the
corresponding data in Tables I and II shows that the tensors of
the pairs with methyl and phenyl as partner have almost the same
magnitude and the same orientation in the molecular frame,
although the crystalline environments are completely different.
This increases our confidence in the experimental tensors. More
importantly it requires that whatever motion methyl undergoes in
acetyl benzoyl peroxide, phenyl must undergo in dibenzoyl
peroxide.

Far from resolving our dilemma about the ground state of
benzoyloxy radical, this additional information has made it more
acute. If we favor the Σ radical with spin on the original peroxy
oxygen, which in its initial orientation is consistent with the g
tensor, the D tensor forces us to accept that in their different
crystals M-B and P-B pairs undergo identical large (about 3 Å)
displacements relative to the peroxide geometry, although the M-P
and P-P pairs have geometries very similar to those in the
peroxide. If we favor the Π radical, the g tensor is anomalous
and no longer attributable to experimental error, and the methyl
or phenyl excursion is still substantial. Again the Σ radical
with spin on the original carbonyl carbon allows minimal radical
motion but seems unlikely to have the observed g tensor.

The B-B Pairs. While measuring the spectra of P-B we noticed
weak doublets with greater electron-electron splitting than P-B
and no hfs, even in undeuterated samples. Their strong g
anisotropy, resistance to saturation at low temperature, and rapid
decay at 35 K (P-B can be observed at 63 K) confirmed our
inference that they were due to benzoyloxy-benzoyloxy pairs (B-B).
Box has mentioned observing B-B after subjecting a solid solution
of dibenzoyl peroxide in dibenzoyl disulfide to ionizing
radiation. His B-B recombined at "a few degrees above 4.2 K."(16)
After brief photolysis at 20 K and cooling to 6 K we observed
eight doublets rather that the four required by symmetry for a
single species. Thus there are two different metastable
arrangements of the B-B pair. One pair, which we denote B-B',
converts rapidly to the other on warming to 25 K. We determined D
and g tensors for both B-B' and B-B in hopes of answering two
questions. What can these pairs tell us about the ground state
of benzoyloxy? How can radical pairs be stable when they are not
separated by an intervening molecule, such as CO_2 or N_2?

Since the crystalline unit cell contains four symmetry-
related molecules, spectra of four B-B orientations are measured
simultaneously in one crystal mounting. However this experimental
efficiency comes at a high price, because there is no direct
method for determining which of four symmetry-related tensors goes

with a particular molecular orientation. The previous radical pairs were sufficiently separated that the point dipole approximation permitted a unique choice of the correct tensor. For B-B (and B-B') two of the four choices may be excluded as giving spin-spin extension in directions nearly orthogonal to reasonable ones. The remaining two possibilities are plotted in Figures 4 and 5 for B-B. Plots for B-B' analogous to Figures 4 and 5 would be practically indistinguishable from 4 and 5, except in the end-on view of 5, where the B-B' vectors are included and indicated by squares. In both figures we have started the spin-spin arrow from the carbonyl oxygen, which seemed to be the best origin for the M-B and the P-B pairs in terms of predicting minimal motion. As before the arrow should terminate near the center of spin density of the other radical in the pair, which in this case is also benzoyloxy. The g tensors of the second benzoyloxy in B-B and B-B' were calculated as before by subtracting the g tensor of the first (assumed equivalent to that found from the P-B pair) from twice the observed tensor. The next to last row of Table II corresponds to the tensor assignment of Figure 4, and the last row to that of Figure 5.

The arrow of Figure 4D suggests that the spin density in the second benzoyloxy radical, as in the first, resides mostly near the original carbonyl oxygen. It does require some motion of this atom, which could be achieved by twisting the carboxy group about the C-phenyl bond. The same sense of rotation is suggested by the orientation of the g tensor for this radical. The magnitudes of the principal g shifts (Table II) are similar to those of the previous benzoyloxy radicals, but differ from them by much more than they differ from one another. If this is the correct tensor assignment, the stability of the B-B pair could perhaps be attributed to the twist, which separates the peroxy oxygens enough that their bonded attraction cannot overcome a lattice barrier to the twist back. The attraction may be particularly weak if most of the spin resides on the original carbonyl oxygens. B-B and B-B' are remarkably similar, differing by less than 0.4 Å in point dipole vectors and 6° in g orientation. The source of a barrier between two such similar geometries is not obvious. As for the M-B and P-B pairs, the D and g tensors suggest spin localization on different oxygens.

The tensor assignment of Figure 5 suggests that the average spin position of the second benzoyloxy radical is very near the carbonyl carbon, as would be expected for a C_{2v} radical. Furthermore the g shifts for this radical in B-B are precisely those that would be predicted for an almost unmoved Π state, both in magnitude and in orientation. The g orientation of the corresponding radical in B-B' differs from this by rotation of 11° about the C-phenyl bond, as shown in the end-on view of Figure 5. It is remarkable that the principal values of these g tensors are so similar to those of the other radical, if that radical is in a Σ state. If this assignment is correct, the barrier to recombi-

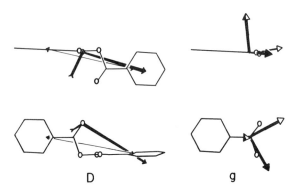

Figure 3. *P–B and P–P tensors relative to dibenzoyl peroxide. Presentation analogous to that of Figure 2, except that the double-headed arrow representing the spin–spin vector of P–P is centered between the substituted ring carbons of the peroxide.*

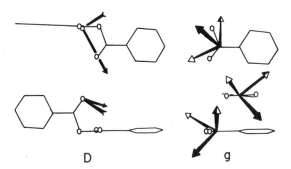

Figure 4. *B–B tensors relative to dibenzoyl peroxide. Tensor assignment of the next-to-last row in Table II. Presentation analogous to Figure 2, except that the middle diagram in the g column is an end-on view obtained by rotating the lower diagram by 90° to the right about a vertical axis in the page.*

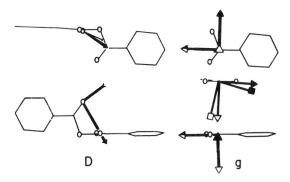

Figure 5. B–B tensors relative to dibenzoyl peroxide. Tensor assignment of the last row in Table II. Presentation analogous to Figure 4, except that the end-on view includes the g tensor of B–B' denoted by squares instead of triangles.

nation of the radicals may be due to a failure of a $\Sigma-\Pi$ radical pair state to correlate with the ground state of peroxide.

Summary and Conclusions

Although we have now measured D and g tensors for four different radical pairs containing the benzoyloxy radical, we cannot yet be certain of the radical's electronic state. If we consider only the g tensors, we would feel confident that the radical in the M-B and P-B pairs is in a Σ state with its spin on the oxygen which was originally peroxidic. However, the D tensors would then indicate that the other radical, methyl or phenyl, must undergo surprisingly large (and similar) motion after the first decarboxylation but return to its initial location after the second.

If we considered only the D tensors of these pairs, we would conclude that the benzoyloxy is in a Σ state with spin on the original carbonyl oxygen, and that very little radical motion occurs during the decarboxylations. The g tensors do not seem consistent with this interpretation.

An unmoved Π benzoyloxy radical would not give such flagrant disagreement with one of the tensors as one or the other of the Σ radicals does. Neither would it give such good agreement with the other tensor.

We could cease trying to understand the nature of the benzoyloxy radical in M-B and P-B and assume only that its spin is near the original site of the carbonyl oxygen and that it has the observed g tensor. This might result from some unanticipated motion or distortion of the radical. If we then risk supposing that one of the benzoyloxy radicals of B-B has these same properties, we may determine the g tensor and spin location of the other. The same is true of B-B'. In either instance one of two plausible interpretations is incorrect. Either 1) the second radical has the same anomalous properties as the first and has its carboxyl group twisted about the C-phenyl bond, or 2) the second radical is in a Π electronic state and is essentially unmoved in B-B but twisted by 11° about the C-phenyl bond in B-B'.

Obviously we must do much more work before claiming that we understand acyloxy radicals and their behavior in solids. Still the work described above shows that single-crystal epr spectroscopy of radical pairs can provide unique, if not always unambiguous, information about the structure and chemistry of free radicals.

Acknowledgments

This work has been supported by the National Science Foundation (DMR 76-01996) and by a Camille and Henry Dreyfus Teacher-Scholar Grant. We thank Dr. Marshall D. Newton and Professor D. E. Wood for helpful discussions of their unpublished work.

Abstract

Low-temperature epr spectra yield A, g, and D tensors for radical pairs generated by uv photolysis of single crystals of diacyl peroxides. The precise, but sometimes ambiguous, information these tensors contain about the structure of the radicals and about their arrangement in pairs is discussed with examples. Pairs including a benzoyloxy radical together with methyl, phenyl, or another benzoyloxy radical are perplexing in that D tensors seem to indicate that electron spin resides mostly on the original carbonyl oxygen, while g tensors seem to indicate that most of the spin resides on the original peroxy oxygen. The radicals in the benzoyloxy-benzoyloxy pair may be in different electronic states.

Literature Cited

1. Nelsen, S. F., and Bartlett, P. D., J. Am. Chem. Soc., (1966), 88, 137.
2. Bartlett, P. D., and McBride, J. M., Pure Appl. Chem., (1967), 15, 89.
3. Karch, N. J., Koh, E. T., Whitsel, B. L., and McBride, J. M., J. Am. Chem. Soc., (1975), 97, 6729.
4. Jaffe, A. B., Skinner, K. J., and McBride, J. M., J. Am. Chem. Soc., (1972), 94, 8510.
5. Wertz, J. E., and Bolton, J. R., "Electron Spin Resonance," McGraw Hill, New York (1972), Table C.
6. Reichel, C. L., and McBride, J. M., J. Am. Chem. Soc., (1977), 99, 6758.
7. Carrington, A., and McLachlan, A. D., "Introduction to Magnetic Resonance," Harper and Row, New York, 1967 Ch. 9.
8. Ibid., Ch. 8.
9. Kikuchi, O., Tetrahedron Lett., (1977), 2421, and references therein.
10. Newton, M. D., personal communication and work in progress.
11. Karch, N. J., and McBride, J. M., J. Am. Chem. Soc., (1972), 94, 5092.
12. Whitsel, B. L., Ph. D. Dissertation, Yale Univ., (1977).
13. Kasai, P. H., Clark, P. A., and Whipple, E. B., J. Am. Chem. Soc., (1970), 92, 2640.
14. Yim, M. B., Kikuchi, O., and Wood, D. E., J. Am. Chem. Soc., in press.
15. Barchuk, V. I., Dubinsky, A. A., Grinberg, O. Ya., and Lebedev, Ya. S., Chem. Phys. Lett., (1975), 34, 476
16. Box, H. C., Budzinski, E. E., and Freund, H. G., J. Am. Chem. Soc., (1970), 92, 5305.

RECEIVED December 23, 1977.

Propagation Reactions of Free Radicals

Molecular Orbital Correlations of Some Simple Radical Reactions

GERALD JAY GLEICHER

Department of Chemistry, Orgeon State University, Corvallis, OR 97331

The application of relatively simple molecular orbital theory to problems involving the formation and reactivity of organic radicals has lagged behind related studies on charged trigonal species. This is somewhat surprising for although radical forming reactions are often judged by the uninitiated to be much less sensitive to structural changes, post hoc evidence indicates that a judicious choice of system can obviate such problems. There also exists a compensating factor in that the uncharged radical system may be simpler to treat theoretically than its charged counterparts due to the absence of a pronounced electric field (1). A too sanguine overview, however, can be adopted. Through the early sixties surprisingly few investigations were planned to make use of potential molecular orbital calculations. Even more disturbing was the observation that in certain studies where such correlations were attempted, overall agreement was unsatisfactory.

The best of the early experimental work is a series of investigations involving radical aromatic substitution by alkyl radicals studied by Szwarc and co-workers (2-6). This and still earlier work by Kooyman and Farenhorst with the trichloromethyl radical (7) have allowed for systematic variation in the steric and electronic properties of the attacking species. Experiments indicate that there is rather little variation in relative rate trends as a function of attacking radical. Theoretically, it can also be pointed out that there is little difference among correlations utilizing parameters determined from ground state calculations and those utilizing calculated energy differences (see below).

A second type of investigation can be typified by the work of Kooyman in the generation of arylmethyl and similarly delocalized radicals and the attempted correlation with calculated molecular orbital parameters (8). Studies of this type represent but part of the overall subject of arylmethyl reactivity which includes not only radical generating hydrogen abstraction processes, Equation 1, but also the corresponding carbonium ion and carbanion reactions, Equations 2 and 3.

$$\underset{CH_3}{\bigcirc} + X\cdot \longrightarrow \underset{\overset{\bullet}{C}H_2}{\bigcirc} + HX \qquad 1$$

$$\underset{CH_2Y}{\bigcirc} \longrightarrow \underset{\overset{\oplus}{C}H_2}{\bigcirc} + :Y^{\ominus} \qquad 2$$

$$\underset{CH_3}{\bigcirc} + :B \longrightarrow \underset{\overset{\ominus}{\overset{\bullet\bullet}{C}H_2}}{\bigcirc} + BH^{\oplus} \qquad 3$$

Kooyman's investigation utilized the relative rates of hydrogen abstraction from a series of aralkyl compounds (and one olefin) by the trichloromethyl radical (8). The experimental results are shown in Table I.

Table I. Relative rates of Hydrogen Abstraction from Selected Hydrocarbons by the Trichloromethyl Radical.

Hydrocarbon	Relative Rate
Toluene	1.0
Diphenylmethane	7.98
2-Octene	11.67
Triphenylmethane	17.90
3-Phenylpropene	28.57
Indene	111.9

A linear correlation between logs of the relative rate constants and calculated HMO pi energy differences between the delocalized product radical (assumed equivalent to the transition state) and the initial unsaturated compound was attempted. The results were quite poor. A correlation coefficient of only 0.47 was obtained. (A formalism strongly analogous to the Hammett equation will be employed throughout the discussion. The slopes of the correlations, unlike rho values, have little recognized physical meaning and are not even comparable if energy differences are calculated by different methods. The numerical values of these slopes well therefore be ignored. However, correlation coefficients shall be given and considered as a measure of the reliability of the calculation employed. All such coefficients are obtained from linear least squares analyses of the correlations.)

Although this poor correlation may be attributable in part to utilization of the overly simple HMO approach and to experi-

mental problems in evaluation of rate constants, there also
exist underlying difficulties concerning the choice of systems.
Benzylic radicals derived from diphenylmethane, triphenylmethane
and 3-phenylpropene cannot achieve complete planarity and
arguments based on delocalization in planar systems cannot be
safely applied. As shall be seen most subsequent workers in the
field of benzylic reactivity have opted to study series of
compounds of overall greater similarity than those utilized by
Kooyman.

Benzylic Reactivity and Molecular Orbital Correlations.

The formation of unsubstituted polycyclic benzylic carbonium
ions has been studied for over twenty years (9, 10). While much
of the earlier work was apparently planned with application of
Hückel molecular orbital theory in mind, the problem continues
to attract workers and more recent publications have combined
experimentally more precise results with advanced theoretical
techniques (11).
The corresponding carbanions have also been generated
utilizing the reaction of polycyclic arylmethanes with strong
bases (12, 13). Initial theoretical calculations were again
with HMO theory.
Irrespective of the complexity of the molecular orbital
approach employed, it is possible to attempt correlations using
some calculated energy difference, e.g. Kooyman's results (8),
or some parameter derived from the isolated aromatic starting
material. While the latter approach has worked well in correlating the results of electrophilic and radical aromatic substitution (14), it has been less frequently applied to the generation of charged benzylic species. Mechanistically this neglect
is reasonable. The transition states for generation of the
benzylic ions must show a strong resemblance to the charged
intermediate. Any approach which ignores this will probably be
incorrect. The utilization of energy differences should also be
applicable to relatively endergonic radical forming reactions as
well (15).
The HMO correlations of relative rates of formation of the
polycyclic benzylic ions lead to some interesting conclusions.
Compounds of an α-naphthyl type (i.e. those having the exo-cyclic
methylene unit attached to a carbon adjacent to a point of
annelation) always show a reactivity less than that expected
based on compounds of the β-naphthyl type. A tendency to treat
data in terms of dual correlations has been thus developed. The
infrequent inclusion of nonalternant (i.e. non-benzenoid) systems
would usually generate data which could not be accommodated by
either the α-naphthyl or β-naphthyl correlations. These findings
for the generation of arylmethyl anions (12, 13) are illustrated
in Figure 1. A single HMO correlation including data from both
α-naphthyl and β-naphthyl type systems yields a correlation

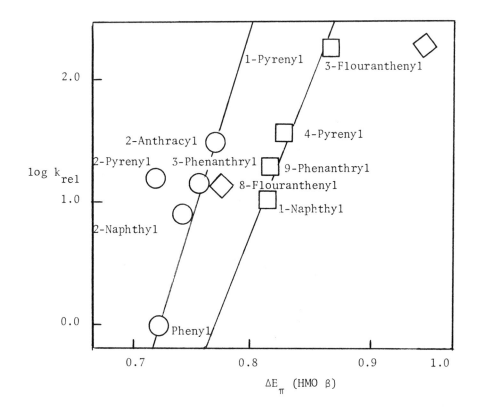

Figure 1. Correlation of the relative rates of formation of arylmethyl anions with calculated HMO energy differences: (○), β-naphthyl compounds; (□), α-napthyl compounds; (◇), non-alternant compounds.

coefficient of only 0.82. If a dual correlation is employed
correlation coefficients of 0.97 (α-naphthyl) and 0.99
(β-naphthyl, excluding one point) are found. Inclusion of the
data from non-alternant systems, as pointed out, decreases the
reliability.

In order to explain this duality, recourse was made to
special steric factors operative in the α-naphthyl series which
could increase the energy difference between ground and transition states. This peri effect is a specific result of non-bonded
interaction between the methylene unit and the atom or group on
the "other side" of the point of annelation as is shown for the
α-naphthylmethyl carbanion in Figure 2A. A rotation of ninety
degrees around the exocyclic carbon-carbon bond shown in Figure
2B can remove this unfavorable interaction but only with complete
loss of conjugation between the exocyclic carbon and the
remainder of the system. A semi-quantitative estimate of these
interactions may be gleaned if recourse is made to models of the
systems and standard values for hydrogen-hydrogen and carbon-
hydrogen repulsion terms (16). Initially models were constructed
using an average aromatic carbon-carbon bond length of 1.40 Å
and a carbon-hydrogen bond length of 1.085 Å. The pertinent
internuclear distances and repulsive interactions giving rise to
the peri effect are shown in Figure 2. It is assumed that some
practical compromise is reached with the resulting lessened
delocalization causing a decreased reactivity. In the above
carbanion case, for example, a rotation of seventeen degrees from
planarity is estimated (13).

It is possible, however, that intuitively attractive as the
peri effect is, there is no need to invoke it for simple,
unsubstituted, polycyclic arylmethyl intermediates. The complete
neglect of electron repulsion which underlies HMO calculations
cannot be justified. While this approach can yield reasonable
results for alternant hydrocarbons because of the uniform
electron distribution, there is less reason to believe that it
will be satisfactory for odd-alternant ions (17) as, in fact,
is observed above. Dewar and Thompson (18) have applied the
results of SCF calculations to the problem of carbanion formation.
These workers were able to incorporate α-naphthyl, β-naphthyl and
non-alternant derivatives into a single relationship with a correlation coefficient of 0.97. This is shown in Figure 3. The SCF
approach can also make use of a bond order-bond length relationship to determine two center resonance and repulsions integrals
(19). It is thus possible to arrive at consistent molecular
structures as well as energies. Such structures for α-naphthyl
species also tend to predict a slightly decreased importance for
the peri interaction, relative to that obtained from the usual
HMO approximation. This is due to greater calculated internuclear separations being obtained for those atoms causing the
repulsive peri effect.

Although it is clear that some of the discrepancy between

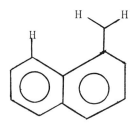

A) Planar, standard bond lengths

| C---H | 2.44 Å |
| H---H | 1.71 Å |

E_{C--H}	0.725 kcal/mole
E_{H--H}	4.175 kcal/mole
strain	4.900 kcal/mole

B) As A, CH_2 rotated 90°

| C---H | 2.44 Å |
| H---H | 2.74 Å (2) |

E_{C--H}	0.725 kcal/mole
E_{H--H}	0.050 kcal/mole
strain	0.725 kcal/mole

C) Planar, SCF bond lengths

| C---H | 2.49 Å |
| H---H | 1.76 Å |

E_{C--H}	0.529 kcal/mole
E_{H--H}	3.520 kcal/mole
strain	4.049 kcal/mole

D) As C, 3° in-plane bends

| C---H | 2.62 Å |
| H---H | 1.97 Å |

E_{C--H}	0.261 kcal/mole
E_{H--H}	1.484 kcal/mole
strain	1.745 kcal/mole
angle strain	0.544 kcal/mole
total strain	2.289 kcal/mole

Figure 2. Various geometries of the α-naphthylmethyl system with accompanying peri interactions and calculated strains

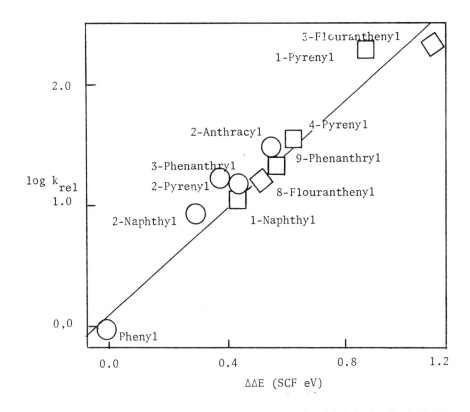

Figure 3. Correlation of the relative rates of formation of arylmethyl anions with calculated SCF energy differences

α-naphthyl and β-naphthyl type systems hitherto assigned to peri effects is not steric but rather electronic in origin, such an effect might still be important in arylmethyl reactivity. A quantitative evaluation of the peri effect has been undertaken (20). Based upon bond lengths determined from SCF calculations, it is shown in Figure 2C that the non-bonded interactions involving the exo-cyclic atoms were smaller than thought. These interactions could be further diminished by slight (three degree) in-plane deformations without any loss of delocalization. This is shown in Figure 2D. Out of plane bending of hydrogen atoms may further reduce peri interactions. Inclusion of atoms or groups larger than hydrogen within the potential peri interactions, however, may well necessitate a rotation from co-planarity of the exo-cyclic group.

Formation of Arylmethyl Radicals by Hydrogen Abstraction.

Unlike the corresponding ions, benzylic radicals are true alternant hydrocarbon species. It was deemed of interest to see whether rates of formation of such species could be related to changes in delocalization and also whether a similar dichotomy of results between HMO and SCF correlations would be observed. To examine these questions a series of arylmethanes was reacted with bromotrichloromethane at 70° (21, 22). The reactions were initiated with trace amounts of benzoyl peroxide, Equation 4.

$$ArCH_3 + BrCCl_3 \xrightarrow{\ \ peroxide\ \ } ArCH_2Br + HCCl_3 \qquad 4$$

Although subsequent research in other laboratories have led to claims that bromine atom is the chain carrying species in hydrogen abstractions utilizing bromotrichloromethane (28), our original presentation of results was based on the rate determining step shown in Equation 5.

$$ArCH_3 + {}^{\cdot}CCl_3 \longrightarrow ArCH_2{}^{\cdot} + HCCl_3 \qquad 5$$

The relatively endothermic nature of this process should allow for utilization of of a calculated energy difference as a structural parameter with which to correlate reaction rates.

This particular reaction is not free from complications. The trichloromethyl radical is known to attack the rings of polycyclic aromatics (7). In order to correct for this, it was assumed that any enhanced reactivity of the methylarene relative to the parent arene reflected only hydrogen abstractions. A competition between these two species can thus be used to determine (allowing for statistical correction) the amounts of hydrogen abstraction and ring substitution for each arylmethane. Such an approach presumes that the methyl substituent does not electronically influence attack in the aromatic unit. At the time, however, both experimental (24) and theoretical (25)

results seemed to justify this view for the trichloromethyl
radical.
    Table II presents the originally obtained percent of re-
action at the methyl group and corrected relative rates of
hydrogen abstraction. Also shown are the calculated relative
energy differences between the arylmethyl radical and the initial
arene. It can be seen that a large range of experimental
reactivities, nearly three powers of ten, is encountered.

Table II. Relative Rates of Hydrogen Abstraction from a Series of
           Unsubstituted Arylmethanes by Trichloromethyl Radical
           at 70°.

Arylmethane	% methyl Hydrogen abstraction	(k_X/k_0) H abstr		$\Delta\Delta E$ (SCF)
Toluene	100	0.172	± 0.008	(0.000)
1-Methyltriphenylene	84.6	0.362	± 0.027	0.076
2-Methyltriphenylene	69.1	0.395	± 0.014	0.068
3-Methylphenanthrene	86.9	0.547	± 0.027	0.060
1-Methylphenanthrene	69.5	0.569	± 0.035	0.139
2-Methylnaphthalene	68.4	0.682	± 0.057	0.076
9-Methylphenanthrene	91.0	0.845	± 0.056	0.178
1-Methylnaphthalene	94.3	(1.00)		0.175
6-Methylchrysene	77.0	2.19	± 0.04	0.214
2-Methylanthracene	74.5	5.89	± 0.47	0.215
1-Methylanthracene	54.3	13.1	± 1.4	0.340
1-Methylpyrene	76.5	18.7	± 1.0	0.352
9-Methylanthracene	86.4	112.0	± 5.0	0.519

Reprinted with permission of the Journal of Organic Chemistry, <u>93</u>,
2008, (1971). Copyright by the American Chemical Society.

    Although the arylmethyl radicals are alternant hydrocarbons,
the Hückel method again proved inadequate in correlating the data.
The compounds fell into the usual two sets. While the separate
correlation coefficients of 0.948 for the α-naphthyl points and
0.904 for the β-naphthyl points were good and fair, the overall
correlation coefficient of 0.86 was unacceptable. The single
SCF correlation is shown in Figure 4. The overall correlation
coefficient is now 0.977. Some slight improvement can still be
made if a dual correlation is utilized with the two new correlation coefficients now being greater than 0.99. Perhaps this
does reflect a small <u>peri</u> effect.
    Why does the Hückel method still fail in treating arylmethyl
reactivity even in neutral systems? This must be attributable
to the neglect of electronic interaction terms in general and
specifically neglect of spin polarization. The latter is
particularly important in odd electron systems and is due to
differential interactions between electron pairs of the same and
opposite spin. Hückel calculations cannot account for fine

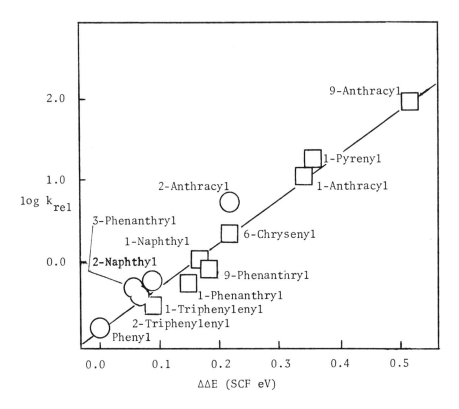

Journal of the American Chemical Society

Figure 4. Correlation of the relative rates of hydrogen abstraction from polycyclic arylmethanes by the trichloromethyl radical with calculated SCF energy differences

structure in the esr spectra of allyl and benzyl radicals caused
by spin polarization while SCF calculations can (26).
   Although it was felt that a correlation with a calculated
energy difference was in keeping with the view that a large
degree of carbon-hydrogen bond breaking occurs in hydrogen
abstraction by trichloromethyl radical, correlation with several
ground state parameters (14) was also undertaken. In all cases
the resulting correlations were much poorer than that discussed
above. The use of negative evidence is suspect of course.
However, it is felt that these results give some support to the
view that a late transition state is encountered in this process.
(22)
   It was at this point that concern over one of the initial
assumptions developed. The problem that a methylarene might
undergo ring substitution at a significantly different rate than
the parent system, could not be ignored. This question seemed to
be particularly important for 9-methylanthracene. This compound
should not only be most susceptible to ring substitution, but,
because of its high reactivity in the hydrogen abstraction
reaction, can disproportionately influence that· overall correlation. In order to evaluate the importance of this possible
effect the trichloromethylation of a series of 9-substituted
anthracenes was studied in accord with Equation 6.

$$\text{(anthracene with X, H)} + \,^{\bullet}CC1_3 \longrightarrow \text{(product with X, H, CCl}_3) \qquad 6$$

The meso position(s) of anthracenes is known to be most prone to
reaction and has been shown to account for about ninety percent
of radical substitution (27).
   The results are given in Table III. As can be seen there is
a substituent effect with electron donating groups favoring, as
might be expected, the trichloromethylation of the ring (28).
This data may be treated within a standard Hammett type correlation to produce a rho value of -0.83 (correlation coefficient
of 0.970) when plotted against σ_p^+. The data for 9-methylanthra-
cene, however, shows a significant upwards deviation from the
correlation. This is indicative of an additional mode of
disappearance for this compound which must be hydrogen abstraction
from the methyl group. The total reactivity of 9-methylanthracene
is due mostly (65%) to this latter process. The original amount
of reaction at this site was believed to be 86% of the total
reactivity of this compound. Most serendipitously, however, the
original correlation for hydrogen abstraction from arylmethanes is
only slightly changed. There is only a two percent decrease in
the slope of the correlation and a change in the correlation
coefficient from 0.977 to 0.973.

238   ORGANIC FREE RADICALS

Table III. Relative Reactivities of 9-Substituted Anthracenes
toward Trichloromethyl Radical Addition at 70.0°.

Substituent	k_X/k_H
NO_2	0.71 ± 0.10
CN	0.34 ± 0.04
CO_2CH_3	0.74 ± 0.07
Br	1.00 ± 0.04
Cl	1.02 ± 0.04
H	1.00
C_6H_5	2.28 ± 0.43
$i-C_3H_7$	2.38 ± 0.30
C_2H_5	3.39 ± 0.44
CH_3	6.85 ± 0.89
OCH_3	5.52 ± 0.60

Reprinted with permission of the Journal of the American Chemical
Society, 96, 787, (1974). Copyright by the American Chemical
Society.

It is of interest that 9-ethyl and 9-isopropylanthracene do
not show the same large deviation from the correlation for ring
substitution. The simplest conclusion is that exocyclic hydrogen
abstraction is less likely here because of peri effects caused by
replacing one or both of the methylene hydrogens by methyl groups.
Two other studies involving hydrogen abstraction from
polycyclic arylmethanes have also been reported in the literature.
Gilliom and coworkers have studied many of the same compounds
discussed above in reaction with bromine atom (29). It was
observed that methylanthracenes and methylpyrenes underwent
exclusive ring reactions. The other compounds employed showed
reactivities in this reaction which paralleled those observed in
the trichloromethyl radical reaction. One must again conclude
that appreciable radical character is again developed in the
trasition state based upon the good correlation with SCF calculated energy differences. The reaction of the arylmethanes with
t-butyl hypochlorite was also undertaken (22). It was hoped
that the less selective t-butoxy radical might cause a reactivity
trend which could be correlated with ground state properties.
Although chlorine atom traps were employed, exclusive ring
chlorination was observed for several of these molecules. The
remaining systems showed marginal selectivity. All attempt at
correlation with either calculated energy differences or ground
state parameters were unsuccessful.

Formation of Arylmethyl Radicals by Addition.

In an attempt to determine whether ground state molecular
orbital parameters could correlate a relatively exothermic

radical reaction, attention was directed away from atom abstraction and toward an addition process. The addition of a thiyl radical to vinylarenes is shown in Equation 7 (30). No measurable attack by the thiyl radical in the polycyclic portion

$$ArCH=CH_2 + \cdot S \!-\!\!\!\bigcirc \!\!\!\bigcirc \longrightarrow Ar\dot{C}HCH_2S \!-\!\!\!\bigcirc \!\!\!\bigcirc \qquad 7$$

of the olefin takes place; thus making the reaction in theory a simple one to study. The results are presented in Table IV. The overall range of reactivities is much smaller than that

Table IV. Relative Reactivities of Vinylarenes toward Thiophenol at 70°.

Compound	k_X/K_{std}	$\Delta\Delta E$ (SCF)
Styrene	1	0
2-Vinylnaphthalene	2.8 ± 0.3	0.061
9-Vinylphenanthrene	2.0 ± 0.2	0.125
1-Vinylnaphthalene	4.0 ± 0.5	0.140
6-Vinylchrysene	3.8 ± 0.2	0.142
2-Vinylanthracene	5.9 ± 1.2	0.167
1-Vinylanthracene	6.5 ± 1.5	0.209
1-Vinylpyrene	11.5 ± 3.0	0.234
9-Vinylanthracene	1.8 ± 1.2	0.424

found in the corresponding hydrogen abstraction study and is indicative of the expected more exothermic process. Failure was again encountered, however, in the attempted utilization of ground state properties to correlate the experimental data. The two most likely of such parameters, free valence on the terminal carbon and bond order of the exocyclic double bond, both yielded correlation coefficients of less than 0.5. A much better correlation could again be obtained with a calculated SCF energy difference. This is shown in Figure 5. The correlation coefficient, while only 0.932, is still fair. Improvement is noted if the data are separated into the traditional two sets. Correlation coefficients of 0.95 (α-naphthyl) and 0.97 (β-naphthyl) are now found. Peri interactions must be very real in these systems as the exocyclic portion of the supposed transition state is no longer a simple methylene group. Indeed, as can be seen, it is impossible to incorporate the results for 9-vinylanthracene within any of the above correlations as no planar conformation is possible for the intermediate radical without a particularly severe peri interaction. The remaining

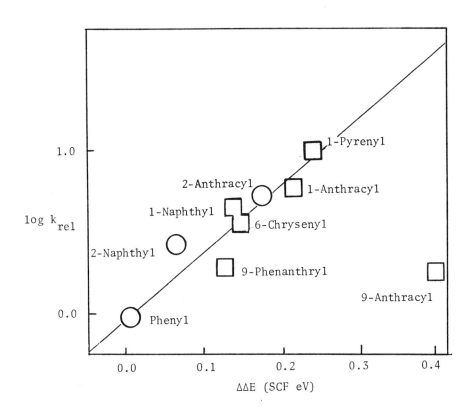

Figure 5. Correlation of the relative rates of thiyl radical addition to polycyclic vinylarenes with calculated SCF energy differences

α-naphthyl compounds may adopt a conformation with the bulky
exocyclic group directed away from the peri hydrogen. Support
for this view is obtained from a consideration of similar thiyl
radical addition to a homologous series of isopropenylarenes
as shown in Equation 8. (30). The data in Table V show   all

$$ArC = CH_2 \quad + \quad \cdot S \bigcirc \longrightarrow Ar\overset{\cdot}{C}-CH_2 S \bigcirc \qquad 8$$
$$\underset{CH_3}{|} \qquad\qquad\qquad\qquad \underset{CH_3}{|}$$

Table V.   Relative Reactivities of Isopropenyl Arenes toward
Thiophenol at 70.0°.

Compound	k_X/k_{std}
α-Methylstyrene	1.0
2-Isopropenylnaphthalene	1.55 ± 0.12
1-Isopropenylnaphthalene	0.14 ± 0.01
9-Isopropenylphenanthrene	0.05 ± 0.002
9-Isopropenylanthracene	0.02 ± 0.005

members of the α-naphthyl series to now possess greatly reduced
reactivity attributable to steric effects as it is now impossible
for all large groups on the exocyclic carbon to "point away from"
the peri hydrogen.

Conclusions.

    Certain distinct conclusions may be drawn from these
studies. Firstly, as might have been expected, is the obvious
fact that delocalization can have a tremendous effect on the
ease of benzylic radical formation. The peri effects, invoked
by earlier mechanicians, seem to be less important than claimed
for non-substituted systems, but, may exert a large effect if
substituents are attached to either the exocyclic atom or the
peri site. (Although only rate retarding peri effects have been
treated above, accelerative counterparts can be easily imagined.)
Finally, from the theoretical point of view, all of the work on
formation of benzylic like radicals cited enforce the view that
calculations which do not include interelectron terms are inadequate and that ground state parameters cannot successfully

correlate the data.

Abstract

The correlation of the rates of formation of arylmethyl
free-radicals by molecular orbital theory will be discussed.
Different levels of sophistication among pi-electron methods
lead to conflicting conclusions concerning the degree of
possible electron localization and the importance of non-bonded
interactions with the principal radical site. The arylmethyl
radical systems have been generated both by hydrogen abstraction
(1) and addition (2) reactions.

$$ArCH_3 \;+\; X\cdot \;\;\rightarrow\;\; ArCH_2{}^\bullet \;+\; HX \quad (1)$$

$$\underset{\overset{|}{R}}{ArC} = CH_2 \;+\; Z\cdot \;\rightarrow\; \underset{\overset{|}{R}}{Ar\overset{\bullet}{C}} - CH_2Z \quad (2)$$

The sensitivity toward steric factors varies and is much more
pronounced in the latter process.

Literature Cited

(1) Williams, G. H., "Homolytic Aromatic Substitution", p. 7,
     Pergamon Press, New York, 1960.
(2) Levy, M. and Szwarc, M., J. Chem. Phys., (1954), 22, 1621.
(3) Smid, M. and Szwarc, M., J. Am. Chem. Soc., (1956), 78,
     3322.
(4) Smid, M. and Szwarc, M., J. Am. Chem. Soc., (1957), 79,
     1534.
(5) Smid, M. and Szwarc, M., J. Chem. Phys.,   (1962), 29,
     432.
(6) Stefani, A. P. and Szwarc, M., J. Am. Chem. Soc., (1962),
     84, 3661.
(7) Kooyman, E. C. and Farenhorst, E., Trans. Faraday Soc.,
     (1953), 49, 58.
(8) Kooyman, E. C., Disc. Faraday Soc., (1951), 10, 163.
(9) Dewar, M. J. S. and Sampson, R. J., J. Chem. Soc., (1956),
     2789.
(10) Dewar, M. J. S. and Sampson, R. J., J. Chem. Soc., (1957),
     2946, 2952.
(11) Streitwieser, A., Jr., Hammond, H. A., Jagow, R. H.,
     Williams, R. M., Jesuitis, R. G., Chang, C. J., and Wolf,
     R., J. Am. Chem. Soc., (1970), 92, 5141.
(12) Streitwieser, A., Jr., and Langworthy, W. C., J. Am. Chem.
     Soc., (1963), 85, 1757.
(13) Streitwieser, A., Jr., Langworthy, W. C. and Brauman, J.I.,
     J. Am. Chem. Soc., (1963), 85, 1761.
(14) Greenwood, H. H. and McWeeny, R., Adv. Phys. Org. Chem.,
     (1966), 4, 73.

(15) Hammond, G. S., J. Amer. Chem. Soc., (1955), 77, 334.
(16) Bartell, L. S., J. Chem. Phys., (1960), 32, 827.
(17) Dewar, M. J. S., Rev. Mod. Phys., (1963), 35, 586.
(18) Dewar, M. J. S., and Thompson, C. C., Jr., J. Am. Chem. Soc., (1965), 87, 4414.
(19) Dewar, M. J. S. and Gleicher, G. J., J. Am. Chem. Soc., (1965), 87, 685.
(20) Gleicher, G. J., J. Am. Chem. Soc., (1968), 90, 3397.
(21) Unruh, J. D. and Gleicher, G. J., J. Am. Chem. Soc., (1969) 91, 6211.
(22) Unruh, J. D. and Gleicher, G. J., J. Am. Chem. Soc., (1971) 93, 2008.
(23) Tanner, D. D., Arhart, R. J., Blackburn, E. V., Das, N.C., and Wada, N., J. Am. Chem. Soc., (1974), 96, 829.
(24) Farenhorst, E. and Kooyman, E. C., Recl. Trav. Chim. Pays-Bas, (1962), 81, 816.
(25) Greenwood, H. H., Nature (London), (1955), 176, 1024.
(26) Fessenden, R. W., and Schuler, R. H., J. Chem. Phys., (1963), 39, 2147.
(27) Iwamura, H., Iwamura, M., Sato, S. and Kushida, K., Bull. Chem. Soc. Jap., (1971), 44, 876.
(28) Arnold, J. C., Gleicher, G. J., and Unruh, J. D., J. Am. Chem. Soc., (1974) 96, 787.
(29) Roark, R. B., Roberts, J. M., Croom, D. W. and Gilliom, R. D., J. Org. Chem. (1972), 37, 2042.
(30) Church, D. F., and Gleicher, G. J., J. Org. Chem., (1976), 41, 2327.

RECEIVED December 23, 1977.

14

Nucleophilic Reactions of Superoxide Anion Radical

WAYNE C. DANEN,* R. JAY WARNER, and RAVINDRA L. ARUDI

Department of Chemistry, Kansas State University, Manhattan, KS   66506

Superoxide ion, $O_2^{\bar{\cdot}}$, is the anion radical derived by addition of an electron to molecular oxygen. It is one of the simplest anion radicals and, undoubtedly, the most important. This species is capable of reacting with a variety of substrates owing to its anionic, radical, and redox nature although the nuceophilic and reducing electron transfer processes appear to be the predominate reaction pathways. This report will attempt to review the important features of the nucleophilic behavior of $O_2^{\bar{\cdot}}$ with well-defined chemical reactants. No attempt will be made to cover the vast biochemical literature involving or implicating $O_2^{\bar{\cdot}}$.

Although $O_2^{\bar{\cdot}}$ is a stable anion radical, easily generated or even available commercially as the potassium salt, its chemical behavior has received significant attention only within the last several years. Chemical interest was spawned by the discovery of Fridovich and coworkers (1) that superoxide dismutase, an enzyme present in all aerobic organisms studied, has as its function the dismutation of $O_2^{\bar{\cdot}}$ into H_2O_2 and O_2 (equation 1). Since

$$2\ O_2^{\bar{\cdot}}\ +\ 2\ H^+\ \longrightarrow\ H_2O_2\ +\ O_2 \tag{1}$$

superoxide dimutase is so common among respiring organisms, $O_2^{\bar{\cdot}}$ is thought to be a deleterious species whose cytotoxicity has promoted the evolution of such defenses. In fact, superoxide has been suggested to be involved in various biological disorders such as radiation damage to tissue, cancer, aging processes and oxygen toxicity; it may have beneficial effects in biological defense mechanisms.

Chemical studies have employed either electrogenerated $O_2^{\bar{\cdot}}$ or solutions or suspensions of KO_2. Dipolar aprotic solvents are usually used in either case as $O_2^{\bar{\cdot}}$ is quite sensitive to proton sources. The electrogenerated tetraalkylammonium superoxides are both soluble and stable in a variety of dipolar aprotic solvents (DMSO, DMF, CH_3CN, pyridine, acetone) (2). The solubility of KO_2 in such solvents is not particularly high, e.g., the estimated concentration of saturated KO_2 in DMSO is only about 0.02 M (3). However, the use of crown ethers such as 18-crown-6 greatly enhances the solubility of KO_2 (4) and the commercial availability of both KO_2 and crown ethers has resulted in their rather common usage for chemical studies of O_2^{-}.

Superoxide ion exhibits spectral properties in the ultraviolet region. The λ_{max} for O_2^{-} is 250 nm in acetonitrile with a reported extinction coefficient of 2580 ± 300 $M^{-1}cm^{-1}$ (5). The reduction potential for oxygen in DMSO is -0.77 V versus SCE (2) and the pK_a for $HOO\cdot$, the conjugate acid of O_2^{-}, has been determined to be 4.8 (6).

I. Reactions with Alkyl Halides and Sulfonate Esters

A. Mechanism. Alkyl halides undoubtedly represent the most well-tested functional group for nucleophilic reactivity. That superoxide ion reacts with alkyl haldies by an S_N2 mechanism has been demonstrated. Dietz, et al., (7), observed a relative reactivity which fell in the series n-BuBr > sec-BuBr > i-BuBr > t-BuBr for variation of alkyl group structure and in the series n-BuBr > n-BuOTs > n-BuCl for variation of leaving group. The former order is consistent with a S_N2 reaction mechanism and the latter order suggests that superoxide anion radical is a strong nucleophile.

Results obtained by San Filippo and co-workers (8) paralleled those of Dietz concerning substrate reactivity but indicated further that substitution was predominant with primary halides, whereas substantial elimination occurred with secondary and tertiary systems. These workers, as well as Johnson and Nidy (9) reported the essentially complete inversion of configuration at the chiral center in the carbon–oxygen bond formation which is typical of a S_N2 Walden inversion mechanism (eq. 2).

$$O_2^{-} + \underset{\underset{CH_3}{|}}{\overset{\overset{C_6H_{13}}{|}}{H\text{---}C\text{---}X}} \xrightarrow[\text{2. } H_2O]{\text{1. } KO_2,\text{DMSO}} \underset{\underset{CH_3}{|}}{HO\text{---}\overset{C_6H_{13}}{C}\cdots H} + X^{-} \quad (2)$$

In a similar manner, Corey, et al., (10) have reported that the p-toluenesulfonate of trans-4-t-butylcyclohexanol was converted into pure cis-4-t-butylcyclohexanol in 95% yield (eq. 3) in DMSO-DME. Likewise, the cis-methanesulfonate was transformed into pure trans-4-t-butylcyclohexanol in 96% yield.

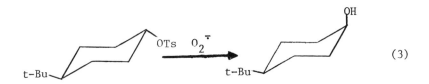

(3)

These workers ($\underline{10}$,$\underline{11}$) have also utilized $O_2^{\bar{\cdot}}$ to achieve an important synthetic objective in the prostaglandin field, namely the efficient conversion of 15-$\underline{R}$ (unnatural) prostaglandins into the 15-$\underline{S}$ (natural) isomers by nucleophilic displacement. These data all demonstrate the S_N2 inversion character of $O_2^{\bar{\cdot}}$ reacting with alkyl halides and sulfonate esters.

As noted by equation 2, the alcohol is the ultimate reaction product in DMSO solvent. However, there are several reactions that take place when $O_2^{\bar{\cdot}}$ is mixed with an alkyl halide. Equations 4-10 list the overall mechanism originally proposed by

$$RX \;+\; O_2^{\bar{\cdot}} \quad\begin{array}{c}\nearrow\\\searrow\end{array}\quad \begin{array}{ll} RO_2\cdot \;+\; X^- & (4)\\[1em] R(-H) \;+\; HO_2\cdot \;+\; X^- & (5)\end{array}$$

$$HO_2\cdot \;\rightleftharpoons\; H^+ \;+\; O_2^{\bar{\cdot}} \tag{6}$$

$$HO_2\cdot \;+\; O_2^{\bar{\cdot}} \;\longrightarrow\; HO_2^- \;+\; O_2 \tag{7}$$

$$RO_2\cdot \;+\; O_2^{\bar{\cdot}} \;\longrightarrow\; RO_2^- \;+\; O_2 \tag{8}$$

$$RO_2^- \;+\; RX \;\longrightarrow\; RO_2R \;+\; X^- \tag{9}$$

$$RO_2^- \;+\; Me_2SO \;\longrightarrow\; RO^- \;+\; Me_2SO_2 \tag{10}$$

Dietz, $\underline{et\ al}$., ($\underline{7}$) which was subsequently expanded by Gibian and Ungermann ($\underline{12}$). The course of the reaction is dependent upon both the alkyl halide structure and the reaction solvent. For a primary alkyl halide, the displacement of halide ion (eq. 4) is dominant over the elimination reaction (eq. 5) as determined by product analysis; e.g., for the reaction of KO_2 with 1-bromo-octane and 2-bromooctane in DMSO, olefins were isolated in yields of 1% and 34%, respectively ($\underline{8}$). It is apparent that secondary halides give significant olefin yields under these experimental conditions. The $HO_2\cdot$ formed in equation 5 may ionize to super-oxide (eq. 6) or be reduced (eq. 7). The bulk reaction of the alkyl peroxy radical with $O_2^{\bar{\cdot}}$ (eq. 8) is the analogue to equation

7 and is equally exothermic. Finally, the peroxy anion can undergo either reaction 9 or 10 to form products. The displacement
reaction of RO_2^- with an alkyl halide is quite facile but equation 10 is favored by at least a factor of four when DMSO is the
solvent (12).

Equation 9 can be made the principal reaction pathway and
Johnson and Nidy (9) have reported a convenient and synthetically
useful preparation of dialkyl peroxides using benzene as solvent
and solubilizing the KO_2 with crown ethers. Alcohols were
frequently formed as by-products, probably resulting from reduction of the dialkyl peroxides with O_2^- (13).

Interestingly, Corey, *et al.*, (10) produced a cyclic peroxide in DMSO. The peroxide shown in equation 11 was formed

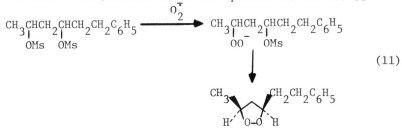

$$\underset{\substack{|\\ OMs}}{CH_3CHCH_2}\underset{\substack{|\\ OMs}}{CHCH_2}CH_2C_6H_5 \xrightarrow{O_2^-} \underset{\substack{|\\ OO}}{CH_3CHCH_2}\underset{\substack{|\\ OMs}}{CHCH_2}CH_2C_6H_5$$

(11)

in 35% yield in DMSO. Apparently, the close proximity of the
second mesylate group in the proposed peroxy mesylate allowed
the intramolecular cyclization process to compete favorably with
reaction with DMSO. Such a process may be of value in the
synthesis of biologically important prostaglandin endoperoxides.

B. __Kinetics of Reaction with Alkyl Halides__ (14). Prior
to the work of Danen and Warner, (14,15) few rate constants for
the reaction of O_2^- with alkyl halides had been reported. In an
electrochemical study, Merritt and Sawyer (16) had determined the
pseudo-first-order rate constants at 28°C for three butyl chlorides in DMSO solvent. In a similar manner, Dietz, *et al.*, (7)
had reported a pseudo-first-order rate constant for 1-bromobutane
reacting with electrogenerated O_2^- in DMF containing tetra-n-
butylammonium perchlorate. San Filippo and coworkers (8) had
determined the relative reactivity of several alkyl halides but
had not reported any absolute rate constants.

Danen and Warner (15) have reported the rate constants in
Table I for reaction of KO_2 in DMSO with a series of alkyl
bromides. The rates were determined by stopped-flow spectrophotometry under pseudo-first-order conditions. The usual reactivity
order characteristic of a S_N2 process was evident in the series
MeBr > EtBr > n-BuBr > i-PrBr >> 1-bromoadamantane, which reflects the increasing inaccessibility of the reaction center.
However, the 10-fold difference in reactivity between CH_3Br and
i-PrBr is smaller than frequently observed; Streitwieser (17)
noted a 1000-fold difference in average relative reactivity

Table I. Rate Constants for the Reaction of KO_2 with Alkyl
Bromides in DMSO at 25.0°

Alkyl Bromide	k_2 ($M^{-1}s^{-1}$)
CH_3Br	$(6.7 \pm 0.2) \times 10^2$
CH_3CH_2Br	$(3.5 \pm 0.2) \times 10^2$
$CH_3(CH_2)_3Br$	$(1.5 \pm 0.1) \times 10^2$
$(CH_3)_2CHBr$	$(6.5 \pm 0.1) \times 10^1$
1-bromoadamantane	$<<1.0^a$

[a]Too slow to measure by stopped-flow spectrophotometry.

towards nucleophiles for these two alkyl groups.

The unreactivity of 1-bromoadamantane with O_2^{-} demonstrated
a lack of any appreciable electron transfer type substitution
process (eq. 12). Although an electron transfer type mechanism
has been demonstrated for nitroaromatic halides (see Section
II. C), the difference in reduction potentials between O_2^{-} and
aliphatic bromides is apparently too great to allow such a
transfer to occur.

$$RBr \xrightarrow[-O_2]{O_2^{-}} [RBr]^{-} \xrightarrow{-Br^-} R\cdot \xrightarrow{O_2} RO_2\cdot \longrightarrow Products \quad (12)$$

The rate constants reported in Table I are very large for a
nucleophilic reaction at 25°. Danen and Warner (15) compared
O_2^{-} with other nucleophiles under comparable conditions and
noted that O_2^{-} reacts several orders of magnitude faster than
many nucleophiles indicating that O_2^{-} is a very potent nucleophile. Although there are numberous factors which determine
the nucleophilicity of various anions toward organic substrates,
it was suggested that a possible important factor in determining
O_2^{-} nucleophilicity is a significant electron-transfer contribution in the transition state (eq. 13). Since similar

$$O_2^{-} + RBr \longrightarrow \left[O_2^{-}\text{---}R\text{---}Br \leftrightarrow \overset{\cdot}{O}_2\text{---}R\text{---}Br\right]^{\ddagger} \longrightarrow RO_2\cdot + Br^- \quad (13)$$

contributions can be drawn for all nucleophiles, the key factor
for O_2^{-} is the inherent stability of molecular oxygen.

The Arrhenius parameters, E_a and $\Delta S^{\ddagger}$, have been determined
to be 11.0 kcal/mole and -15.6 e.u., respectively, for the
reaction of O_2^{-} with 1-bromobutane (18). For comparison purposes,
the activation parameters for various nucleophiles reacting with

alkyl bromides in dipolar aprotic solvents are presented in
Table II along with the values for superoxide. The low E_a for
$O_2^{\cdot-}$ is reflected in its nucleophilic reactivity. The entropy

Table II. Energies and Entropies of Activation of Some
           Displacement Reactions (14)

Nucleophile	Substrate	Solvent	E_a	$\Delta S^{\ddagger}$
$O_2^{\cdot-}$	n-BuBr	DMSO	11.0	-15.6
I^-	n-BuBr	acetone	20.5	-13.9
Cl^-	n-BuBr	acetone	17.5	-21.1
N_3^-	n-BuBr	DMSO	16.91	+ 1.62
$C_6H_5S^-$	n-BuBr	DMF	13.3	- 9.4
N_3^-	MeBr	DMF	16.7	-12.9
SCN^-	MeBr	DMF	16.0	-17.5

of activation for $O_2^{\cdot-}$ is indicative of a bimolecular displacement reaction. Since entropy is measure of randomness or
disorder, a negative $\Delta S^{\ddagger}$ implies that the transition state is
more ordered than the reactants. This is the normal trend
observed in most bimolecular reactions, since for reaction to
take place two molecules in random motion must come together.
For S_N2 Walden inversion reactions, where specific backside
attack is required, a significantly negative $\Delta S^{\ddagger}$ value is
usually observed as evident from the data in Table II.
    The effects of halogen atom substituents on the reactivity
of alkyl bromides toward $O_2^{\cdot-}$ have been determined for a series of
$\alpha-$, $\beta-$, and $\gamma-$halogen substituted alkyl bromides. Rate constants
were determined by stopped-flow techniques and are listed in
Table III. Since displacement of chloride by $O_2^{\cdot-}$ is much slower

Table III. Rate Constants for the Reaction of KO_2 with Halogen.
            Substituted Alkyl Bromides in DMSO at 25.0° (14)

Dihalide	$k_2 (M^{-1}s^{-1})$	Relative Rate[b]
$BrCH_2Br$[a]	$(1.28\pm0.02) \times 10^1$	0.037
$ClCH_2Br$	$(4.0\pm0.2) \times 10^1$	0.11
$BrCH_2CH_2Br$[a]	$(2.42\pm0.02) \times 10^2$	0.69
$ClCH_2CH_2Br$	$(2.67\pm0.02) \times 10^2$	0.76
$ClCH_2CH_2CH_2Br$	$(4.71\pm0.04) \times 10^2$	1.35

[a]Corrected for statistical factor of two.
[b]Relative to ethyl bromide (350 $M^{-1}s^{-1}$).

than that of bromide, the rate constants given are assumed to be for bromide displacement only.

The largest effect on reactivity was caused by α-halogen substitution as indicated by the relative rates shown in Table III. The relative rates for the first two entries are characteristic of an S_N2 process. For example, the nucleophilic displacement reaction of KI in acetone with the same alkyl bromide series (Et-, ClCH$_2$-, BrCH$_2$-) gave the relative rates 1.0:0.13:0.041 (19).

The decrease in reactivity upon halogen substitution in the α-position can be accounted for as follows. Halogen atoms, which are more electronegative than carbon, tend to destabilize the developing positive charge on carbon and, in turn, increase the energy of the transition state. Since reactivity is decreased by introduction of either an α-methyl (electron donating) or an α-halogen (electron withdrawing), it is likely that the decrease is at least partially due to steric hindrance. Finally, and closely related to steric hindrance, is the effect called neighboring orbital overlap, which is due to electron repulsion between the incoming nucleophile and the α-position halogen.

The effect of β-halogen substituents on the reactivity was in the anticipated direction as a result of the electronegativity effect but of less magnitude than usually observed, i.e., the rates of reaction of β-haloethyl bromides with thiophenoxide ion in methanol are a factor of about seven less than that of ethyl bromide (20).

In contrast to the decrease in reactivity exhibited by the other entries in Table III. the γ-halogen substituted example showed increased reactivity. This was a trend that had been previously observed (21) for S_N2 reactions with γ-oxygen substituents but a satisfactory explanation for this behavior has not been advanced.

In summary of this section, it has been conclusively demonstrated that the reaction of $O_2^{\overline{\cdot}}$ with primary alkyl halides and tosylates occurs by an initial S_N2 displacement process. Secondary and tertiary substrates give significant amounts of elimination products. In DMSO solvent, alcohols are the ultimate reaction product while dialkyl peroxides are formed in synthetically useful yields in benzene using crown ethers to solubize the KO_2. The available rate data indicate that $O_2^{\overline{\cdot}}$ is a very potent nucleophile.

II.   Other Nucleophilic Reactions of Superoxide.

A.   Reaction with Acyl Chlorides and Esters.   In view of the facile reaction of $O_2^{\overline{\cdot}}$ with alkyl halides, it is not unexpected that acyl halides should react rapidly with $O_2^{\overline{\cdot}}$. Johnson (22) has reported that diacyl peroxides are conveniently produced in the reaction of KO_2 with acyl chlorides according to equation 14. The reaction proceeds readily in benzene even

$$2 \ \underset{\displaystyle \overset{\text{O}}{\|}}{R\text{-}C}\text{-Cl} \ + \ 2 \ KO_2 \ \longrightarrow \ \underset{\displaystyle \overset{\text{O}}{\|}}{R\text{-}C}\text{-O-O-}\underset{\displaystyle \overset{\text{O}}{\|}}{C}\text{-R} \ + \ 2 \ KCl \ + \ O_2 \quad (14)$$

without crown ether. In fact, the present authors (3) have demonstrated that diacyl peroxides are rapidly consumed by KO_2 solubilized in benzene with 18-crown-6 generating the anion of the corresponding acid. Even without crown ether, Johnson has noted that it is advantageous to work-up the acyl chloride-KO_2 reactions as soon as the acyl chloride has been consumed to avoid production of the acid. The mechanism of formation of diacyl peroxides almost certainly involves steps 15-17 analogous to the

$$\underset{\displaystyle \overset{\text{O}}{\|}}{R\text{-}C}\text{-Cl} \ + \ O_2^{\cdot -} \ \longrightarrow \ \underset{\displaystyle \overset{\text{O}}{\|}}{R\text{-}C}\text{-OO}^{\cdot} \ + \ Cl^- \quad (15)$$

$$\underset{\displaystyle \overset{\text{O}}{\|}}{R\text{-}C}\text{-OO}^{\cdot} \ + \ O_2^{\cdot -} \ \longrightarrow \ \underset{\displaystyle \overset{\text{O}}{\|}}{R\text{-}C}\text{-OO}^- \ + \ O_2 \quad (16)$$

$$\underset{\displaystyle \overset{\text{O}}{\|}}{R\text{-}C}\text{-OO}^- \ + \ \underset{\displaystyle \overset{\text{O}}{\|}}{R\text{-}C}\text{-Cl} \ \longrightarrow \ \underset{\displaystyle \overset{\text{O}}{\|}}{R\text{-}C}\text{-OO-}\underset{\displaystyle \overset{\text{O}}{\|}}{C}\text{-R} \ + \ Cl^- \quad (17)$$

production of dialkyl peroxides as discussed above.

San Filippo and coworkers (23) have shown that a variety of carboxylic esters are cleaved by KO_2 in benzene with crown ether to produce after acidic work-up the corresponding carboxylic acid and alcohol in good to excellent yields (eq. 18). A mechanism involving alkyl-oxygen cleavage was ruled out by stereochemical studies involving esters derived from chiral alcohols. A

$$\underset{\displaystyle \overset{\text{O}}{\|}}{R\text{-}C}\text{-OR'} \ + \ KO_2 \ \longrightarrow \ \underset{\displaystyle \overset{\text{O}}{\|}}{R\text{-}C}\text{-OH} \ + \ R'OH \quad (18)$$

mechanism involving initial additon of $O_2^{\cdot -}$ to the carbonyl carbon was favored although the subsequent steps were only speculated; peroxy intermediates were implicated. An initial addition-elimination mechanism is consistent with the influence the departing alkoxide ion exerts on the rate of reaction; i.e., $R^1 =$ phenyl > primary > secondary > tertiary.

Triphenyl phosphate was readily cleaved by $O_2^{\overline{\cdot}}$ but tri-n-octyl phosphate showed no appreciable reactivity. Simple amides and nitriles were also largely unaffected by $O_2^{\overline{\cdot}}$ under conditions equivalent to those employed for ester cleavage.

Magno and Bontempelli (24) have reported on the reaction kinetics of electrogenerated $O_2^{\overline{\cdot}}$ in DMF with phenyl benzoate and p-chlorophenyl benzoate; second-order rate constants of 3.0 ± 0.3 and 25 ± 5 $M^{-1}sec^{-1}$, respectively, were determined by electrochemical techniques. An e.c.e. (electron transfer-chemical reaction-electron transfer) mechanism was shown to be operative and the following mechanism was suggested (equations 19-23). The

$$O_2 \ + \ e^- \ \longrightarrow \ O_2^{\overline{\cdot}} \tag{19}$$

$$\underset{\quad\quad\;\;O}{C_6H_5-\overset{\displaystyle O}{\overset{\|}{C}}-OAr} \ + \ O_2^{\overline{\cdot}} \ \longrightarrow \ C_6H_5-\overset{\displaystyle O}{\overset{\|}{C}}-OO\cdot \ + \ ArO^- \tag{20}$$

$$C_6H_5-\overset{\displaystyle O}{\overset{\|}{C}}-OO\cdot \ + \ e^- \ \longrightarrow \ C_6H_5-\overset{\displaystyle O}{\overset{\|}{C}}-OO^- \tag{21}$$

$$C_6H_5-\overset{\displaystyle O}{\overset{\|}{C}}-OO^- \ + \ C_6H_5-\overset{\displaystyle O}{\overset{\|}{C}}-OAr \ \longrightarrow \ C_6H_5\overset{\displaystyle O}{\overset{\|}{C}}-OO-\overset{\displaystyle O}{\overset{\|}{C}}C_6H_5 \ + \ ArO^- \tag{22}$$

$$C_6H_5-\overset{\displaystyle O}{\overset{\|}{C}}-OO-\overset{\displaystyle O}{\overset{\|}{C}}-C_6H_5 \ + \ 2\ e^- \ \longrightarrow \ 2\ C_6H_5-\overset{\displaystyle O}{\overset{\|}{C}}-O^- \tag{23}$$

greater reactivity of the p-chlorophenyl ester is consistent with reaction 20 being the rate-limiting step.

B. <u>Nucleophilic Addition to Double Bonds</u>. There are at least two reports of $O_2^{\overline{\cdot}}$ adding in nucleophilic fashion to a carbon-carbon double bond. Although the reaction mechanisms were not elucidated in detail, in both reports the double bond was activated for nucleophilic addition.

Benzylidenefluorene, a hydrocarbon susceptible to nucleophilic attack, was shown by Dietz, <u>et al</u>. (7) to react with electrogenerated $O_2^{\overline{\cdot}}$ in the presence of O_2 to produce fluorenone and benzoate. It was proposed that $O_2^{\overline{\cdot}}$ initiated an autoxidation reaction (equation 24); one equivalent of $O_2^{\overline{\cdot}}$ was necessary to remove the hydrocarbon completely.

$$(24)$$

These authors also showed that electrogenerated O_2^{-} reacted with cyclohexen-3-one to give the corresponding epoxide in 30% yield (equation 25). Cyclohexene, in contrast, gave no epoxide implying that a double bond activated towards a Michael-type of addition is required.

$$(25)$$

C. Electron Transfer Reactions Mimicking Nucleophilic Attack. There are several reports in which O_2^{-} nominally appears to react as a nucleophile but, in fact, involve an initial electron-transfer from O_2^{-} followed by reaction of O_2 with the radical anion generated from the organic substrate. These results appear appropriate for this review. There are certainly other reactions which occur by such a process instead of a simple nucleophilic displacement but isotopic labeling studies or chiral reactants must be utilized to distinguish between the two mechanisms. Because of the relatively low reduction potential for O_2, -0.77 V versus SCE (2), any substrate with a more negative reduction potential will likely be reduced by O_2^{-} via an electron transfer process. Depending upon the substrate, the ultimate product(s) of such a reaction may or may not resemble those expected from a simple nucleophilic reaction.

Frimer and Rosenthal (25) have demonstrated that the reaction of O_2^{-} with nitro substituted aromatic halides occurs via an electron transfer from O_2^{-} to the substituted benzene to yield the anion radical which is subsequently scavenged by molecular oxygen (equation 26). They were able to distinguish this reaction pathway from direct addition of O_2^{-} to the aromatic ring as in normal nucleophilic aromatic substitution by utilizing

$$\text{(diagram)} \qquad (26)$$

^{18}O-enriched KO_2. By conducting the reaction with enriched KO_2 in benzene saturated with unlabeled O_2, mass spectroscopic analysis of the resulting phenol revealed the presence of an ^{18}O tag of less than 10%. This result implied that the phenolic oxygen was incorporated in large part after the equilibration with molecular oxygen dissolved in solution.

Dougherty, et al., (26) have observed a similar reaction in the gas phase. The negative chemical ionization spectra of 4-bromobenzophenone, 4-nitrochlorobenzene, and 2,4-dinitrochloro-benzene were all replaced by spectra of the corresponding phenolate anions when $O_2^{\bar{\cdot}}$ was generated in the system.

Another nominally nucleophilic reaction of $O_2^{\bar{\cdot}}$ is the efficient production of carboxylic acids from chalcones (equation 27). Rosenthal and Frimer (27) have also studied this reaction

$$\text{(diagram)} \qquad (27)$$

utilizing ^{18}O-enriched KO_2 and determined that oxygen is not incorporated into the product carboxylic acids by direct nucleo-philic attack of $O_2^{\bar{\cdot}}$. Instead, the reaction proceeds by a preliminary electron transfer from $O_2^{\bar{\cdot}}$ to the enone system and the resulting anion radical then reacts with the surrounding molecular oxygen as depicted in equations 28-32. A similar mechanism enabled the understanding of the surprising formation of 2-hydroxy-2,4,5-triphenylfuranone-3 in the reaction of KO_2 with tetracyclone (28)(equation 33).

$$R\text{-}\overset{\overset{O}{\|}}{C}\text{-}CH\text{=}CH\text{-}R \; + \; O_2^{\cdot -} \longrightarrow \; R\text{-}\overset{\overset{O^-}{\|}}{\underset{\cdot}{C}}\text{-}CH\text{=}CH\text{-}R \; + \; O_2 \qquad (28)$$

$$R\text{-}\overset{\overset{O^-}{\|}}{\underset{\cdot}{C}}\text{-}CH\text{=}CH\text{-}R \; + \; O_2 \longrightarrow \; R\text{-}\overset{\overset{O^-}{\|}}{C}\text{-}\underset{\underset{O\text{-}O}{|}}{CH}\text{-}\overset{\cdot}{CH}\text{-}R \qquad (29)$$

$$R\text{-}\overset{\overset{O^-}{\|}}{C}\text{-}\underset{\underset{O\text{-}O}{|}}{CH}\text{-}\overset{\cdot}{CH}\text{-}R \longrightarrow \; R\text{-}\overset{\overset{O}{\|}}{\underset{\cdot}{C}}\text{-}O^- \; + \; R\text{-}\overset{\cdot}{CH}\text{-}\overset{\overset{O}{\|}}{CH} \qquad (30)$$

$$R\text{-}\overset{\cdot}{CH}\text{-}\overset{\overset{O}{\|}}{CH} \quad
\begin{cases}
\xrightarrow{SH} \; RCH_2\overset{\overset{O}{\|}}{CH} \; \xrightarrow{O_2^{\cdot -}} \; RCH_2CO_2^- & (31) \\[2ex]
\xrightarrow{O_2^{\cdot -}} \; R\text{-}\underset{\underset{O\text{—}O}{|}}{\overset{\overset{O^-}{|}}{CH}}\text{-}CH \; \longrightarrow \; R\overset{\overset{O}{\|}}{CH} \; \xrightarrow{O_2^{\cdot -}} \; RCO_2^- & (32)
\end{cases}$$

$$\underset{Ph}{\overset{Ph}{\diagdown}}\hspace{-0.3em}\begin{array}{c}\overset{O}{\|}\\[-0.3em]\end{array}\hspace{-0.3em}\underset{Ph}{\overset{Ph}{\diagup}} \; + \; O_2^{\cdot -} \longrightarrow \longrightarrow \; \underset{HO}{\overset{Ph}{\diagdown}}\hspace{-0.3em}\begin{array}{c}\overset{O}{\|}\\[-0.3em]O\end{array}\hspace{-0.3em}\underset{Ph}{\overset{Ph}{\diagup}} \; + \; PhCO_2^- \qquad (33)$$

In conclusion of this review of the nucleophilic properties of $O_2^{\cdot -}$, the versatile nature of this unique anion radical should be emphasized. This chapter attempted to cover only the main features of the nucleophilic reactions of $O_2^{\cdot -}$ with well-defined chemical substrates; no attempt was made to treat any of the biochemical reactions. Moreover, in addition to nucleophilic properties, $O_2^{\cdot -}$ is capable of reacting as a free radical as well as an electron transfer agent or electron acceptor. Thus, the understanding of this ubiquitous anion radical is probably only in its late infancy even though a computer search of the 1972-mid 1977 Chemical Abstracts revealed 850 references to "super-oxide".

III. Acknowledgement

This work was written while W. C. Danen was on sabbatical

leave at the Los Alamos Scientific Laboratory, Los Alamos, New
Mexico. This author would like to express his appreciation to
the personnel of AP Division, in particular John H. Birely and
Samuel M. Freund, for their hospitality during that time.

IV. Literature Cited

(1)   Fridovich, I., Acct. Chem. Res., (1972), 5, 321, and
      references therein.
(2)   Peover, M. E., and White, B. S., Electrochim. Acta, (1966),
      11, 1061.
(3)   Danen, W. C., and Arudi, R. L., unpublished results.
(4)   Valentine, J. S., and Curtis, A. B., J. Amer. Chem. Soc.,
      (1975), 97, 224.
(5)   Fee, J. A., and Hildenbrand, P. G., FEBS Lett., (1974), 39,
      79.
(6)   Behar, D., Czapski, G., Rabani, J., Dorfman, L. M., and
      Schwarz, H. A., J. Phys. Chem., (1970), 74, 3209.
(7)   Dietz, R., Forno, A. E. J., Larcombe, B. E., and Peover,
      M. E., J. Chem. Soc. (B), (1970), 816.
(8)   San Filippo, Jr., J., Chern, C-I., and Valentine, J. S.,
      J. Org. Chem., (1975), 40, 1678.
(9)   Johnson, R. A., and Nidy, E. G., J. Org. Chem., (1975), 40,
      1680.
(10) Corey, E. J., Nicolaou, K. C., Shibasaki, M., Machida, Y.,
      and Shiner, C. S., Tetrahedron Lett., (1975), 3183.
(11) Corey, E. J., Nicolaou, K. C., and Shibasaki, M., J.C.S.
      Chem. Comm., (1975), 658.
(12) Gibian, M. J., and Ungermann, T., J. Org. Chem., (1976), 41,
      2500.
(13) Peters, J. W., and Foote, C. S., J. Amer. Chem. Soc., (1976)
      98, 873. These authors report the reaction of O_2^- with
      hydroperoxides. Di-t-butyl peroxide was reported to be
      unreactive but the primary and secondary dialkyl peroxides
      generated by Johnson and Nidy (9) may be more susceptible
      to reduction.
(14) This section taken largely from the M.S. thesis of R. Jay
      Warner, Kansas State University, 1976.
(15) Danen, W. C., and Warner, R. J., Tetrahedron Lett., (1977),
      989.
(16) Merritt, M. V., and Sawyer, D. T., J. Org. Chem., (1970),
      35, 2157.
(17) Streitwieser, Jr., A., "Solvolytic Displacement Reactions,"
      p. 13, McGraw-Hill Book Co., New York, N.Y., (1962).
(18) Danen, W. C., and Warner, R. J., unpublished results.
(19) Hine, J., Thomas, C. H., and Ehrenson, S. J., J. Amer. Chem.
      Soc., (1955), 77, 3886.
(20) Hine, J., and Brader, W. H., J. Amer. Chem. Soc., (1953),
      75, 3964.
(21) Reference 17, p. 16.
(22) Johnson, R. A., Tetrahedron Lett., (1976), 331.

(23) San Filippo, Jr., J., Romano, L. T., Chern, C-I. and Valentine, J. S., J. Org. Chem., (1976), 41, 586.

(24) Magno, F., and Bontempelli, G., J. Electroanal. Chem., (1976), 68, 337.

(25) Frimer, A., and Rosenthal, I., Tetrahedron Lett., (1976), 2809.

(26) Levonowich, P. F., Tannenbaum, H. P., and Dougherty, R. C., J.C.S. Chem. Comm., (1975), 597.

(27) Rosenthal, I., and Frimer, A., Tetrahedron Lett., (1976), 2805.

(28) Rosenthal, I., and Frimer, A., Tetrahedron Lett., (1975), 3731.

RECEIVED December 23, 1977.

15

Kinetic Analysis of the Methylene Blue Oxidations of Thiols

EARL S. HUYSER and HSIAO–NEIN TANG

Department of Chemistry, University of Kansas, Lawrence, KS 66044

Kinetics traditionally have been among the more powerful tools available for the study of organic reation mechanisms. Much of our present understanding of the behavior of free radicals as reaction intermediates can be attributed to data accumulated by various competitive kinetic investigations (e.g. the determination of transfer and copolymerization constants in vinyl polymerization reactions). Somewhat less exploited as a means of gaining insight into the mechanistic details of free radical chain reactions has been the examination of the rate laws of such reactions. Neglect of this area of investigation is surprising because those reactions that have been subjected to such studies (e.g., vinyl polymerizations, brominations, autoxidations) have yielded valuable information concerning the overall mechanisms of free radical chain reactions. The lack of extensive effort in this area may be due, at least in part, to the fact that the experimentally determined rate constants are combinations of the rate constants of the several individual steps in the overall reaction. Generally, these observed rate constants are not readily resolvable into the rate constants of the individual steps of the reaction and thereby available for the classical examination the activation parameters of these reactions. Furthermore, the observed rate laws often involve fractional kinetic orders of the reactants (and sometimes the products). Further, these kinetic orders often depend on experimental conditions, in particular the relative concentrations of the reactants (and sometimes the products). While such factors as inconsistent kinetic orders may appear at the outset to be detrimental to the study of a reaction mechanism by kinetic analysis, these same factors may, in some instances, prove to be valuable probes for investigation of the mechanisms of free radical chain reactions. The work presented here describes such an investigation.

Methylene Blue Oxidations of Thiols

The oxidations of thiols to disulfides, reactions
encountered in a variety of biological processes, have been
accomplished in the laboratory by a variety of oxidizing agents

$$2 \ RSH + Ox \longrightarrow RSSR + OxH_2$$

(e.g. molecular oxygen, ferricyanide, quinones, flavins, and
azocompounds) (1). In this study, methylene blue (MB^+) was used
as the oxidizing agent (2). The thiols subjected to oxidation by

MB^+ were mercaptoethanol (RSH) and dithioerythritol ($R(SH)_2$)
which were oxidized to the corresponding disulfides (RSSR and
RS_2, respectively) with concurrent reduction of MB^+ to
leucomethylene blue (LMB).

The reaction rates of the oxidations of mercaptoethanol and
dithioerythritol were determined at various concentration ratios
of the thiols with respect to MB^+ over a range of pH's and ionic
strengths in both water and D_2O by following the decrease in the

absorption maximum of MB^+ at 650 nm. The high molar extinction
coefficient of MB^+ (ε = 51,200) allows for rate determinations at
very low concentrations of this reagent. That the reactions
investigated actually do occur was established by isolation of
both leucomethylene blue and the corresponding disulfides as
reaction products.

$\underline{MB^+ \text{ Oxidations of Mercaptoethanol}}$. A mechanism for the
oxidation of mercaptoethanol with MB^+ that is consistent with our
data for the reaction is shown in the sequence 1-10. This
mechanism evolved for the most part from our kinetic studies of
the reaction. Steps 2-4 of the reaction scheme comprise a free

$$RS^- + MB^+ \xrightarrow{k_1} RS\cdot + MB\cdot \tag{1}$$

$$RS\cdot + RS^- \underset{k_{-2}}{\overset{k_2}{\rightleftharpoons}} RS\dot{\overline{S}}R \tag{2}$$

$$RS\dot{\overline{S}}R + MB^+ \xrightarrow{k_3} MB\cdot + RSSR \tag{3}$$

$$MB\cdot + RSH \xrightarrow{k_4} LMB + RS\cdot \tag{4}$$

$$2\ RS\cdot \xrightarrow{k_5} RSSR \tag{5}$$

$$RS\cdot + MB\cdot \xrightarrow{k_6} RS^- + MB^+ \tag{6}$$

$$RS\cdot + RS\dot{\overline{S}}R \xrightarrow{k_7} RS^- + RSSR \tag{7}$$

$$MB\cdot + RS\dot{\overline{S}}R \xrightarrow{k_8} MB^+ + 2\ RS^- \tag{8}$$

$$2\ MB\cdot + H^+ \xrightarrow{k_9} MB^+ + LMB \tag{9}$$

$$2\ RS\dot{\overline{S}}R \xrightarrow{k_{10}} 2\ RS^- + RSSR \tag{10}$$

radical chain sequence that accounts of the stoichiometry of the
oxidation of mercaptoethanol by MB^+. The sulfur-sulfur linkage
of the disulfide is effected in the formation of the radical
anion $RS\dot{\overline{S}}R$, a radical which is partitioned between fragmenting to
the species from which it was formed and transfering of an
electron to MB^+ to yield the disulfide as a reaction product and
the methylene blue derived radical $MB\cdot$. The latter propagates
the chain by abstracting the sulfur-bonded hydrogen from the
thiol yielding leucomethylene blue (LMB) and a chain-carrying
thiyl radical ($RS\cdot$). The overall reaction, which occurs at room
temperature, is initiated (step 1) by an interaction (possibly an
electron transfer) between a sulfide ion and MB^+. There are six
termination reactions possible (steps 5-10) for a chain sequence
involving three different chain-carrying radicals. Our rate data
indicate that only those shown in 5-8 are likely termination
processes for the reaction.

Examination of this mechanism reveals that both the thiol and the sulfide ion are required as reactants. The pH of the medium, which determines the relative amounts of these components, therefore, has a significant influence on the rate of the oxidation reaction. Table I shows the oxidation rates of mercaptoethanol by MB^+ are pH dependent in that the reaction rate

Table I

Effect of pH on MB^+-Oxidation

Rates of Mercaptoethanol

pH^a	Rate x 10^6 M sec^{-1}
8.20	1.01 (0.05)
8.60	2.14 (0.08)
9.00	4.89 (0.08)
9.30	6.51 (0.26)
9.60	7.40 (0.08)
9.90	7.66 (0.13)
10.30	8.36 (0.09)
10.70	7.66 (0.32)

a. Buffered with borate. All solution adjusted to ionic strength = 0.4 with KCl.

is greatest at a pH roughly one pH unit greater than that required for about equal distribution of the thiol and sulfide ion as dictated by the ionization constant of mercaptoethanol (pK_A = 9.34) (3). The lower oxidation rates above pH 10.3 indicate involvement of the thiol in a rate-limiting step when the thiol concentration is low.

The observed rate laws for the oxidation of mercaptoethanol by methylene blue under different reaction conditions are consistent with the steady-state rate laws derived from the proposed mechanism. If step 2 of the reaction sequence, namely formation of the disulfide radical anion from a thiyl radical and a sulfide ion, is the rate limiting step of the chain sequence, and therefore only termination by 6 (coupling of thiyl radicals) occurs, the derived rate law for the reaction is Eq. 11. This rate law, which takes into account the distribution of the thiol and sulfide as determined by the K_A of the thiol and the acidity of the medium, predicts that the observed rate law would be half order in MB^+ and three halves order in mercaptoethanol. This is the rate law expected, however, only if the concentration of MB^+ is sufficiently high so that the second term in the denominator is neglible. At pH's below the pH maximum, and at a sufficiently

$$\text{Rate} = \frac{(k_1/2k_5)^{1/2} k_2 [MB^+]^{1/2} (K_A [RSH]/[H^+])^{3/2}}{1 + k_{-2}/k_3 [MB^+]} \tag{11}$$

high concentration of MB^+, the formation of the disulfide radical anion may well be expected to be the only limiting reaction in the chain sequence since the reactants for the other two steps (MB^+ and RSH) are present in sufficient concentrations to obviate their participation in rate limiting steps. Table II shows the kinetic orders of MB^+ and RSH for the oxidation reaction at pH 9.9.[*] The rate law for the reaction at pH 9.9 very closely follows the derived rate law shown in Eq. 11 in that at the higher concentrations of MB^+, the reaction is three-halves order in RSH and half order in MB^+. However, the derived rate law indicates that at lower concentrations of MB^+ the second term in the denominator could be significant and the observed kinetic order of MB^+ would be greater than one half. The increase in kinetic order of MB^+ observed at lower concentrations of MB^+ is therefore likely a reflection of the less effective partioning of the RSSR to the MB^+.

At pH > 10.3, the concentration of RSH relative to RS^- is small and reaction 4, the hydrogen atom abstraction from RSH by the methylene blue derived radical $MB\cdot$, is a rate limiting factor. If reaction 4, along with the disulfide radical anion formation in reaction 2, are the rate limiting steps in the chain sequence, the reaction would follow the derived rate law shown in Eq. 12. Note that, in terms of the measurable kinetic

$$\text{Rate} = \left(\frac{k_1 k_2 k_4 [MB^+]([RSH] - K_A [RSH]/[H^+])}{k_6 (1 + k_{-2}/k_3 [MB^+])}\right)^{1/2} K_A [RSH]/[H^+] \tag{12}$$

orders of MB^+ and RSH, the expected observed rate law at pH 10.7 is not distinguishable from the observed rate law for the reaction at pH 9.9, namely that the reaction is half order in MB^+ and three-halves order in mercaptoethanol. Table III shows that at the higher concentrations of MB^+, the reaction at pH 10.7

[*]A word concerning these kinetic orders and those in subsequent Tables is warranted. All kinetic orders were determined by the method of initial rates. The kinetic order listed at each concentration of MB^+ is the average calculated from at least three (generally four or five) determinations of the reaction rate at that concentration of MB^+. The value in parenthesis following the kinetic order indicates the outside limits of certainty based on the calculation of the kinetic order from the fastest and slowest rates at that particular concentration.

Table II

Kinetic Data for MB$^+$ Oxidations

of Mercaptoethanol at pH 9.9^a

Rate = $k_{obs}[MB^+]^{\alpha}[RSH]^{\beta}$

Kinetic Orders

[MB$^+$] x 10^6	$\underline{\alpha}$	$\underline{\beta}^b$
		1.48 (.11)^c
7.84	.47 (.12)^c	
		1.49 (.13)
5.92	.50 (.12)	
		1.48 (.11)
4.36	.48 (.10)	
		1.50 (.12)
3.05	.50 (.10)	
		1.48 (.13)
1.66	.74 (.05)	
		1.50 (.16)
.78	.89 (.18)	
		1.52 (.10)
.59	1.07 (.18)	
		1.49 (.11)

a. Borate buffered, μ = 0.40.
b. [RSH] = 8.15 x 10^{-3} and 5.78 x 10^{-3}.
c. Extreme limits of reliability.

is indeed three halves order in RSH and half order in MB$^+$.

At lower concentrations of MB$^+$, the increase in kinetic order of this reagent reflects less favorable partitioning the disulfide radical anion RSSR to MB$^+$ as noted at pH 9.9. The decrease in the kinetic order of RSH at lower MB$^+$ concentrations indicates that RSSR is possibly involved in a termination reaction.

The derived rate law for the reaction if steps 3 and 4 are rate limiting is shown in Eq. 13 and indicates that if reaction

$$\text{Rate} = \left(\frac{k_1 k_3 k_4 (K_A[RSH]/[H^+])}{k_8(1 + k_{-2}/k_3[MB^+])}\right)^{1/2} [MB^+][RSH]^{1/2} \qquad (13)$$

3 is a rate limiting step, the kinetic order of RSH is unity.

<u>MB$^+$ Oxidations of Dithioerythritol.</u> Although the kinetic rate laws observed for the mercaptoethanol oxidations are

Table III

Rate Data for Mercaptoethanol

Oxidations at pH 10.7[a]

$$Rate = k_{obs}[MB^+]^\alpha[RSH]^\beta$$

[MB$^+$] x 10^6	α	β[b]
		1.50 (.10)
7.84	0.50 (.16)	
		1.49 (.09)
5.92	0.48 (.13)	
		1.50 (.08)
4.36	0.55 (.11)	
		1.47 (.08)
3.05	0.51 (.09)	
		1.45 (.08)
1.66	0.72 (.05)	
		1.38 (.10)
.78	0.81 (.16)	
		1.35 (.06)
.59	0.93 (.04)	
		1.28 (.04)
.44	1.02 (.12)	
		1.25 (.13)
.30	1.17 (.16)	
		1.18 (.13)

a. Borate buffered, $\mu = 0.40$.
b. [RSH] = 6.38 x 10^{-3} and 3.89 x 10^{-3}.

consistent with those expected for the proposed mechanism, these studies were somewhat less than satisfying because the derived rate laws predict the same observed rate laws at pH's above and below the pH of the rate maximum. The oxidations of dithioerythritol are more informative because, owing to the unimolecularity of the disulfide radical anion formation (in contrast to the bimolecularity of the disulfide radical anion formation in the case of mercaptoethanol), the observed and derived rate laws are pH-dependent. The mechanism proposed for the dithioerythritol oxidations by MB$^+$ is shown in the sequence 14-23.

As in the case of the mercaptoethanol oxidations, the oxidation rates of dithioerythritol are pH-dependent showing a maximum at pH $\sim$ 9.6 (see Table IV). The lower pH observed for the maximum rate relative to that observed for mercaptoethanol (see Table I) parallels the lower pK_A for the first ionization

$$MB^+ + R\underset{S^-}{\overset{SH}{<}} \xrightarrow{k_{14}} MB\cdot + R\underset{S\cdot}{\overset{SH}{<}} \qquad (14)$$

$$R\underset{S\cdot}{\overset{SH}{<}} \underset{k_{-15}}{\overset{k_{15}}{\rightleftarrows}} R\underset{S}{\overset{S^-}{<}}| + H^+ \qquad (15)$$

$$R\underset{S}{\overset{S^-}{<}}| + MB^+ \xrightarrow{k_{16}} R\underset{S}{\overset{S}{<}}| + MB\cdot \qquad (16)$$

$$MB\cdot + R\underset{SH}{\overset{SH}{<}} \xrightarrow{k_{17}} LMB + R\underset{S\cdot}{\overset{SH}{<}} \qquad (17)$$

$$2 R\underset{SH}{\overset{S\cdot}{<}} \xrightarrow{k_{15}} R\underset{SH\ HS}{\overset{S-S}{<}}R \qquad (18)$$

$$R\underset{SH}{\overset{S\cdot}{<}} + MB\cdot \xrightarrow{k_{19}} R\underset{SH}{\overset{S^-}{<}} + MB^+ \qquad (19)$$

$$R\underset{S}{\overset{S^-}{<}}| + R\underset{SH}{\overset{S\cdot}{<}} \xrightarrow{k_{20}} R\underset{S}{\overset{S}{<}}| + R\underset{SH}{\overset{S^-}{<}} \qquad (20)$$

$$R\underset{S}{\overset{S^-}{<}}| + MB\cdot + H^+ \xrightarrow{k_{21}} R\underset{S^-}{\overset{SH}{<}} + MB^+ \qquad (21)$$

$$2 R\underset{S}{\overset{S^-}{<}}| + H^+ \xrightarrow{k_{22}} R\underset{S}{\overset{S}{<}}| + R\underset{S^-}{\overset{SH}{<}} \qquad (22)$$

$$2 MB\cdot + H^+ \xrightarrow{k_{23}} LMB + MB^+ \qquad (23)$$

constant of dithioerythritol (9.0) (**4**) relative to that of mercaptoethanol. Not shown is a second rate maximum observed at pH $\sim$ 11 reflecting the dibasic character of dithioerythritol (second pK_A = 9.9) (**4**). For the present study, only the rate maximum associated with the first ionization is pertinent.

If the dithioerythritol oxidations do indeed follow the same mechanistic path as the mercaptoethanol oxidations with respect to relative reagent concentrations, at pH's below that observed for the maximum rate and at the high concentrations of MB^+, only the unimolecular disulfide radical anion formation (reaction 15) would be rate limiting. The derived rate law based on only

Table IV

Effect of pH on MB^+-Oxidation

Rates of Dithioerythritol[a]

pH	Rate x 10^6 M sec^{-1}
8.21	0.11 (0.01)
8.44	0.28 (0.03)
8.86	0.62 (0.02)
9.28	1.08 (0.09)
9.58	1.37 (0.13)
9.88	1.12 (0.09)
10.24	1.03 (0.02)
10.56	0.80 (0.02)

a. Ionic strength = 0.40.

termination by reaction 18, the chain-carrying radical in the rate limiting step, is Eq. 24. This derived rate law predicts that at a pH below 9.6 and relatively high concentrations MB^+ the

$$\text{Rate} = \frac{(k_{14}/2k_{18})^{1/2}k_{15}([MB^+]K_A[R(SH)_2]/[H^+])^{1/2}}{1 + k_{-15}/k_{16}[MB^+]} \qquad (24)$$

observed rate law would be half order in both MB^+ and dithioerythritol, a prediction that is consistent with rate law obtained at pH 9.0 (Table V) when the MB^+ is sufficiently high.

However, the derived rate law indicates that the kinetic order of MB^+ should increase with decreasing concentration of MB^+. The latter effect, as in the mercaptoethanol reactions, reflects the less effective partitioning of the disulfide radical anion toward the MB^+ at the lower concentrations of MB^+.

In the pH range above that observed for the rate maximum, hydrogen atom abstraction by MB· from the thiol (reaction 17) is a rate limiting step in the chain sequence. If termination occurs by reaction 19 and the derived rate law for the reaction (Eq. 25) predicts that the reaction would be half order in MB^+

$$\text{Rate} = \left(\frac{k_{14}k_{15}k_{17}[MB^+]([RSH]-K_A[R(SH)_2]/[H^+])(K_A[R(SH)_2]/[H^+])}{k_{19}(1 + k_{-15}/k_{16}[MB^+])} \right)^{1/2} \qquad (25)$$

Table V

Kinetic Data for MB^+ Oxidations

of Dithioerythritol at pH 9.0^a

$$\text{Rate} = k_{obs}[MB^+]^\alpha[R(SH)_2]^\beta$$

Kinetic Orders

$[MB^+] \times 10^6$	α	β^b
		.52 (.04)
7.84	.52 (.14)	
		.51 (.03)
5.92	.74 (.13)	
		.52 (.04)
4.36	.86 (.12)	
		.51 (.04)
3.05	1.08 (.10)	
		.51 (.05)
1.66	1.25 (.07)	
		.49 (.05)
1.13	1.38 (.08)	
		.45 (.12)

a. Borate buffer, $\mu = 0.40$.
b. $[RSH] = 2.59 \times 10^{-3}$ and 1.01×10^{-3}.

and first order in dithioerythritol. The kinetic orders observed for the dithioerythritol oxidation by MB^+ at pH 10.3 (Table VI) show the reaction is indeed first order in dithioerythritol and half order in MB^+ at the higher concentrations of the latter reagent. The kinetic order of MB^+ at lower concentrations of MB^+ is greater than half order and indicates the involvement of the fractional term in the denominator of derived rate law resulting from less effective partitioning of the disulfide radical anion toward the MB^+ at low concentrations of this reagent.

The observed rate laws for the methylene blue oxidations of both mercaptoethanol and dithioerythritol at the different pH's are consistent with the mechanisms proposed for these oxidations. The most significant effect of the pH on the reaction is that of establishing the relative amounts of thiol and sulfide ion. Initiation of the chain sequence by the reaction of a sulfide ion with MB^+ would be expected to be more rapid, and therefore the overall oxidation faster, if the pH of the medium is increased. However, at the higher pH's where the relative amount of undissociated thiol is small, the reaction rate diminishes indicating not only that the thiol is likely a reactant in the overall reaction but that it is involved in a rate limiting step of the reaction at higher pH's. While not as supportive as the

Table VI

Kinetic Data for MB^+ Oxidations

of Dithioerythritol at pH 10.3[a]

$$Rate = k_{obs}[MB^+]^\alpha[R(SH)_2]^\beta$$

Kinetic Orders

$[MB^+] \times 10^6$	α	β[b]
		.99 (.04)
7.84	.49 (.12)	
		.98 (.03)
5.92	.50 (.14)	
		.98 (.04)
4.36	.53 (.14)	
		.99 (.05)
3.05	.67 (.13)	
		1.01 (.06)
1.66	.79 (.09)	
		.98 (.03)
1.13	.89 (.13)	
		.99 (.07)

a.   Borate buffer, μ = 0.40.
b.   [RSH] = 1.56 x 10^{-3} and 5.58 x 10^{-4}.

rate laws for the dithioerythritol oxidations, the observed rate laws for the mercaptoethanol oxidation are consistent with the hypothesis that hydrogen atom abstraction from the thiol is a rate limiting step at higher pH's where the concentration of thiol is small.

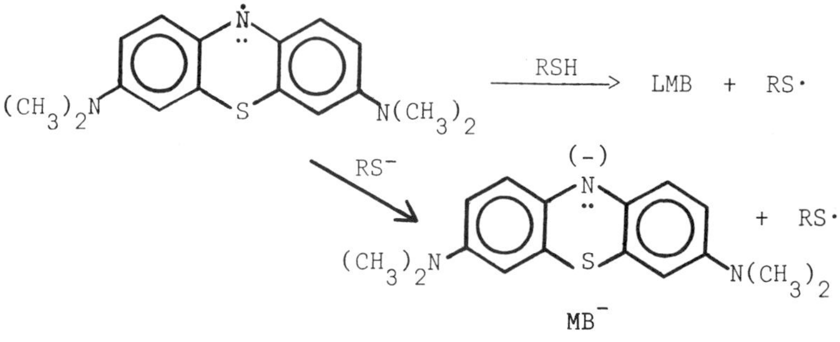

   Finding hydrogen atom abstraction from the thiol is the chain propagating step that produces the chain-carrying thiyl radical rather than electron transfer from a sulfide ion is not unexpected in terms of the expected chemical behavior of the methylene blue derived radical MB·. The most probable structure of MB· is the hybrid radical having as a major contributor to the hybrid the structure with the unpaired electron localized on heterocyclic nitrogen. If electron transfer from RS⁻ to this species did occur, it would yield MB⁻, the conjugate base of the weakly acidic leucomethylene blue. On the other hand, hydrogen atom abstraction from the thiol not only yields a more stable MB· derived reaction product, namely LMB, than electron transfer but the reaction itself involves a favorable polar effect in that the thiol is a good electron acceptor substrate and the MB· a good electron donor radical.

Solvent Isotope Effects. The involvement of the abstraction of the hydrogen atom from the thiol by MB· as a rate limiting factor at the higher pH's is supported by comparison of the oxidation reaction rates in H_2O and D_2O at different pH's. Table VII lists the observed rate constants for the oxidations of both mercaptoethanol and dithioerythritol at different acidities.

Examination of the solvent isotope effects as measured by k'_H/k'_D (where, in the case of mercaptoethanol at each pH, k' is the observed reaction rate constant calculated from the rate law $v = k_{obs}[RSH]^{3/2}[MB^+]^{1/2}$ and in the case of dithioerythritol, k' is calculated from the rate law $v = k_{obs}[R(SH)_2]^{1/2}[MB^+]^{1/2}$ at the lower pH's and from $v = k_{obs}[R(SH)_2][MB^+]^{1/2}$ at the highest pH) shows that an isotope effect is observed only under those conditions where breaking and S-H bond is a rate limiting factor, namely at the higher pH's where the free thiol concentrations are low. The inverse isotope effects at the lower pH's in both cases can be ascribed to the difficulties encountered in comparing the sulfide to thiol ratios in H_2O and D_2O at a designated pH (5). It is possible that at the lower pH's the higher observed rate constant in D_2O reflects an increased rate of initiation owing to a higher concentration of sulfide ion in D_2O relative to that in the H_2O solution with which it is compared. At the higher pH, this factor is outweighed by the positive isotope effect resulting from the involvement of the

Table VII

Solvent Isotope Effects

pH	k_H'	"pD"	k_D'	k_H'/k_D'
		Mercaptoethanol		
9.90	1.69	10.6	1.88	0.92
10.30*	2.08	11.0*	2.44	0.85
10.80	2.05	11.2	1.83	1.12
		Dithioerythritol		
9.30	1.06	9.90	1.17	0.91
9.60*	1.32	10.20*	1.43	0.92
9.90	1.20	10.60	0.91	1.31

* pH or "pD" at which maximum rate is observed.

breaking of the S-H bond in a rate limiting step. Thus, the observed isotope effect at the higher pH's as measured by the k_H/k_D values in Table VII likely are considerably smaller than the actual isotope effect operative in the hydrogen atom abstraction from the thiols by the methylene blue derived radical.

Ionic Strength Considerations. Reactions involving charged reactants may be expected to be influenced by the ionic strength of the medium in which the reaction occurs. The most profound effects are observed in reactions between two charged species, either both with the same charge or each oppositely charged. The effect of a change in the ionic strength of the medium on the rate constant of a reaction between ionic species can be expressed in the Eq. 26 which is a combination of Brønsted equation and the Debye-Hückel rate limiting law (6). This equation, where k_o and k are the rate constants at infinite

$$\ln k = \ln k_o + 2 Z_A Z_B \alpha \sqrt{\mu} \qquad (26)$$

dilution of any charged species in solution and at ionic strength μ, respectively, α is a constant related to the solvent and temperature ($\alpha = 0.509$ in water at 25°) and Z_A and Z_B are the charges of the ionic species A and B, respectively, predicts that the rate constant for a reaction between species of like charges will increase with increasing ionic strength of the medium. Conversely, the rate constant of a reaction between species of opposite charge should be expected to decrease with increasing ionic strength. Examination of the proposed mechanisms for the methylene blue oxidations of mercaptoethanol and dithioerythritol shows that in each case there are two steps in the reaction sequence that involve interactions between species that are oppositely charged. These are the initiation reactions (the reaction between MB^+ and the sulfide ion, reactions 1 and 14) and the chain propagating electron transfer reactions between the disulfide radical anions and MB^+ (reactions 3 and 16).

That interactions between charge species are rate limiting factors in these oxidations was evident in the early stages of our work in that reliable rate data could not be obtained for these oxidations unless precautions were taken to maintain a constant ionic strength by addition of a strong electrolyte (KCl was used in this work). All of the data presented in the work covered up to this point was obtained at an ionic strength of 0.4. It should be noted that at ionic strengths greater than about 0.01, the Debye-Hückel limiting law is no longer strictly valid and corrections must be made to account for the changes in the activities of the charged species in media of high ionic strength. The inclusion of the correction factor "b" is necessary to accommodate these changes in the activities of the charged species at high ionic strengths as shown in Eq. 27.

$$\ln k = \ln k_o + 2 Z_A Z_B \, \alpha \, \sqrt{\mu} \, - \, b\mu \qquad (27)$$

Not having quantitative values for "b" for the reactions of MB^+ with either the sulfide ions or the disulfide radical anions, we can evaluate the effects of ionic strength in these reactions at high ionic strength only in a semi-quantitative manner.

Table VIII lists the rates of oxidation of mercaptoethanol by methylene blue at various ionic strengths. Note that at any given concentration of MB^+ the rate of the reaction decreases with increasing ionic strength. At the higher concentrations of MB^+, only the initiation reaction rate constant would be influenced by the change in the ionic strength of the solution. The observed decrease in the reaction rate at the higher ionic strengths is consistent with a decrease in the value of the rate constant of the initiation reaction.

As pointed out earlier, the electron transfer reaction of the disulfide radical anion with MB^+ is a rate limiting factor at low concentrations of MB^+. The extent to which this reaction is a rate limiting factor, as evidenced by the change in the kinetic order of MB^+, depends not only on the concentration of MB^+ but also on the reaction rate constant for this chain propagating reaction. Equation 28, which relates the expected change in the kinetic order of MB^+ ($\Delta\alpha$) and the change in ionic strength of two solutions in which the rate law is determined for a given concentration of MB^+, can be derived from the Eq. 26 and the assumption that the rate law changes can be ascribed only to a change of the value of the reaction rate constant for the reaction

Table VIII

Effect on Ionic Strength on

Mercaptoethanol[a] Oxidation Rates[b]

μ	Rate x 10^8 M sec^{-1} = [MB$^+$] x 10^6			
	5.88	4.33	3.02	1.89
0.1	7.88	6.78	5.64	4.46
0.4	5.41	4.63	3.83	2.91
1.0	5.14	4.40	3.61	2.68
2.0	3.00	2.57	2.07	1.47

a. [RSH] = 1.57×10^{-2}.
b. pH 10.3.

of the disulfide radical anion with MB$^+$. The equation predicts
that the magnitude of the change in the kinetic order of MB$^+$ is

$$\Delta\alpha = -1.08 \left| \mu_{(1)}^{1/2} - \mu_{(2)}^{1/2} \right| /\log [MB^+] \qquad (28)$$

inversely related to the concentration of MB$^+$ and that changes in
the kinetic order of MB$^+$ may be observable experimentally only if
the rate laws are measured in solutions with markedly different
ionic strengths.

Table IX shows the rate laws for the oxidation of
mercaptoethanol by MB$^+$ at pH 10.3 at different ionic strengths.
At the highest concentration of MB$^+$, the rate law, as expected,
is not influenced by the ionic strength of the medium because the
reaction of the disulfide radical anion is not a rate limiting
factor. At the lower concentrations of MB$^+$, however, there is
evidence that the rate law for the oxidation is dependent on the
ionic strength of the medium. The rate law changes at the high
ionic strength and low MB$^+$ concentrations are consistent with the
changes that would be predicted if the rate constant for the
electron transfer reaction from the disulfide radical anion to
MB$^+$ were decreased in value, namely an increase in the kinetic
order of MB$^+$ and a decrease in the kinetic order of the thiol.
These are the same rate law changes observed at constant ionic
strength when the concentration of MB$^+$ is very low (see Table
III).

Table IX

Effect on Ionic Strength on

Kinetic Orders of Reactants

$$\text{Rate} = k_{obs}[MB^+]^{\alpha}[RSH]^{\beta}$$

μ	α	β
	$[MB^+] = 5.11 \times 10^{-6}$	
.1	.49 (.13)	1.51 (.13)
.4	.51 (.09)	1.51 (.17)
1.0	.51 (.09)	1.50 (.07)
2.0	.51 (.10)	1.42 (.08)
	$[MB^+] = 3.67 \times 10^{-6}$	
.1	.51 (.05)	1.52 (.10)
.4	.55 (.06)	1.51 (.09)
1.0	.55 (.05)	1.45 (.11)
2.0	.60 (.12)	1.35 (.10)
	$[MB^+] = 2.46 \times 10^{-6}$	
.1	.50 (.11)	1.48 (.04)
.4	.59 (.11)	1.42 (.18)
1.0	.64 (.11)	1.40 (.15)
2.0	.73 (.17)	1.27 (.10)

Summary

The agreement between the changes in the observed rate laws of the MB^+ oxidations of thiols with changes in the reactant concentrations, pH and ionic strength and the changes predicted from the steady-state derived rate laws based on the proposed mechanism for disulfide formation support this mechanism. Of particular interest to us is that it is possible to extract information concerning the behavior of a free radical chain reaction by kinetic analysis of the reaction. Such kinetic analysis of other redox reactions may be expected to be of value in not only establishing the possible intermediacy of free radicals but also the details of steps in the chain sequence in which free radicals are reaction intermediates.

Abstract

The effects of pH, ionic strength, reagent concentration ratios and deuterium substitution of the sulfur-bonded hydrogens on both the rates and rate laws for the methylene blue oxidations of mercaptoethanol and dithioerythritol have been determined. A free radical chain mechanism consistent with the observed kinetic behavior of the oxidation reactions is proposed. A key feature of the proposed mechanism is the formation of the sulfur-sulfur linkage of the disulfide in the reversible formation of a disulfide radical anion (RSŚR) as a chain propagating step in the chain sequence.

Literature Cited

(1) For general references, see Capozzi, G., and Modena, G., "The Chemistry of the Thiol Group", Chap. 17, Patai, S., Ed., Wiley, New York, N.Y., 1974; Jocelyn, P. C., "Biochemistry of the SH Group", Chap. 14, Academic Press, London, 1972.
(2) Schrauzer, G. W., and Sibert, J. W., Archs. Biochem. Biophys., (1969) 126, 257.
(3) Kreevoy, M. M., Harper, E. T., Durall, R. E., Wilgus, H. S. and Ditsch, L. T., J. Amer. Chem. Soc., (1960), 82, 4899.
(4) Zahler, W. L. and Cleland, W. W., J. Biol. Chem., (1968), 243, 716.
(5) For discussion see Laughton, P. M. and Robertson, R. S., "Solute-Solvent Interactions", pp. 406-415, Coetzee, J. F., and Ritchie, C. D., Editors, Marcel Dekker, New York, 1969.
(6) See Hammett, L. P., "Physical Organic Chemistry", Chap. 7, McGraw-Hill, New York, 1970.

RECEIVED December 23, 1977.

The Geometry of Free Radical Displacements at Di- and Trivalent Sulfur

J. A. KAMPMEIER, R. B. JORDAN, M. S. LIU,
H. YAMANAKA, and D. J. BISHOP

Department of Chemistry, University of Rochester, Rochester, NY 14627

The displacement reaction (eq. 1) is one of the fundamental unit processes in free radical chemistry.

$$X\cdot + Y - Z \rightarrow X - Y + Z\cdot \qquad (1)$$

This process, with suitable variation in X, Y, and Z, serves to organize a very large number of free radical reactions.(1,2) The determination of rate constants (or relative rate constants) for these reactions has been a major preoccupation of free radical chemists. (1,2) It is striking, however, that we have very little understanding of the geometrical characteristics of these reactions. The historical reason for this is easy to see; free radical displacements at potentially chiral reaction centers are not common, thereby prohibiting the classical experimental approaches to stereochemical questions. For example, free radical displacements at carbon are rare and seem, in general, to be confined to the reactions of strained ring systems.(3-5) A few reaction centers (e.g., P, Si, some metals) can be prepared in chiral form and do invite the use of standard techniques.(6) However, many important displacement reactions take place at mono- or divalent reaction centers, where there is no possibility of using a chiral probe of mechanism. In this paper, we report the results of experiments designed to probe the preferred geometry and mechanism of radical displacements at divalent sulfur, a reaction center characterized by a plane of symmetry. Further, the technique is general and can presumably be used to study the geometries of other displacement reactions at centers which are inherently achiral or cannot be easily prepared in chiral form.(7)

Photolysis of methyl 2-(o-iodophenyl)ethyl sulfide (I, R=CH₃) in a cyclohexane solution gives a

reduction product, methyl 2-phenylethyl sulfide
(II, R=CH$_3$) and two displacement products, dihydro-
benzothiophene (III) and benzothiophene (IV) (eq. 2).
A control experiment involving the photolysis of

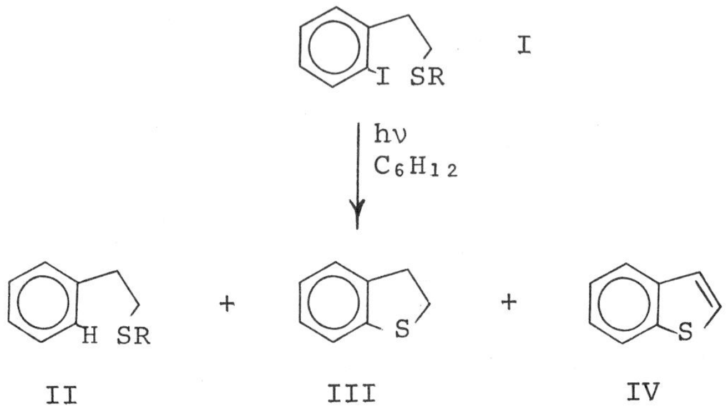

$$\text{I}$$

hv
C$_6$H$_{12}$

(2)

II   III   IV

iodobenzene in cyclohexane containing dihydrobenzothiophene shows smooth conversion of dihydrobenzothiophene to benzothiophene. Benzothiophene is a secondary product and the yield of displacement product from
the photolysis of the iodosulfide (I) is, therefore,
taken as the sum of the yields of III and IV. Typical
results for the photolysis of the methyl sulfide (I)
are shown in Table I. Mass balances are good, showing
that the important reaction products are identified;
conversions are relatively low to minimize secondary
reactions.

Table I.
Products of Photolysis of Methyl 2-(o-Iodophenyl)ethyl
Sulfide[a](I)

Solvent	$\dfrac{10^2 \text{ Red'n}^{b}}{\text{Red'n + Disp}^{\intercal}}$	% Con-version	Mass Balance
CH$_3$CN	10	23	96
Cyclohexane	17	20	94
Cyclohexane/0.5M BuSH	75	–	–
Cyclohexane/3M BuSH	87	11	100
BuSH (NEAT)	97	12	101

a. [I]~0.1M; 254 nm   b. 10^2 [II]/([II]+[III]+[IV])

The key point to be derived from the data in
Table I is that the displacement reaction is progress-
ively interrupted by the reduction reaction as the
hydrogen donor ability of the medium increases. This

is clear evidence that displacement and reduction are competitive fates of a common precursor. Since the reduction reaction involves capture of the aryl radical (V) from the photolysis of the iodo sulfide precursor, (8,9) radical V is implicated as the source of the displacement product (eq. 3). Table II reports data

$$(3)$$

showing that this displacement reaction proceeds with a number of leaving groups covering a wide range of "stability" as free radicals. We have investigated the fate of the leaving group (R·) in some of these displacement reactions and find typical free radical products (e.g., I, R=C_6H_5 gives C_6H_6 and C_6H_5I).

Table II

Products of Photolysis of Various 2-(o-Iodophenyl)ethyl Sulfides (I) in Media of Different H-Donor Ability[a]

$$\underline{\quad\quad} 10^2 \text{ Red'n/(Red'n + Disp')}\underline{\quad\quad}^b$$

R	C_6H_{12}[c]	$C_6H_{12}/BuSH$[d]	$BuSH$[e]
C_6H_5[f]	28-40	–	97
CF_3	35	63	96
CH_3	17	75	97
C_2H_5	12	62	96
t-C_4H_9	9	71	95
$CH_2C_6H_5$[f]·	9	52	92

a. [I]~0.1 M; 254 nm   b. 10^2 [II]/[II]+[III]+[IV])
c. cyclohexane solvent d. 0.5 M n-BuSH
e. n-BuSH solvent
f. Mass balances are poor for these two cases due to competitive photochemical homolysis of $ArCH_2CH_2$--SC_6H_5 and $ArCH_2CH_2S$--$CH_2C_6H_5$. Mass balances are >95% in the other cases.

Any concern that the displacement product origi-
nates in an excited state reaction or in competitive
photolysis of the S-R bond is mitigated by the observation that reaction of iodosulfides (I) with tri-n-
butyltin radicals, via the $Bu_3SnH/Bu_3Sn\cdot$ chain, also
gives the reduction and displacement products (II and
III); the distribution of products is a function of
the concentration of the hydrogen donor, tri-n-butyltin hydride (Table III).

Table III

Distribution of Products in the Reaction of 2-(o-
Iodophenyl)ethyl Sulfides (I) with Tri-n-Butyltin
Hydride

R	10^2 Red'n/(Red'n + Disp')[a]	
	0.1 M Bu_3SnH[b]	~2.2 M Bu_3SnH[c]
C_6H_5	57	86
CH_3	49	–
$CH_2C_6H_5$	16	44

a. 10^2 [II]/([II]+[III])   b. In cyclohexane at 105°-110°,
AIBN as initiator   c. No solvent; 105°-110°

We conclude that we are dealing with a free radical
displacement at divalent sulfur (eq. 3). We arrived
at a similar conclusion in a study of the prototype
of the system described in this paper.(10)   Other
free radical displacements at the divalent sulfur of
sulfides are known.(11-17)
        The salient point in this study is that radical V
has, in principle, two displacement opportunities. One
gives dihydrobenzothiophene, as described; the alternative would give radical VI, which should be reduced
to an o-ethylphenyl sulfide (VII)(eq. 4). We have
searched carefully for the product (VII) of this alternative displacement in every reaction described in
this paper; VII is stable to the reaction conditions
and is never observed.(18) Authentic radical VI(R=C_6H_5,
CH_3, $CH_2C_6H_5$) is available in cyclohexane from the thermal or photochemical decomposition of the corresponding
peresters (VIII)and gives rise, inter alia, to dihydrobenzothiophene and the expected reduction product VII;
the product of the alternative displacement, II, is not
formed. Since it is experimentally established that
the radicals V and VI are captured by cyclohexane to

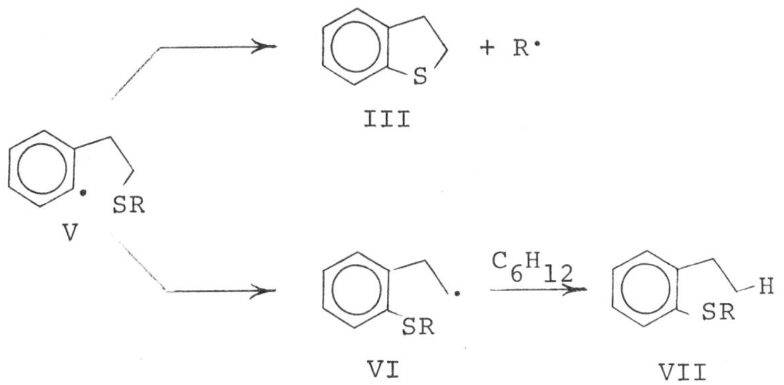

(4)

give stable products II and VII, respectively, it is certain that alternative displacement paths are less favorable than those giving dihydrobenzothiophene. These observations are summarized in Scheme I.

Scheme I

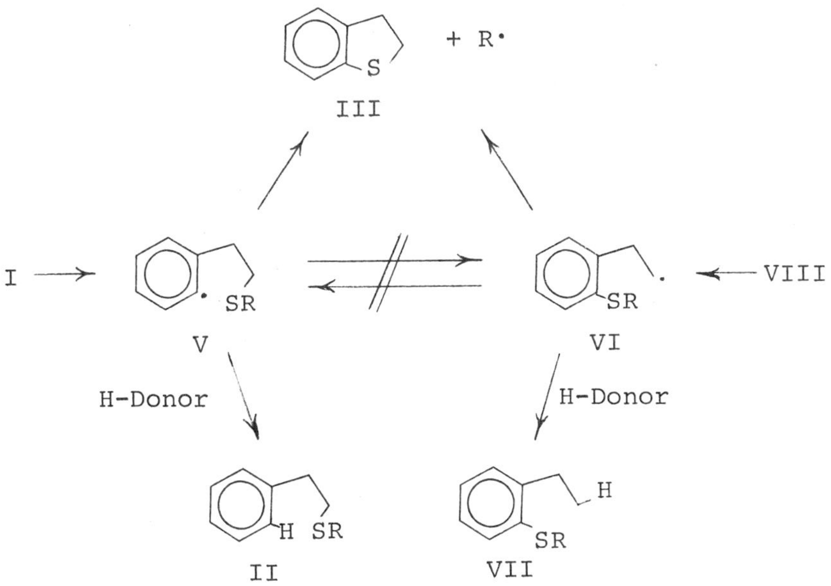

The displacement reactions which give dihydrobenzothiophene are exocyclic pathways; the alternative
displacements are forced, and refuse, to follow an
endocyclic pathway. The clear preference for the
exocyclic pathway cannot, in the case of radical V, be
assigned to differences in leaving group ability. For
radical V, R=C_2H_5, the leaving groups in the two competing displacements are both $\cdot CH_2\sim$; for R=C_6H_5,
CF_3 and CH_3, an argument based on the relative stability of the leaving group favors the endocyclic
path, contrary to observation. In fact, the distinction between the two displacement reactions resides in
the different geometries available to the competing
displacement paths. In particular, the endocyclic
path is constrained to a non-linear geometry by the
cyclic nature of the reaction, whereas the exocyclic
path can attain a linear arrangement of the three centers involved in the reaction. We conclude, therefore,
that these free radical displacements at divalent
sulfur involve a linear, back-side, "inversion" path;
the alternative non-linear path has the character of
a front-side, "retention" reaction. A three-center,
three-electron description (IX) of the reaction path
(Scheme II) is consistent with the observed geometrical
preference.(19)

Scheme II

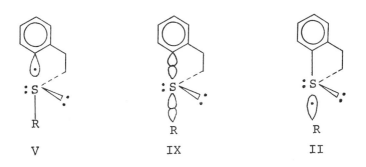

V   IX   II

An analogous free radical displacement reaction
is observed at sulfoxide centers, but fails for the
corresponding sulfone. Thus, photolysis of methyl
2-(o-iodophenyl)ethyl sulfoxide (X, R=CH_3) in cyclohexane gives methyl 2-phenylethyl sulfoxide (XI,
R=CH_3) and dihydrobenzothiophene oxide (XII)(eq. 5).
Photolysis of ethyl(2-o-iodophenyl)ethyl sulfone in

cyclohexane gives only the reduction product, ethyl-

$$\text{X} \quad \xrightarrow[\text{C}_6\text{H}_{12}]{h\nu} \quad \text{XI} \quad + \quad \text{XII} \qquad (5)$$

2-phenylethyl sulfone. The results of the photolysis
of the sulfoxides are given in Table IV.

Table IV

PRODUCTS OF PHOTOLYSIS OF ALKYL 2-(o-IODOPHENYL)
ETHYL SULFOXIDES[a](X) IN CYCLOHEXANE

R	$\dfrac{10^2\ \text{Red'n}}{\text{Red'n + Disp'}}$
CH_3[c]	47
CH_3[d]	54
$\text{C(CH}_3)_3$[c]	16

a. 254 nm in cyclohexane
b. $10^2[\text{XI}]/([\text{XI}]+[\text{XII}])$.
c. $[\text{X}] \sim 10^{-4}$ M.
d. $[\text{X}] \sim 10^{-2}$ M.

The course of these radical displacements at the
trivalent sulfur center of the sulfoxides is analogous to that observed for the sulfides; the exocyclic
displacement path obtains. Methyl (or t-butyl)
o-ethylphenyl sulfoxide, the ultimate product of the
alternative, endocyclic, displacement path is not
observed.   Our conclusion is also analogous; the
free radical displacements at the sulfoxide sulfur
involve a linear, back-side, "inversion" path.   A
three-center, three-electron description (analogous
to IX) of the reaction path is consistent with these
observations.   It is useful to note that sulfoxide
centers can be prepared in chiral form.   These conclusions are, therefore, subject to test by classical
techniques.   Other free radical displacements at
sulfoxide centers are known.(20,21)

In summary, we note that the preference for the
linear displacement geometry observed in this work
probably obtains to free radical displacement at many
other centers such as hydrogen, oxygen, halogen, and
phosphorous. In other cases, suggestive results are
already available.(22,23) For example, the failure
(24) of the displacement at chlorine in eq. 6 is
reasonably attributed to the non-linear, endocyclic
geometry of the reaction path. In any case, the approach we have used to define the preferred geometry
of the displacement at divalent and trivalent sulfur
is, in principle, applicable to a variety of other
systems where classical chiral techniques are not
readily accessible.

(6)

It is important to consider the detailed nature
of these displacement reactions at di- and trivalent
sulfur centers; i.e., is the three-center, three-
electron species (IX) a transition state or an intermediate on the displacement path? In fact, the present system was specifically designed to provide
information on this question. As we have previously
described, radical V partitions competitively between
reduction and displacement paths. In Scheme III,
mechanisms A and B involve the three-center, three-
electron species as transition state and intermediate,
respectively. If the concentration and the nature of
the reducing agent (H-donor) are held constant, then
the rate of formation of reduction product should be
a constant, independent of the nature of substituent
on sulfur, since the S-R bond is not involved in the
rate-determining step of the reduction reaction. This
reaction, therefore, can serve as a kinetic reference
point. If the displacement reaction involves a simple
one-step reaction (path A), the rate of formation of

Scheme III

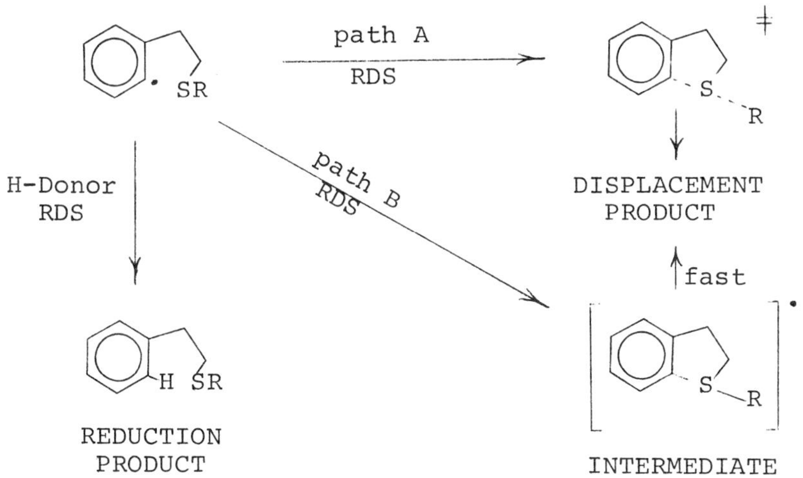

displacement product will vary with the nature of the leaving group (R). In the alternative mechanism involving rate-determining formation of an intermediate followed by fast product formation (path B), the rate of the displacement will, to a first approximation, not depend on substituent R, since the S-R bond is broken in a subsequent, fast step. Thus, the influence of the substituent R on the ratio of the yields of reduction to displacement product in a given solvent provides a potential distinction between mechanisms A and B.

In actual fact, a control experiment shows that this simple analysis cannot be strictly correct. Thus, the photolysis of 0.1M ethyl 2-(o-iodophenyl)ethyl sulfide in cyclohexane-d_{12} ($\overline{I}$, R=C_2H_5) gives ethyl 2-(o-deuteriophenyl)ethyl sulfide (~16%) and ethyl 2-phenylethyl sulfide (~84%). Thus, there are sources of hydrogen, other than solvent, participating in the reduction reaction. In fact, the formation of benzothiophene as a side product in these reactions implicates dihydrobenzothiophene as a reductant. In addition, the starting material (I) has a variety of reactive C-H bonds α to sulfur and to the aromatic ring. As the nature of the substituent (R) is varied, the reducing ability of the starting material will vary and, therefore, the assumption that the rate of reduction is independent of $\dot{R}$ cannot be strictly true. On the other hand the 16/84 competition observed between reaction with C_6D_{12} and other H-donors clearly demonstrates that these different reductions have comparable rates. This is confirmed, in a related case, by the observation that the experimental reduction/displacement ratio is independent of the absolute concentration of the starting material and reaction products. Thus, photolysis of I (R=Me) at 0.1 M concentration gives reduction/displacement = 17/83, while photolysis at 10^{-4} M concentration of I gives a product ratio of 19/81. Thus, although the precise details of the reduction reaction may vary with the specific system, the total rate of reduction is approximately constant.

Tables II and III present the results of the determination of the distribution of products between reduction and displacement paths in cyclohexane solution as the nature of the substituent (R) is varied through a series of sulfides. The striking result is that big changes in the nature of the leaving group have little effect on the distribution of products. The relative rates of displacement by phenyl radical on the hydrogens of various alkanes is a model for the leaving group effect in a direct, one-step, displacement reaction. As shown in Table V, these displacements are very sensitive to the nature of the leaving group.

Table V

$\phi\cdot + H-R \rightarrow \phi H + R\cdot$

R	Rel. Rate(25)
CH_3	1
C_2H_5	40
$CH(CH_3)_2$	300
$C(CH_3)_3$	1600

The relative rates of displacement by phenyl at sulfur can be obtained from the data in Table II by assuming that the rate of reduction is independent of substituent (R). These relative rates are 1.0 (R=CH_3), ~1.5 (R=C_2H_5), ~2 (R=$\underline{t}$-C_4H_9). The contrast is striking and clearly indicates that the S-R bond is not significantly involved in the rate-determining step for the formation of displacement product. A similar lack of sensitivity to leaving group is seen for the displacements at sulfoxide centers (Table IV). The displacement at hydrogen and sulfur are comparably exothermic

and so it does not seem reasonable to dismiss the absence of a leaving group effect by an "early transition
state" argument. Rather, these results would seem to
argue clearly for the rate-determining formation of an
intermediate, followed by a fast cleavage of the S-R
bond (mechanism B).

However, the results of the competition between
reduction and displacement at a sulfoxide center con-
tradict this simple interpretation. Thus, radical V
(R=CH₃) in the sulfide series gives a reduction/dis-
placement ratio of 17/83 in cyclohexane. The corresponding radical derived from the methyl sulfoxide
(X, R=CH₃) distributes approximately 50/50 between
reduction and displacement paths in cyclohexane. This
change in product distribution reflects a trivial
change in the rate of displacement as the reaction
center is changed from divalent sulfide to trivalent
sulfoxide. Unfortunately, we do not have suitable
models for the expected variations in rate with variations in the reaction center in the addition-fragmen-
tation mechanism (mechanism B). The rates ·of addition
of phenyl radical to sulfide and sulfoxide centers
might just be coincidentally identical, although this
seems very unlikely. Rather, the comparison of sulfide and sulfoxide reactions centers indicates that
the formation of the new aryl-sulfur bond is also not
involved in the rate determining step of the displacement reaction.

We are confronted, therefore, with a displacement
reaction in which a new bond is formed to sulfur and an
old bond to sulfur is broken and, yet, neither of these
processes seems to be involved in the rate determining
step. A clue to dissecting this dilemma is found in a
comparison of the behavior of the prototype radical
(XIII) with the radical involved in the present studies

CH₃S

XIII   XIV

(XIV). In the presence of a very good reducing agent
(n-butanethiol) as solvent, radical XIII still gives
>95% displacement to give dibenzothiophene. In con-
trast, radical XIV is trapped under comparable conditions to give >95% reduction product. This structural
change from XIII to XIV is, in fact, the only way in
which we have been able to vary the rate of the displacement reaction. The significant difference between the two radicals is a conformational difference.
Radical XIII needs only to "rock" about the central
bond to bring the radical and the sulfur centers into
juxtaposition; in XIV, significant rotation around
the central "ethane" bond is required. As argued earlier, the distribution of the radicals between reduction
and displacement products does not seem to be determined by the relative rates of radical attack on H-donor
and the sulfur centers. Rather, the distribution of
products is consistent with a competition between the
rate of reduction and the rate of rotation around C-C
bonds required to convert radicals XIII and XIV into
conformations suitable for the displacement reactions
at sulfur. This competition between reduction and conformational change has nothing to do with the sulfur
and, therefore, changes in the ease of bond formation
and/or bond breaking are irrelevant. This analysis is
summarized in Scheme IV. When k_D is large, the product
composition is determined by the ratio, $k_R[SH]/k_1$.

Scheme IV

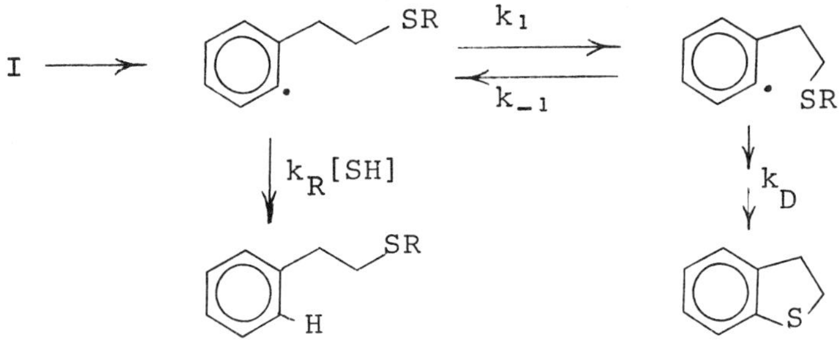

$$\frac{[II]}{[III]} = \frac{k_R[SH]k_{-1}}{k_D k_1} + \frac{k_R[SH]}{k_1}$$

$$k_D > k_1, \quad k_{-1} \sim k_R[SH]$$

Trotman-Dickenson reported an activation energy of 6.7 kcal/mol for the reaction of phenyl radicals with isobutane in the gas phase.(26) On the basis of this, we were confident that conformational interconversion (k_1, k_{-1}) would be rapid with respect to other reactions and that the product composition would be determined by the ratio $k_R[SH]k_{-1}/k_D k_1$. Recent experimental work by J. P. Lorand,(24) however, shows that the rate of reaction of phenyl radicals with the tertiary hydrogens of alkanes is about 10^2 times faster than that reported by Trotman-Dickenson. A competition between $k_R[SH]$ and k_1 is, therefore, plausible.

In summary, these experiments give a clear insight into the geometric course of the displacement reaction at sulfide and sulfoxide centers. The kinetic analyses do not permit a distinction between concerted one-step displacements and two-step mechanisms involving an intermediate because the processes in competition are probably reduction and conformational change. By implication, the rates of reaction at sulfur must be very rapid; they must be faster than either reduction or conformational change. The rate of reaction of hydroxyl radical with dimethylsulfoxide is reported to be diffusion controlled.(28) B. P. Roberts and co-workers have estimated the rate constant for displacement by t-butoxy radical at the sulfur of thietan at be 6×10^6 ℓ/m/sec at 243°K.(17) The addition of alkoxy radicals to dialkylsulfoxylates is competitive with the addition of these alkoxy radicals to phosphites, a reaction that proceeds at close to the diffusion controlled rate at room temperature.(29)

Finally, if the three-center, three-electron species is an intermediate, the observed geometric selectivity of the displacement requires that formation of products must be faster than equilibration of various isomeric three-center, three-electron species. Trialkoxy sulfuranyl radicals have been observed by e.s.r. These species are assigned a T-shape (like IX); equilibration of apical and equatorial substituents is slow on the e.s.r. time scale.(30,31)

Literature Cited

1. Ingold, K. U. and Roberts, B. P., "Free Radical Substitution Reactions," Wiley-Interscience, New York, 1971.
2. Kochi, J. K., Ed., "Free Radicals," Vol. I and II, Wiley-Interscience, New York, 1973.

3. Incremona, J. H. and Upton, C. J., J. Amer. Chem. Soc., 94, 301 (1972).
4. Maynes, G. G. and Applequist, D. E., ibid., 95, 856 (1973).
5. Shea, K. J. and Skell, P. S., ibid., 95, 6728 (1973).
6. For the stereochemistry of radical displacements at P, see W. G. Bentrude, this Volume.
7. For a related study of free radical displacement at oxygen, see N. Porter, this Volume.
8. Blair, J. M. and Bryce-Smith, D., J. Chem. Soc., 1788 (1960).
9. Wolf, W. and Kharasch, N., J. Org. Chem., 26, 283 (1961).
10. Kampmeier, J. A. and Evans, T. R., J. Amer. Chem. Soc., 88, 4096 (1966).
11. Benati, L., Montevecchi, P. C., Tundo, A., Zanardi, G., J. Chem. Soc., Perkin I, 1272 (1974); 1276 (1974).
12. Schmidt, U., Hochrainer, A., and Nikiforov, A., Tetrahedron Letters, 42, 3677 (1970).
13. Haszeldine, R. N., Higginbottom, B., Rigby, R. B., and Tipping, A. E., J. Chem. Soc., Perkin I, 155 (1972).
14. Jakubowski, E., Ahmed, M. G., Lown, E. M., Sandhu, H. S., Gosavi, R. K., and Strausz, O. P., ibid., 94, 4094 (1972).
15. Brazen, W. R., Cripps, H. N., Bottomley, C. G., Farlow, M. W., and Krespan, C. G., J. Org. Chem., 30, 4188 (1965).
16. Weseman, J. K., Williamson, R., Greene, J. L., Jr., and Shevlin, P. B., Chemical Communications, 901 (1973).
17. Chapman, J. S., Cooper, J. W., and Roberts, B. P., Chemical Communications, 407 (1976).
18. Control experiments show that approximately 1% conversion of I to VII is detectable by our analytical techniques.
19. Bonacic-Koutecky, V., Koutecky, J., and Salem, L., J. Amer. Chem. Soc., 99, 842 (1977).
20. Gilbert, B. C., Norman, R. O. C. and Sealy, R. C., J. Chem. Soc., Perkin II, 303 (1975).
21. Lagercrantz, C. and Forshult, S., Acta Chem. Scand. 23, 811 (1969).
22. Ref. 2, Wilt, J. W., Vol. I, Ch. 8, p. 378 ff.
23. Drury, R. F. and Kaplan, L., J. Amer. Chem. Soc., 94, 3982 (1972).
24. Chen, K. S., Tang, Y. H., Montgomery, L. K., Kochi, J. K., J. Amer. Chem. Soc., 96, 2201 (1974).

25. Russell, G. A., "Reactivity, Selectivity and Polar Effects in Hydrogen Atom Transfer Reactions," in "Free Radicals," J. K. Kochi, Ed., Vol. I, Ch. 7, Wiley-Interscience, New York, 1973.
26. Duncan, F. J. and Trotman-Dickenson, A. F., J. Chem. Soc. 4672 (1962).
27. Lorand, J. P., Private Communication, 1976.
28. Meissner, G., Henglein, A., and Beck, G., Z. Naturforsch., 1967, 22b, 13.
29. Bentrude, W. G., "Phosphorous Radicals," in "Free Radicals," J. K. Kochi, ed., Vol. II, Ch. 22, Wiley Interscience, New York, 1973.
30. Morton, J. R. and Preston, K. F., J. Phys. Chem. 77, 2646 (1973).
31. Chapman, J. S., Cooper, J. W. and Roberts, B. P., Chemical Communications, 835 (1976).

RECEIVED December 23, 1977.

17

Succinimidyl and Related Radicals

PHILIP S. SKELL and JAMES C. DAY

Department of Chemistry, The Pennsylvania State University,
University Park, PA 16802

The chemical properties of succinimidyl radicals remained a mystery until the simultaneous recognition in the laboratories of Professor J. G. Traynham (1) and ours (2) that the radical could be a chain carrier in the sense first suggested by Bloomfield (3), albeit incorrectly (4), for the allylic bromination reaction.

For substrates which react readily with bromine atoms, succinimidyl mediated brominations are carried out in the presence of bromine scavengers (ethylene or t-butylethylene), employing a good solvent for NBS such as methylene chloride. In these circumstances bromination occurs with a selectivity which closely resembles Cl· mediated chlorinations.

Additions to alkenes and arenes also occur readily.

In all of these reactions the selectivities are remarkable for radical reactions, being characteristic of no-stabilization of transition states by contributions from product structures. Each of these reactions appears to be an encounter-controlled reaction. As an illustration, succinimidyl radical adds to styrene as follows: to the double bond twice as rapidly as to the phenyl nucleus.

With substrates which do not react with bromine atoms, a mixture of Br_2 and NBS is effective in bromination, succinimidyl abstracting the H-atom, the product radical reacting with Br_2, and the Br· reacting with NBS to regenerate the succinimidyl radical.

These two methods for generating succinimidyl radicals (in the presence of olefins or bromine) produce different succinimidyl radicals, ground state (π) in the presence of Br_2 and excited state (σ_N) in the presence of olefins (the evidence is described elsewhere).

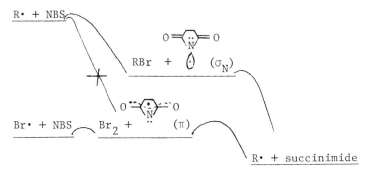

The ring opening reaction, NBS → β-bromopropionyl isocyanate, is a reaction of σ, but not π.

Financial support came from the Air Force Office of Scientific Research (2748C).

Literature Cited

(1) Traynham, J. G., and Lee, Y. S. , J. Am. Chem. Soc., 96, 3590 (1974).

(2) Day, J. C., Lindstrom, M. J., and Skell, P. S., ibid., 96, 5616 (1974).

(3) Bloomfield, G. F., J. Chem. Soc., 114 (1944).

(4) Adam, J., Gosselain, P. A., and Goldfinger, P., Nature, 171, 704 (1953); Gosselain, P. A., Adam, J., and Goldfinger, P., Bull. Soc. Chem. Belg., 65, 533 (1956); Sixma, F. L. J., and Riem, R. H., K. Ned. Akad. Wet. Proc., 61B, 183 (1958); McGrath, B. P., and Tedder, J. M., Proc. Chem. Soc., 80, (1961); Pearson, R. E. and Martin, J. C., J. Am. Chem. Soc., 85, 354, 3142 (1963); Walling, C., Rieger, A. L., and Tanner, D. D., ibid., 85, 3129 (1963); Russell, G. A., Do Boer, C., and Desmond, K. M., ibid., 85, 365, 3139 (1963); Incremona, J. H. and Martin, J. C., ibid., 92, 627 (1970).

RECEIVED December 23, 1977.

18

The Chemistry of Some Nitrogen-Centered Radicals

CHRISTOPHER J. MICHEJDA, DWANE H. CAMPBELL,
DAVID H. SIEH, and STEVEN R. KOEPKE

Department of Chemistry, University of Nebraska, Lincoln, NE 68588

The chemistry of nitrogen-centered radicals has been re-viewed comprehensively by Nelsen (1), and, more recently, by Forrester (2). The present paper deals largely with the chem-istry of a more restricted group of these radicals, the dialkyl-amino radicals.
These radicals are an intriguing class because of their seemingly dichotomous behavior. Besides being normal, electron deficient, and therefore, electrophilic species, they are also basic and hence could exhibit nucleophilic properties. The pK_A of Me_2NH+ has been estimated to be 6.5-7.3 (3). Profound changes in amino radical reactivity are produced by protonation, complexation by metals, and by substitution with electron at-tracting groups. These effects can be seen as a result of changes in electron density around the central atom, i.e. these changes which reduce non-bonding electron density at nitrogen tend to make the radical more electrophilic, and, therefore, more reactive in typical radical reactions. It is convenient to discuss the reactivity of amino radicals in terms of the afore-mentioned types--protonated radicals, metal complexed radicals, those substituted by electron attracting substituents and, finally, the unadulterated, neutral amino radicals.

Protonated Amino Radicals. The reaction of hydroxylamine with $TiCl_3$ in aqueous, acidic methanol results in the formation of the simplest protonated amino radical, NH_3^+. This radical ad-ded readily to butadiene and to simple olefins to form products which were the result of coupling of the intermediate β-amino-alkyl radicals (4). The addition reactions of protonated di-alkylamino radicals were described in a series of elegant papers by Neale and his co-workers (5-10). The radicals were generated from the appropriate N-chloroamines in presence of the unsatu-rated system. The most effective acid solvent combination was found to be 4M sulfuric acid in glacial acetic acid, but other acid/solvent combinations were also used. The reactions pro-ceeded by chain mechanisms, which were initiated by light or by

Fe(II) salts, although spontaneous initiation was also observed.

$$R_2\overset{+\cdot}{N}H + \underset{X}{\overset{/}{\underset{/}{C}}=C\overset{\diagdown}{\diagup} \longrightarrow R_2\overset{+}{N}H-\underset{X}{\overset{|}{C}}-\overset{|}{C}\cdot \overset{R_2\overset{+}{N}HCl}{\longrightarrow} R_2NHC-\underset{X}{\overset{|}{\underset{|}{C}}}-\overset{Cl}{\underset{}{C}}\cdot + R_2\overset{+\cdot}{N}H$$

The reaction was most effective when X was an electron withdrawing substituent, but only because this suppressed the competing
ionic chlorination of the double bond. In some cases, the ionic
chlorination was the predominant reaction. Protonated amino
radical addition to unsubstituted olefins, acetylenes and al-
lenes, however, gave acceptable yields of products, particularily
in the case of the sterically unhindered substrates. The products obtained from the acetylenes were the α-chloroketones or
aldehydes.

$$R_2NCl + R'-C\equiv C-R \longrightarrow \left[R_2N-\underset{R'R}{\overset{|}{C}}=\overset{|}{C}-Cl \right] \overset{H_2O}{\longrightarrow} R'-\overset{O}{\overset{||}{C}}-\overset{Cl}{\underset{}{C}}H-R$$
$$R' = H, \text{ alkyl}$$

One interesting aspect of the reactivity of protonated amino
radicals is that the reactivity toward unsaturation is much
higher than toward allylic abstraction. There is essentially no
competing allylic abstraction even in very favorable cases (7).
Intramolecular addition of protonated amino radicals has
been examined by several groups. The following reaction has
been carried out using silver perchlorate in acetone, ferrous
sulfate or acidic (4M H_2SO_4) conditions (11). The yields varied,

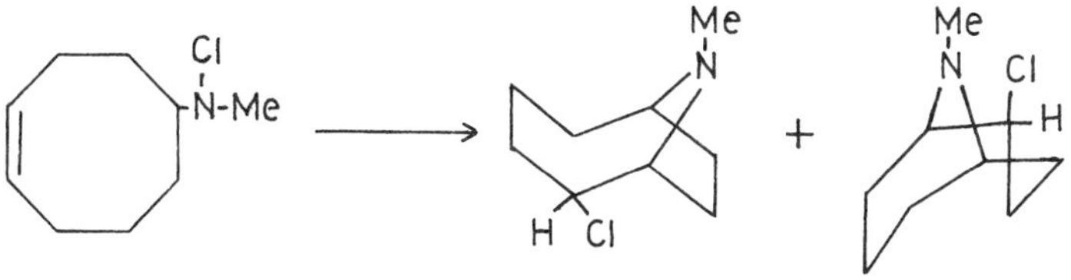

depending on conditions, but the authors concluded that either a
metal-coordinated or a protonated amino radical was the reactive
species. The elegant synthesis of the azatwistane also probably
involved the protonated amino radical (12). Chow (13) has

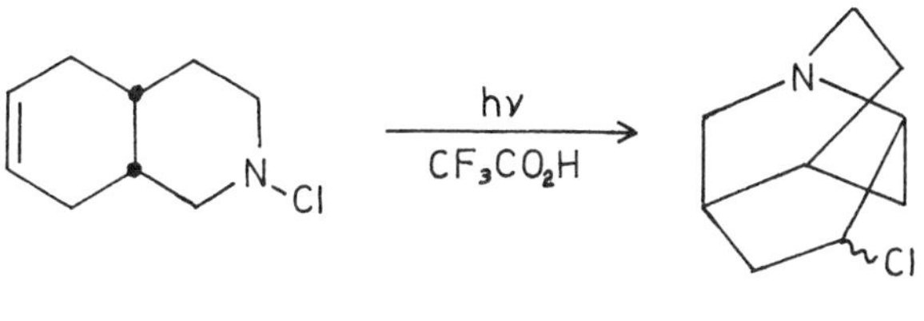

studied acid catalyzed photolysis of N-nitrosamines. The photolysis leads to protonated amino radicals, which add to alkenes
to produce, generally, amino oximes. In the intramolecular case,
efficient ring closure to the five membered ring occurs, as in
the following example (14,15).

The question of amination of aromatics by protonated amino
radicals has been examined extensively by Minisci and co-workers
(16). Benzene is aminated in good yields by a variety of N-
chloroamines in acidic solutions, catalyzed by iron(II) sulfates.
With activated aromatics competing chlorination and sulfonation
complicates the reactions, but in many cases good yields of the
substituted anilines are obtained. In the case of alkylbenzenes,
benzylic chlorination competes with nuclear amination. The
latter is favored by high acid concentrations. Thus, the reaction of toluene with N-chlorodimethylamine (17) gives 95 percent amination and 6 percent benzylic chlorination in neat H_2SO_4,
but 100 percent benzylic chlorination in neat acetic acid. The
amount of nuclear amination increases with the concentration of
H_2SO_4 in acetic acid.

Complexed Amino Radicals. Since protonated amino radicals
perform so well in addition reactions it is not altogether surprising that metal complexed amino radicals also add well to unsaturated substrates. This area of amino radical chemistry has
been studied by Minisci and his co-workers (18-21).

$$R_2NCl + Fe(II) \xrightarrow[\text{MeOH}]{H_2O} (R_2N\cdot)Fe(III)Cl \xrightarrow{\qquad} R_2N\overset{|}{\underset{|}{C}}-\overset{|}{\underset{|}{C}}-Cl$$

For example, the addition of N-chloropiperidine to cyclohexene,
catalyzed by iron(II) sulfate, gives predominately the cis product in good yield (21),

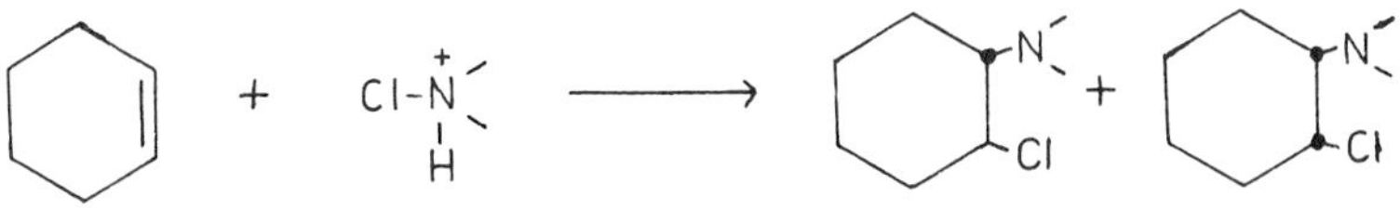

The protonated amino radicals give a mixture of cis and trans products. The stereoselectivity of the complexed radical is pre-sumably due to the intramolecular transfer of the chlorine atom. A decided advantage of the addition of N-chloramines catalyzed by a redox process in neutral media as opposed to the acid cata-lyzed reaction, is that the reaction conditions are much milder and ionic chlorination is not a competing process.

Tetralkyl-2-tetrazenes are excellent, general sources of dialkylamino radicals because they are prepared readily from the corresponding unsym-dialkylhydrazines by oxidation, and can be decomposed to the radicals either by thermal or photochemical methods (see below).

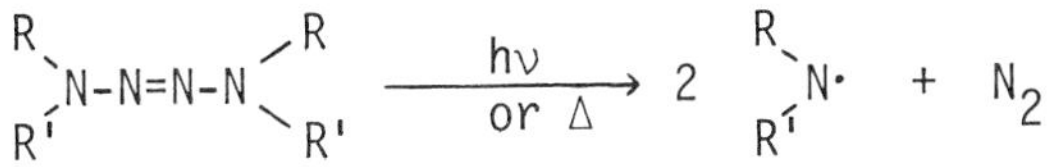

In keeping with their amine origins, tetrazenes are basic and can act as ligands in coordination complexes. Noltes and van den Hurk (22) reported that tetrazenes formed isolatable 1:1 com-plexes with zinc halides, zinc alkyls and aryls. They found that the zinc chloride complex was particularly unstable and lost nitrogen on heating between 40° and 60°. The structure which these workers proposed for the complex and its ready loss of nitrogen suggested to us (23) that the following reaction was occurring with tetramethyl-2-tetrazene (TMT).

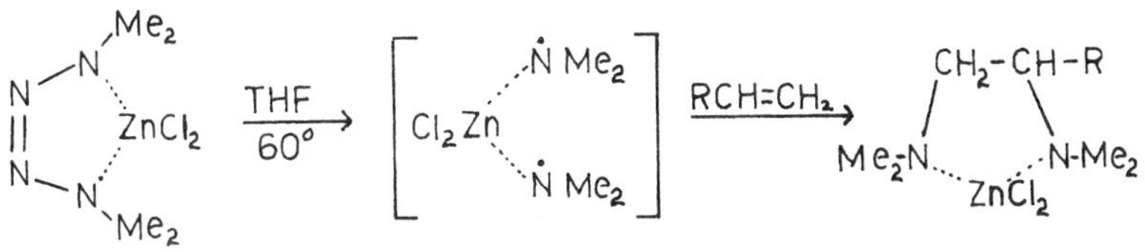

The putative biradical intermediate was trapped by styrene and α-methylstyrene in 30-40 percent yields. That the reaction was not concerted was shown by the fact that cis and trans-β-methyl styrenes gave an identical mixture of threo and erythro diamine adducts (24).

Ph\   /Me
  \C=C/
 H/   \H   Me₂N   H   Ph   Me₂N   H   Ph
                    \··· ◁ /   \··· ◁ /
Ph\   /H   THF   C——C   +   C——C◁
  \C=C/   ————→   /   \ \   /   \
 H/   \Me   60°   Me   H NMe₂   Me   NMe₂

Doubts about the validity of the biradical intermediate arose
when the single crystal x-ray structure of the bis(perfluoro-
phenyl)tetramethyl-2-tetrazenezinc(II) was shown to be the
following (25).

Me₂N\   /N≈N\
      N——N   ···NMe₂
        \ Zn /
         Ar   Ar

The transoid stereochemistry of the tetrazenes seems to be more
stable than the cisoid, as suggested by other x-ray structures of
tetrazenes, carried out in our laboratories (26). It thus became difficult to rationalize the biradical on the basis of the
structure of the final complex. The crucial evidence, which
destroyed the biradical hypothesis, was the result of the following isotope scrambling experiment. A mixture of TMT and TMT-d₁₂
was treated with zinc chloride in refluxing THF in the presence
of excess indene. If the reaction to form the diamine adduct
proceeded via the biradical, the resulting diaminoindane would
have been either fully labelled in methyl groups or not at all.
If, on the other hand, the dimethylamino groups came from different tetrazenes, the product would have contained approximately
half-labelled and half-unlabelled methyl groups. The experiment
showed that the latter case obtained and, therefore, the biradical could not be an intermediate.

TMT
  +   [indene structure]   ZnCl₂
            ────────────────→   THF
TMT-d₁₂

                              N(Me-d₃)₂
                          [indane] N(Me-d₃)₂
                                 +
                              NMe₂
                          [indane] NMe₂

                              N(Me-d₃)₂
                          [indane] NMe₂

Indene was used instead of styrene in this experiment because the
diamine adduct of styrene failed to give a molecular ion in the
mass spectrum.
       The addition of the amino radicals from TMT to styrene cata-
lyzed by zinc chloride is now seen to be a stepwise process.

$$TMT:ZnCl_2 \longrightarrow Me_2\overset{\bullet}{N}:ZnCl_2 + N_2 + \cdot NMe_2$$

$$Me_2\overset{\bullet}{N}:ZnCl_2 + PhCH=CH_2 \longrightarrow Ph\underset{\bullet}{C}H-CH_2-NMe_2$$
$$ZnCl_2$$

$$Ph\underset{\bullet}{C}H-CH_2-NMe_2 + (\cdot NMe_2) \longrightarrow \underset{Me_2N\cdots}{\overset{Ph-CH-CH_2}{\underset{ZnCl_2}{\diagdown NMe_2}}}$$
$$ZnCl_2$$

The product crystallized out of the reaction mixture as the zinc chloride complex. The state of the dimethylamino radical shown in brackets in the last equation is uncertain. It may or may not be coordinated with zinc chloride or it may originate from an S_H2 displacement by styryl radical on the TMT:ZnCl$_2$ complex. Curiously no dimerization or disproportionation of the β-amino-styryl radical was observed. Perhaps the coordination by zinc chloride prevents other reactions from occurring, allowing the combination with the second amino radical to occur reasonably efficiently. The reaction works well only with conjugated alkenes. Only slight reactions were observed with olefins such as 1-octene.

The addition reactions of the zinc chloride-complexed amino radicals, described above were carried out in the absence of oxygen. When oxygen was bubbled through the reaction mixture, the products were not the corresponding diamines but rather amino alcohols were isolated (27). Thus with styrene and α-methylstyrene, the reaction proceeded in the following fashion.

$$TMT:ZnCl_2 + PhCR=CH_2 \xrightarrow[THF]{O_2} Ph\underset{OH}{\overset{|}{C}}R-CH_2NMe_2$$

These products could be accounted for by assuming the addition of an amino radical to the styrene, followed by capture of the intermediate radical by oxygen and the reductive breakdown of the resulting hydroperoxide to the alcohol. A reaction of that type has been discussed by Minisci and Galli (28).

$$PhCR=CH_2 \xrightarrow{\cdot NMe_2} Ph\underset{\bullet}{C}R-CH_2NMe_2 \xrightarrow{O_2} \longrightarrow Ph\underset{OH}{\overset{|}{C}}R-CH_2NMe_2$$

This mechanism, however, does not explain the behavior of the other alkenes studied. Thus indene gave two amino alcohols. Only one of them, trans-2-(dimethylamino)-1-indanol, could

have been formed by the radical addition mechanism above. Likewise, the reaction of cis and trans-β-methylstyrenes gave products which were difficult to rationalize by that mechanism.
Thus, the cis isomer gave exclusively the threo-products, 1-(dimethylamino)-1-phenyl-2-propanol and 2-(dimethylamino)-1-phenyl-
1-propanol, while the trans isomer gave exclusively erythro-1-
(dimethylamino)-1-phenyl-2-propanol. The stereospecificity of
this reaction, as well as the wrong regiospecificity, are sufficient grounds to reject the radical addition mechanism.
    All of these products can be accounted for by the assump-
tion that the initial products of the reaction are the alkene
epoxides, which then react with either dimethylamine or TMT to
form the amino alcohols. Thus, when authentic cis and trans-β-
methylstyrene epoxides were treated with dimethylamine or the
TMT:ZnCl$_2$ complex in THF solution precisely the same amino alcohols were formed as in the TMT:ZnCl$_2$:O$_2$ reaction. The same
was true for the other alkenes studied. The incursion of the
radical reaction, however, can be detected with indene because
the only product formed from indene epoxide and dimethylamine
was trans-1-(dimethylamino)-2-indanol and, hence, the other
regioisomer must have been formed by the radical reaction. The
epoxidation hypothesis was strengthened considerably by the observation that cyclooctene reacted with the TMT:ZnCl$_2$:O$_2$ mixture
to give cyclooctene oxide in 12 percent isolated yield. This
epoxide resists nucleophilic ring opening because the backside
approach to the epoxide function is sterically hindered. In
accord with this, no amino alcohol was formed when cyclooctene
oxide was treated with dimethylamine or with the TMT:ZnCl$_2$
mixture.
    The epoxidation reaction proceeds in the following fashion.

$$TMT:ZnCl_2 \longrightarrow Me_2\overset{\cdot}{N}:ZnCl_2$$

$$Me_2\overset{..}{N}\cdot \ + \ O_2 \ \rightleftharpoons \ Me_2\overset{..}{N}\text{-}O^{\nearrow O\cdot}$$
$$\overset{\mid}{ZnCl_2} \qquad\qquad\qquad \overset{\mid}{ZnCl_2}$$

$$Me_2\overset{..}{N}\text{-}O^{\nearrow O\cdot} \ + \ \rangle C=C\langle \longrightarrow Me_2N\text{-}O\cdot(ZnCl_2) \ + \ \triangle$$
$$\overset{\mid}{ZnCl_2}$$

When the reaction was carried out in the cavity of an esr spectrometer a transient spectrum of a radical was observed, which

could be simulated adequately by using the literature values for the hyperfine coupling constants for dimethylnitroxyl (a_N = 15.2G, a_H = 12.3G). The line-width (3.9G) was broadened, probably by the presence of oxygen. Nitroxyl radicals which have α-hydrogens decay relatively rapidly by disproportionation to the corresponding hydroxylamine and nitrone (29). Our reaction mixtures always contained considerable quantities of dimethylhydroxylamine. The reaction of the dimethylamino radical with oxygen is reversible because no epoxidation was observed when the temperature of the reaction mixture was increased to 100° or when the oxygen supply was restricted. Under these conditions the diaminated products were formed.

The formation of the simplest amino peroxyl radical, NH_2O_2, has been observed in the reaction of potassium ozonide with ammonia under photolytic conditions (30) if Raman spectrum has been determined. This radical has also been postulated as an intermediate in the reaction of NH_2 with oxygen (31).

The concept of a direct oxygen atom transfer from a radical precursor to an alkene to give an epoxide and another radical ought to be a general reaction. Its energetics, however, require that the final product radical be a stable one, at least substantially more stable than the precursor peroxy radical.

$$X-O^{\nearrow O\cdot} \;+\; \underset{\diagdown}{\overset{\diagup}{=}} \;\longrightarrow\; \overset{O}{\diagup\!\diagdown} \;+\; XO\cdot$$

The above conditions are met by the radical formed from nitric oxide and oxygen, the nitrosoperoxyl radical (32). This radical,

$$NO \;+\; O_2 \;\rightleftharpoons\; ON-O^{\nearrow O\cdot}$$

if left alone presumably reacts with another nitric oxide molecule to form two molecules of nitrogen dioxide. In the presence of an alkene however, the ONOO· radical can transfer an oxygen atom to give the epoxide and nitrogen dioxide.

$$ON-O^{\nearrow O\cdot} \;+\; \underset{\diagdown}{\overset{\diagup}{=}} \;\longrightarrow\; NO_2 \;+\; \overset{O}{\diagup\!\diagdown}$$

The reaction is remarkably clean for some alkenes, notably cyclooctene, which is transformed to the epoxide with no side products. Other alkenes, however, give a plethora of products, many containing nitrogen, together with some epoxide. Reactive olefins sometimes undergo rapid reactions but no epoxides are isolated. For example α-methylstyrene is cleaved quantitatively to acetophenone. This is also the fate of α-methylstyrene oxide when that substance is exposed to the NO/O_2 mixture. The cleavage of the styrene epoxide does not make the substance an

obligatory intermediate in the reaction of α-methylstyrene with
NO/O_2, but the general instability of unprotected epoxides in
reaction mixtures containing NO, NO_2 and O_2 makes such a hypothesis reasonable.
An interesting reaction of the NO/O_2 system is with a mixture of α- and β-cedrene. The β-isomer is left intact.

Amino Radicals Substituted by Electron Attracting Sub-
stitutents. The reactivity of these radicals, particularly toward addition reactions, in greatly enhanced, relative to dialkylamino radicals. The photochemically initiated reaction
between N-bromo-bis-trifluoromethylamine and ethylene gave a
95% yield of the addition product (33). The bis-trifluoro-
methylamino radical was the chain carrying species. Although
the authors also reported efficient addition to several poly-
fluoroolefins, no attempt was made to add this interesting
radical to olefins containing allylic hydrogens. Neale and
Marcus have reported good to excellent yields of addition of
N-halocyanamides and N-halosulfonamides to conjugated and non-
conjugated olefins (34). The amidyl radicals generated from N-
chlorourethanes add to unsaturated systems efficiently (35-37).
Both N-monochloro and N,N-dichlorourethanes can be used, although the latter have been studied more frequently. Some competition from allylic abstraction has been observed with the
radicals derived from the N-halourethanes. The additions of
dichlorourethanes are spontaneous but are accelerated by light,
exhibit an induction period, and are retarded by oxygen.

$$Cl_2NCO_2Et + \overset{\shortmid \quad \shortmid}{\diagdown\!\!=\!\!\diagup} \longrightarrow Cl-\underset{\shortmid}{\overset{\shortmid}{C}}-\underset{\underset{Cl}{\shortmid}}{\overset{\shortmid}{C}}-N-CO_2Et \xrightarrow{Na_2S_2O_3} Cl-\underset{\shortmid}{\overset{\shortmid}{C}}-\underset{\shortmid}{\overset{\shortmid}{C}}-NH-CO_2Et$$

In spite of earlier reports to the contrary, the addition of
amidyls derived from N-halocarboxamides has been shown to be an
efficient reaction (38) as long as the temperature of the reaction mixture was kept low. Thus high yields of addition of
N-chloro or N-bromoacetamide to cyclohexene were observed at
-70°. At higher temperatures allylic abstraction predominated.
The addition was also quenched by N-methylation of the amidyl
radical.

Neutral Amino Radicals. It is generally accepted that
dialkylamino radicals add to olefins very reluctantly. This is
a little surprising because thermodynamic calculations, based on
the best thermochemical values (39), indicate that the addition
of NH_2 to ethylene ought to be exothermic by about 17 kcal/mole.
The apparent lack of reactivity suggests therefore that there is
a relatively high activation barrier to this reaction. Indeed,
this was found to be the case in the ab initio SCF-MO-CI calculation of the potential surface of the NH_2 plus ethylene

reaction (40, 41). The best activation energy for that reaction
appears to be in the range of 16-20 kcal/mole (41). However,
Lesclaux and Khe (42) reported that NH_2 radicals in the gas
phase generated from the flash photolysis of ammonia, reacted
with ethylene, propene and 1-butene with an energy of activation
of ~4 kcal/mole, in all three cases. Bamford (43) found, on
the other hand, that dimethylamino radicals gave exclusively
allylic hydrogen abstraction from propene, in the gas phase. No
product studies were carried out by Lesclaux and Khe, and it is
possible that their kinetics might be complicated by allylic abstraction in the propene and butene case. Addition of dimethylamino radicals to ethylene was reported in the gas phase, but
no yields were given (44). Whatever the true value of the
activation energy might be, there is no doubt that amino
radicals do not add to alkenes rapidly.

Dialkylamino radicals are likewise sluggish in H-abstraction
reactions. Nelsen and Heath (45) found that arylmethylamino
radicals, generated from the corresponding diaryldimethyl-2-
tetrazenes, are difficult to scavenge with H-atom donors. These
workers found, however, that the radicals induced the decomposition of the parent tetrazenes, presumably by H-atom abstraction. We (46) also found that induced decomposition was an
important reaction in the thermal decomposition of tetramethyl-
2-tetrazene when the concentration of the tetrazene was greater
than 0.2M. The relative rates of benzylic hydrogen abstraction
from substituted toluenes by the dimethylamino radicals were
correlated by the Hammett equation, using σ^+ values; the ρ-value
of -1.1 indicated that the radical is moderately electrophilic
in abstraction reactions (46). The deuterium isotope effect for
that reaction was found to be 4.0. Interestingly, the ρ-value
for H-atom abstraction from toluenes by a protonated amino
radical, the piperidinium radical, was found to be -1.34 (47).
Thus it appears that protonation does little to change the
electrophilicity of the radical in abstraction reactions.

Although neutral dialkylamino radicals generally do not
add to simple alkenes it was possible to observe some addition
of dimethylamino radicals to styrene and α-methylstyrene (48).
Photolysis of tetramethyl-2-tetrazene (TMT) in cyclohexane, at
room temperature, in the presence of α-methylstyrene resulted in
the formation of small but reproducible yields of addition products. The bulk of the amino radicals were consumed in the formation of tetramethylhydrazine and dimethylamine. The relative
rate of formation of the adducts from substituted α-methylstyrenes
were correlated by the Hammett equation yielding a ρ = +0.69 $\pm$
0.03 (corr. coeff. .99). This result indicates that the dimethylamino radical behaves like a nucleophile in addition reactions i.e. the styrenes substituted by electron attracting
substituents reacted more rapidly than the electron rich styrenes. This result is in marked contrast to that obtained for
dimethylamino radicals coordinated by zinc chloride (48). These

radicals generated from the TMT:ZnCl$_2$ complex (see above) added
reasonably efficiently to α-methylstyrenes. The Hammett correlation yielded a ρ value of -0.98 ± .04 (corr. coeff. .99).
This cross-over from nucleophilicity to electrophilicity illustrates very graphically how the lone pair on nitrogen influences amino radical reactivity.

The intramolecular cyclization of amino radicals, where
the nitrogen is not coordinated by a metal or the proton, has
not been studied extensively. Surzur, Stella and Tordo (49)
reported the photolysis of N-chloro-N-(4-pentenyl)-N-propyl-
amine in various neutral solvents. The data were accounted for
by the following scheme.

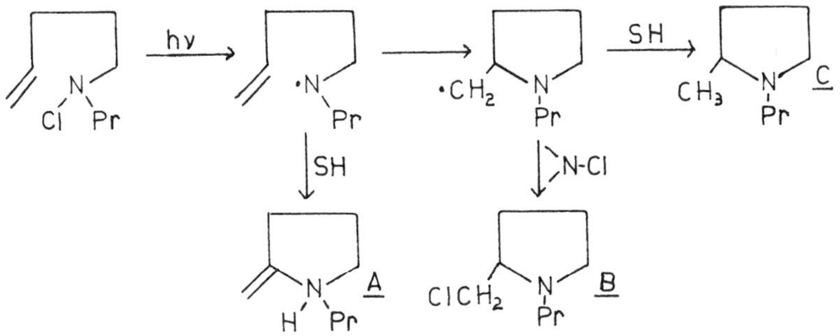

The photolysis was carried out in acetic acid/water, in methanol
and in isopropyl alcohol. In the first solvent B was formed exclusively, but in the alcohols, mixtures of A, B, and C were
obtained. While some of the details of this reaction are not
clear, the evidence for the intramolecular addition of the
neutral amino radical seems to be strong.

We (50) prepared the following tetrazenes in an effort to
provide unequivocal sources of the alkenyl-substituted amino
radicals.

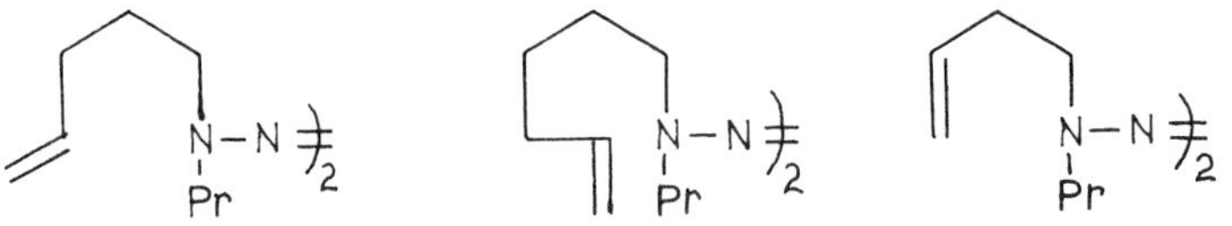

The pentenyl tetrazene D gave on photolysis or thermolysis the
same radical as that reported by Surzur and co-workers. The
tetrazene E led to a radical which could cyclize to a 6 or a
7-membered ring, while the radical from F had only the possibility of cyclization to a 5-membered ring (a 4-membered ring
being unlikely).

Photolysis of the tetrazene D through Pyrex in cyclohexane

gave a large number of products. However, the cyclization products were formed in reasonable yields. It is interesting to

$$\underline{D} \xrightarrow{h\nu} \underset{\underset{\underset{Pr}{|}}{N}}{\bigcirc} + \underset{\underset{\underset{Pr}{|}}{N}}{\bigcirc} + \text{Open chain products}$$

G 19%   H 34%

note that photolysis of D at room temperature leads to almost twice as much of the N-propylpiperidine H as the pyrrolidine G. Surzur et al (49) did not observe any H in their system. Interestingly, however, the thermolysis of D at 143° in cyclohexane (reaction carried out in sealed tubes) led to only 16% of H but 41% of G.

$$\underline{D} \xrightarrow[\bigcirc]{143°} \underline{G} + \underline{H} + \text{Open chain products}$$
41%   16%

Thus the ratio of the cyclic products was inverted at the two temperatures. The reason for this inversion is not at all clear. One possible explanation, however, might be that at higher temperatures the reaction is less selective, i.e. the transition state is less bound, and hence the kinetically favored product G is formed predominatly. At the lower temperature of the photochemically generated radical, the reaction is more selective, the transition state is more product-like and hence the 6-membered ring product H is formed. This argument is really a loose interpretation of the observations in terms of the reactivity-selectivity principle (51).

Thermolysis of the hexenyl tetrazene E in cyclohexane gave the following cyclic products.

$$\underline{E} \xrightarrow[\bigcirc]{143°} \underset{\underset{\underset{Pr}{|}}{N}}{\bigcirc} + \underset{\underset{\underset{Pr}{|}}{N}}{\bigcirc} + \text{Open chain products}$$
3.3%   1.3%

The photolysis gave no detectable cyclized materials. The open chain products were either monomers or dimers of the intermediate N-propyl-N-(5-hexenyl)amino radical, presumably by the Hofmann-Loeffler type of rearrangement.

These various rearranged dimers corresponded to approximately
65 percent of the product. The hydrazine, the product of the
N-N-coupling of the first formed radical, was isolated in less
than one percent yields. Thus, in the hexenyl radical case the
cyclization does not compete well with rearrangement.

No cyclic products were obtained from either the photolysis
or the thermolysis of the butenyl tetrazene, F. The principal
isolated product in the photochemical reaction was the parent
amine, N-propyl-N-(3-butenyl)amine, formed in about 50 percent
yield. In the thermolysis reaction the principal reaction was
the β-scission.

The allyl radical then entered into a variety of cross-combi-
nation reactions. The β-scission of amino radicals had been
observed previously in the case of the methyl-(2-phenylethyl)-
aminium radical (52).

The preferred transoid geometry of tetraalkyl-2-tetrazenes
is in common with other azo compounds. In this case, however,
the cisoid form of the acylic tetrazenes has never been isolated.
In fact Roberts and Ingold (53) suggested that the photochemical decomposition of tetrazenes involved the photochemical
trans to cis isomerization followed by the thermal loss of
nitrogen from the cis-isomer. Some acylated cis-tetrazenes
have been reported (54,55), but the chemical behavior of these
substances differs substantially from the tetraalkyl analogs.

Nelsen and Fibiger (56) prepared the first authentic cis-
tetrazene, 1,4-dimethyl-1,4,5,6-hexahydro-1,2,3,4-tetrazine.

The cis-cyclic tetrazene was formed in 5 percent yield by basic
oxidation of the precursor dihydrazine (R=Me) with sodium
hypochlorite. A general synthesis of tetrazenes of this type
was worked out by Seebach and co-workers (57,58,59). The

method involves the low temperature self-condensation of the lithium derivatives of nitrosamines to give the corresponding tetrazene N-oxides. The oxygen is then readily removed by treatment with trimethylphosphite.

Nelsen and Fibiger studied the thermal decomposition of the cis-tetrazene in tetralin over the temperature range of 130 to $\overline{144}°$. The activation parameters, $\Delta H^{\ddagger}$ = 38 kcal/mole and $\Delta S^{\ddagger}$ = 17 eu, obtained from the kinetic data, were surprisingly high. The activation parameters for tetramethyl-2-tetrazene, for comparison, are $\Delta H^{\ddagger}$ = 36.1 kcal/mole and $\Delta S^{\ddagger}$ + 4.7 eu. These were obtained from a gas phase study (60), but indications are that they do not change significantly in solution (56). This unexpected result was explained by Nelsen on the basis of a conformational effect, i.e. that restricted movement of the cis-cyclic structure prevents the attainment of the most favorable conformation which leads to reaction. Thus, the six-membered ring cis-tetrazene is to be considered as a special case, its stability being the product of the conformationally restricted structure.

It was of interest to prepare and study the next higher homolog. Kreher and Wissman (61) reported the preparation of several 1,4-diaryl- and 1,4-dialkyl-1,4,5,6-tetrahydro-D-tetrazines by oxidation of the corresponding hydrazines.

$$R = C_6H_5, \text{p-}MeC_6H_4, C_6H_{11}$$

They also reported unsuccessful attempts to prepare the higher homologs. We repeated this procedure using N,N'-dimethyl-N,N'-diaminopropylenediamine as the starting hydrazine. Careful, low temperature isolation of the product yielded the corresponding seven membered ring cis-tetrazene in up to 50 percent yield.

The thermolysis of this tetrazene (1,4-dimethyl-1,2,3,4-tetraaza-2-cycloheptene) indicated that the material behaved as

expected for the cis-structure. The rates were measured at
five degree intervals over the temperature range of 65-90°. The
Arhennius parameters calculated from the kinetic data were,
$\Delta H^{\ddagger} = 19 \pm 1$ kcal/mole, $\Delta S^{\ddagger} = -10.5 \pm 1$ eu. Based on differences
in the activation enthalpies between trans and cis azoalkanes,
Nelsen (56) predicted that the difference between the cis and
trans tetrazene activation energies ought to be on the order of
6-8 kcal/mole. It is apparent now that the difference is approximately double of that seen in azoalkanes. The primary
reason that our tetrazene could be isolated at all is that the
entropy of activation is quite negative, suggesting that the
rather loose seven-membered ring must become more rigid in the
transition state. It is therefore entirely possible that
Ingold and Roberts may have been correct in their assumption
that the photochemical decomposition of acyclic tetrazenes
proceeds by photoisomerization to the cis form and the thermal
decomposition of that isomer to the amino radicals.

The products of the decomposition the tetraazacycloheptene
are the five-membered ring hydrazine plus other, as yet un-
identified, materials. When the decomposition was carried out
in a large excess of α-methylstyrene some of the initial biradical was trapped as a radical addition product.

Acknowledgement: We are grateful to NSF for support through
grants MPS 7411792 and CHE 76-24095 and to Dr. H. E. Baumgarten
for his help.

Literature Cited

1. Nelsen, S. F., in "Free Radicals," Kochi, J. K. (ed.), Vol. 2,
    pp. 527-593, John Wiley and Sons, Inc., New York, 1973.
2. Forrester, A. R., in "MTP International Review of Science,
    Organic Chemistry Series Two, Free Radicals," Vol. 10,
    Waters, W. A. (ed.), pp. 87-131, Butterworths, London, 1975.
3. Fessenden, R. W. and Neta, P., J. Phys. Chem., (1972), 76,
    2857.
4. Albissetti, C. J., Coffman, D. D., Hoover, F. W., Jenner,
    E. L., and Mochel, J. Am. Chem. Soc., (1959), 81, 1489.
5. Neale, R. S. and Whipple, E. B., ibid., (1964) 86, 3130.
6. Neale, R. S., Tetrahedron Lett., (1966) 483.
7. Neale, R. S., J. Org. Chem., (1967) 32, 3273.
8. Neale, R. S. and Marcus, N. L., ibid., (1967) 32, 3273.
9. Neale, R. S. and Marcus, N. L., ibid., (1968) 33, 3457.

10. Neale, R. S., Synthesis, (1971) 1.
11. Hobson, J. D. and Riddell, W. D., J. Chem. Soc., Chem. Commun., (1968), 1178.
12. Heusler, K., Tetrahedron Lett., (1970) 97.
13. Chow, Y. L., Accounts Chem. Res., (1973) 6, 354.
14. Chow, Y. L., Perry, R. A., Menou, B. C., and Chen, S. C., Tetrahedron Lett., (1971), 1545.
15. Chow, Y. L., Perry, R. A., and Menou, B. C., ibid (1971), 1549.
16. Minisci, F., Synthesis, (1973) 1.
17. Minisci, F., Galli, R. and Bernardi, R., Chim. Ind. (Milano), (1967) 49, 252.
18. Minisci, F. and Galli, R., ibid, (1963), 45, 1400.
19. Minisci, F. and Galli, R., ibid, (1964), 46, 546.
20. Minisci, F., Galli, R., and Pollina, G., ibid, (1965), 47, 736.
21. Minisci, F., Galli, R., and Cecere, M., ibid, (1966) 48, 347.
22. Noltes, J. G., and van den Hurk, J. W. G., J. Organometallic Chem., (1964), 1, 377.
23. Michejda, C. J., and Campbell, D. H., J. Am. Chem. Soc., (1974), 96, 929.
24. Campbell, D. H., Ph.D. Thesis, University of Nebraska, 1975.
25. Day, V. W., Campbell, D. H., and Michejda, C. J., J. Chem. Soc., Chem. Commun., (1975), 118.
26. Day, V. W., Day, R. O., and Michejda, C. J., submitted for publication.
27. Michejda, C. J., and Campbell, D. H., J. Am. Chem. Soc., (1976), 98, 6728.
28. Minisci, F., and Galli, R., Tetrahedron Lett., (1964)., 3197.
29. Adamic, K., Bowman, D. F., Gillian, J., and Ingold, K. U., J. Am. Chem. Soc., (1971), 93, 902.
30. Giguère, R. A., and Herman, K., Chem. Phys. Lett., (1976), 44, 273.
31. Jayanty, R. K. M., Simonaitis, R., and Heicklen, J., J. Phys. Chem., (1976), 80, 433.
32. Guillory, W. A., and Johnston, H. S., J. Chem. Phys., (1965), 42, 2457.
33. Haszeldine, R. N. and Tipping, A. E., J. Chem. Soc., (1965) 6141.
34. Neale, R. S. and Marcus, N. L., J. Org. Chem., (1969) 34, 1808.
35. Foglia, T. A. and Swern, D., ibid (1966) 31, 3625.
36. Foglia, T. A. and Swern, D., ibid, (1967) 32, 75.
37. Foglia, T. A. and Swern, D., ibid, (1968), 33, 766.
38. Touchard, D. and Lessard, J., Tetrahedron Lett., (1973) 3827.
39. Benson, S. W., J. Chem. Ed., (1965) 42, 502.
40. Shih, S., Buenker, R. J., Peyerimhoff, S. D. and Michejda, C. J., J. Am. Chem. Soc., (1972), 94, 7620.
41. Peyerimhoff, S. D. and Buenker, R. J., Ber. Bunsenges. Phys. Chem., (1974), 78, 119.

42. Lesclaux, R., and Khe, P. V., in "Extended Abstracts of the 12th Informal Conference on Photochemistry," (1976) pp. P3-1 - P3-3, National Bureau of Standards, Gaithersburg, MD.

43. Bumford, C. H., J. Chem. Soc., (1939), 17.

44. Good, A. and Thynne, J. C. J., J. Chem. Soc. (B), (1967), 684.

45. Nelsen, S. F. and Heath, D. H., J. Am. Chem. Soc., (1969), 91, 6452.

46. Michjeda, C. J. and Hoss, W. P., ibid, (1970) 92, 6298.

47. Neale, R. S. and Gross, E., ibid, (1967) 89, 6579.

48. Michejda, C. J. and Campbell, D. H., Tetrahedron Lett., (1977) 577.

49. Surzur, J. M., Stella, L. and Tordo, P., ibid, (1970) 3107.

50. Michejda, C. J. and Sieh, D. H., to be published.

51. Giese, B., Angew. Chemie, Internat. Ed. English, (1977) 16, 125.

52. Minisci, F. and Galli, R., Tetrahedron Lett., (1966) 2532.

53. Roberts, J. R., and Ingold, K. V., J. Am. Chem. Soc., (1973), 95, 3228.

54. Forgione, P. S., Sprague, G. S., and Troffkin, H. J., ibid, (1966), 88, 1079.

55. Jones, D. W., J. Chem. Soc., Chem. Commun., (1970), 1084.

56. Nelsen, S. F., and Fibiger, R., J. Am. Chem. Soc., (1972), 94, 8497.

57. Seebach, D., and Enders, D., Angew. Chemie, Internat. Edit., (1972), 11, 301.

58. Seebach, D., Enders, D., Ranger, B., and Brügel, W., ibid, (1973), 12, 495.

59. Seebach, D., and Enders, D., ibid, (1975), 14, 15.

60. Gowenlock, B. G., Jones, P. P., and Major, J. R., Trans. Faraday Soc., (1961), 57, 23.

61. Kreher, R., and Wissman, H., Chem. Ber., (1973), 106, 3097.

RECEIVED December 23, 1977.

Early Intermediates in Hydrazine Oxidations: Hydrazine Cation Radicals, Hydrazyls, and Diazenium Cations

STEPHEN F. NELSEN

Department of Chemistry, University of Wisconsin, Madison, WI 53706

We have put considerable effort over the past several years into study of the early intermediates expected in hydrazine (I) oxidation. As indicated in Scheme 1, electron and proton loss should alternate in the presence of a one electron oxidant, giving hydrazine radical cations (II), hydrazyl radicals (III), and trisubstituted diazenium cations (IV). Other types

Scheme 1

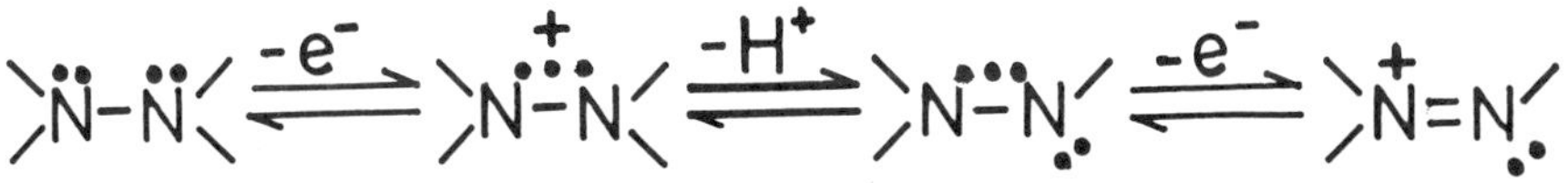

I II III IV

of oxidants would bypass some of these intermediates; a hydride abstraction would convert I directly to IV, and hydrogen atom abstraction would give III. Under most oxidation conditions, II-IV do not build up to observable concentrations. IV deprotonates easily if one of the substituents is hydrogen, giving either azo compounds (often isolable), or 1,1-disubstituted diazeniums (N-aminonitrenes) which have a complex chemistry of their own (1). Our interest in the intermediates of Scheme 1 was sparked by the great conformational charge which must occur between the four unshared electron, two center hydrazine I, well known to be approximately tetrahedral at nitrogen and prefer a gauche orientation of the lone pairs (2), and the diazenium cation IV, which should have sp^2 hybridized, pi bonded nitrogens. We thought the great conformational charge which must occur would make the electron transfer steps unusual, and wished to probe the geometric consequences of the three electron, two center bonding shown for II and III.

© 0-8412-0421-7/78/47-069-309$05.00/0

Analogous bonding occurs in several important classes
of free radicals, including nitroxides, peroxy radicals,
ketone radical anions, and superoxide, all of which
also have a spin-bearing center adjacent to an atom
bearing an unshared pair of electrons.
There had been considerable study of variously
substituted examples of II-IV before our work began.
The most studied example of II was tetraphenylhydrazine
cation radical, whose purple color led Hünig to coin
the term "violenes" (3a) for the series of vinologous
species his group has studied extensively (3). The
solution ESR spectrum of hydrazine radical cation had
been recorded in a flow system (4). Aromatic examples
of III had received extensive study (5), and diphenyl-
picrylhydrazyl was the first example of a neutral
radical stable both to dimerization and reaction with
oxygen. The chemistry of 1,1-dimethyldiazenium cation
had been studied (6), and interesting electrochemical
work on arylhydrazine oxidations had been carried out
by Cauquis and Genies (7).

The Hydrazyine,Hydrazine Radical Cation Redox Equili-
brium

Our work on the first electron transfer, I $\rightleftharpoons$ II,
became possible when it was realized that tetraalkylhydrazine radical cations have a reasonably long lifetime in solution, even at room temperature. This discovery was made completely by accident, when tetramethylhydrazine radical cation, $1^{\ddagger}$ was observed by ESR

$$CH_3 \diagdown \overset{+}{\underset{\displaystyle N}{\cdots}} \diagup CH_3$$
$$CH_3 \diagup N \diagdown CH_3$$

$$1^{\ddagger}$$

during attempted study of tetramethyl-2-tetrazine
maleic anhydride charge-transfer complexes (8a), and
nearly simultaneously by Michejda and coworkers (8b),
who had methylated the tetrazene, and observed $1^{\ddagger}$ as a
decomposition product. The long lifetime of tetraalkyl-
II cation radicals made possible measurements of the
relative free energy differences between I and II by
electrochemical determination of E^{o}, the standard potential for the redox equilibrium (9), and also allowed
convenient ESR study of II with a variety of structural
constraints (10).

The four point substitution of hydrazines provides a unique opportunity to control the geometry at the N-N bond of I, and thus allows placing structural constraints upon II which are not possible with other three electron, two center species. The highest occupied molecular orbitals of a hydrazine are expected to be the symmetric and antisymmetric lone pair combination orbitals, n_+ and n_-. Since the energy separa-

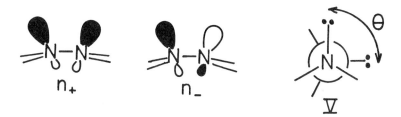

tion between n_+ and n_- should be sensitive to the lone pair,lone pair dihedral angle θ (see V), one would predict that the energy of the highest occupied molecular orbital (homo) for a hydrazine would be sensitive to θ. It was well established that solution redox potentials correlate linearly with the energy of the homo for several aromatic cases (11), and a surprisingly good linear correlation of solution oxidation potential with vapor phase ionization potential was found by Miller and coworkers (12) for a very wide range of structural types of compounds. Photoelectron spectroscopy studies by our group (13) and that of Rademacher (14) have verified the expectation of a great sensitivity of ΔE^O to θ; $\Delta E = 2.3$ eV (53 kcal/mole) for hydrazines with θ near 0° and 180° (13), and the minimum ΔE is near 0.5 eV (11.5 kcal/mole) for gauche, acyclic hydrazines (9b). The photoelectron spectrum for hydrazines provides a convenient method for determining the approximate θ for a hydrazine in the vapor phase, and the short pe timescale allows observation of conformational mixtures in compounds which interconvert too rapidly for investigation by NMR techniques (13e). Nevertheless, E^O is not very sensitive to θ, partially because of compensating energy changes in lower-lying orbitals. An excellent example is provided by the ee and ae conformations of dimethyl hexahydropyridazine 2. Although ΔE differs by a substantial amount, 1.3 eV (30 kcal/mole) because of the large difference in θ, 2ee and 2ae only differ in free energy by about 0.2 kcal/mole (15), requiring that their E^O values are virtually identical, since both give the same cation, 2^+. Their first ionization potentials are also

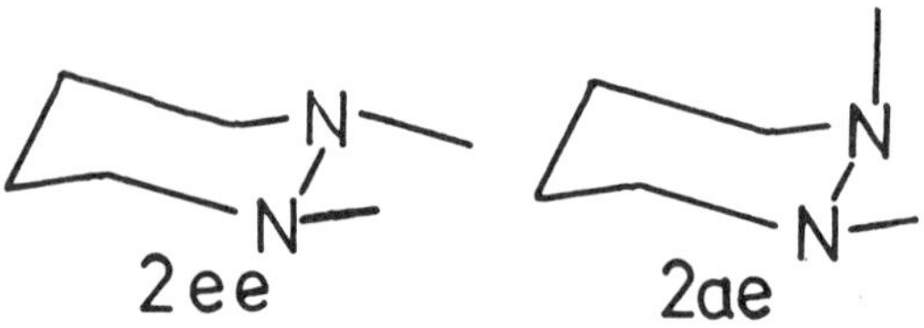

2ee   2ae

virtually the same ($\underline{13},\underline{14}$), showing that the average of
the n_+ and n_- energies is not constant as θ is changed.
    Comparison of E^O values for various tetraalkyl-
hydrazines shows that there is a great increase in
$\underline{R}N,N\underline{R}$ steric interaction in the cation relative to the
neutral species and that six-ring hydrazines have sub-
stantially higher E^O values than five- and seven-ring
compounds, both of which indicate considerable flatten-
ing at the nitrogens of the cation radical ($\underline{9b}$). ESR
studies of the nitrogen splitting constant have demonstrated that although the nitrogens of hydrazine radical cations are considerably flattened compared to
those of the neutral compounds, the equilibrium geometry is not planar for some cases in which planarity
would destabilize cyclic substituents. The double

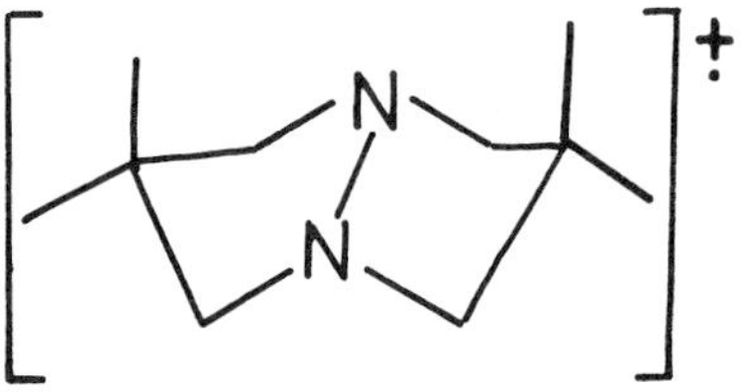

$3^{\ddot{+}}$

nitrogen inversion barrier of $\underset{\sim}{3}^{\ddot{+}}$ is only about 3.4 kcal/
mole ($\underline{10c}$), considerably less than the 12 kcal/mole of
the neutral hydrazine. If relatively unstrained tetraalkylhydrazine cation radicals are not planar at nitrogen in their equilibrium geometry, the barrier to
double nitrogen inversion is quite low. Tetraalkylhydrazine cation radicals clearly are more easily bent
than hydrazine radical cation ($\underline{10c}$); even the lower
nitrogen splitting constant examples such as $\underset{\sim}{1}^{\ddot{+}}$ (a(N) =
13.4 gauss) have larger nitrogen splittings than does
hydrazine radical cation (a(N) = 11.5 gauss ($\underline{4}$)) despite the fact that hyperconjugation to the alkyl
groups must decrease the total spin density at nitrogen.

The ease of twist at the three electron pi bond of II
is being studied by comparing E^O values for 4-6 (16).
Although E^O was found to be the same for 4 and 5, it is
0.38 V (9 kcal/mole) higher for 6 than for 4. We
attribute the lack of detected strain in 5$^+$ to the ease
of bend at the nitrogens, which decreases the amount of
torsional strain applied by the bicyclic system, and a

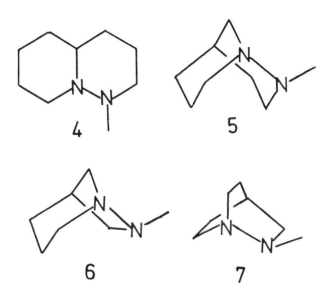

smaller barrier to twisting in a three electron than a
two electron pi system. The corresponding olefin has
been found to have about 12 kcal/mole of strain relative to 3-ethyl-3-hexene by Turner and coworkers (17).
The strain apparent in 6$^+$ demonstrates that substantial
torsional twist of the three electron pi bond is difficult. For 7, where even greater twist is required, the
cation is so destabilized that it is short-lived, and a
thermodynamic E^O value is not currently available.
    In contrast to the small effect of θ on E^O, the
rate of electron transfer is very sensitive to θ, as
has been demonstrated by low temperature cyclic voltammetry experiments. Electron transfer from axial,
equa   al (θ ~ 60°) conformations of six ring hydrazines is far slower than from diequatorial (θ ~ 180°)
conformations, resulting in a kinetic resolution of the
oxidation peaks for these two types of conformations
(18). This phenomenon allows the use of low temperature cyclic voltammetry to determine both equilibrium
and rate constants for conformational change in these

compounds (19).

Many tetraalkylhydrazine radical cations last for
hours to days in solution, but attempts to isolate them
only led to their rapid decomposition. Since the hydrazine dications were known to be very short-lived from
cyclic voltammetry experiments (only an irreversible
second oxidation wave was observed) we thought the problem in isolation was caused by electron transfer disproportionation, followed by very rapid proton transfer. We found that 8, which has Bredt's rule destabilization of the transition state for α-deprotonation,

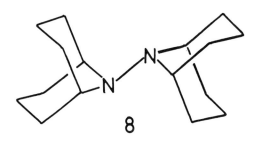

8

gives a dication which is long-lived on the cyclic
voltammetry timescale, and shows E^O values for mono-
and dication formation of -0.01 and +1.18 V vs. sce
(20). These E^O values require a minute value for the
electron transfer disproportion constant K_d =
$[8^{+2}][8]/[8^{\ddagger}]^2$ of 1 x 10^{-20}. It is not surprising that
$8^{\ddagger}$ has the smallest K_d value yet observed for cation
radical, because the small, two atom pi system involved
will maximize electrostatic effects. We find that $8^{\ddagger}$
$PF_6^{\ominus}$ is indefinitely storable, both in solution and as
the solid (20). Preliminary X-ray diffraction data
for $8^{\ddagger}$ $PF_6^{\ominus}$ (21) indicate that the nitrogens are coplanar with their four α-carbons (for neutral 8, the
N-N bond forms an angle of 58° with the CNC plane),
and the N-N distance is surprisingly short 1.28 A°
(neutral 8 has a 1.51 A° N-N distance, unusually long
for a hydrazine (22)).

The Hydrazyl, Diazenium Salt Redox Equilibrium

        Several groups reported esr studies on non-
aromatic hydrazyl radicals during the period 1972-1974
(23), the most detailed work being that of Ingold's
group. Our approach to the problem of studying these
compounds was to devise trialkylhydrazyls which would
have conveniently long lifetimes. Since extensive work

on the isoelectronic nitroxides had shown that α-
hydrogen atom transfer disproportionation is the irreversible decomposition pathway, and that Bredt's rule
effects were sufficient to make norpseudopelletierine-
N-oxyl (24) long-lived, we selected the tert-butyl-
bicyclo[2.2.1]heptyl system 9 as the most easily available long-lived hydrazyl. Two routes were developed
to generate 9 from a bicyclic azo compound 12, as
shown in Scheme 2. Although 9 is easily formed by

Scheme 2.

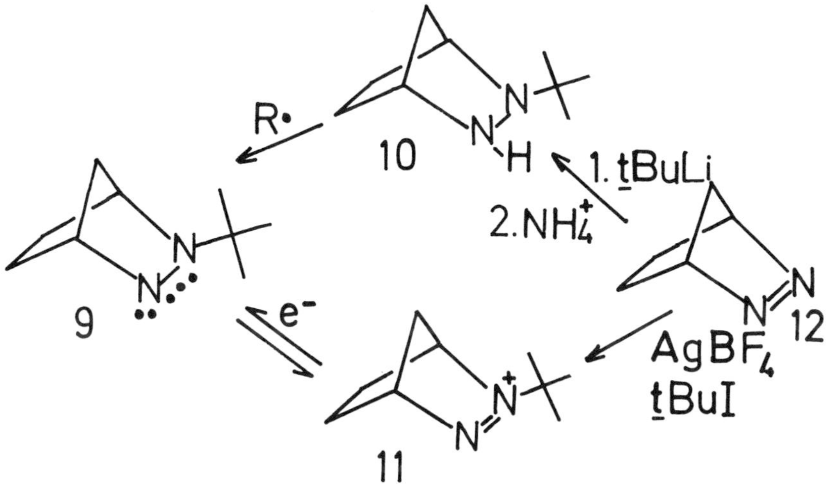

hydrogen abstraction from the trialkylhydrazine 10, the
great sensitivity of 10 to air makes it an inconvenient
precursor. In contrast, the diazenium salt 11 is an
air stable solid which is easy to handle, and 9 is conveniently prepared by electrochemical reduction of 11.
There is no problem with over-reduction of 9 to the
hydrazine anion; indeed, E° for this process is more
than 1.8 V negative of the 9,11 E° in acetonitrile (25),
and 1.5 V more negative in dimethylformamide (26)
facts which require that the hydrazine anion is remarkably basic, presumably because of lone pair-lone pair
interactions. The electron transfer disproportionation
constant is even lower for 9 than for 8†. Preparation

of the 2-methyl and 2-ethyl analogues of 11 was carried
out, and these compounds give only irreversible reduction waves (26), demonstrating that α-hydrogen transfer
is in fact rapid when abstractable α-hydrogens are present in trialkylhydrazyl radicals. The E^O for the 9,11
couple is -0.72 V vs. SCE in acetonitrile, 0.89 V (20
kcal/mole) negative of the 13,13⁺ couple E^O. Removal

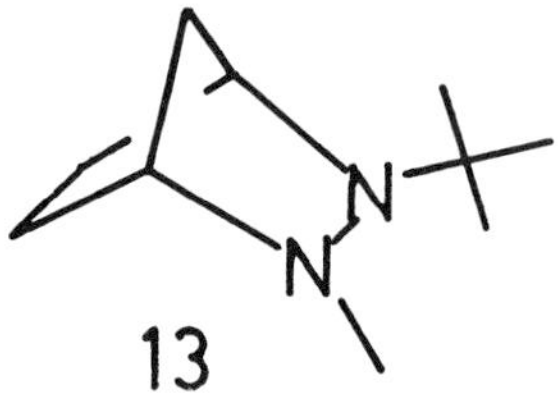

13

of the third pi electron from 9 is thus considerably
more fascile than generation of a three pi electron
system by electron removal from 13. The ESR spectrum
of 9 has nitrogen splittings of 11.03 and 10.25 gauss
(26a), indicating nearly equal spin densities at the
nitrogens. No evidence for dimerization was observed,
even at low temperature, but admission of air to solutions of hydrazyl 9 results in immediate destruction
of its ESR signal, and appearance of the hydrazyloxy
radical 14. Several other hydrazyl radicals also have
been shown to give hydrazyloxy radicals in the presence
of air (27). In the absence of oxygen, however, 9 decomposes in several hours to give 10 and a ca. 10:1
mixture of 15 and 16. This process seems best described as "homodisproportionation", since a β-hydrogen
is removed to give 15 and 16. The β-scission process,
which is rather rapid for 16 (23h), does not occur at
all rapidly for N-tert-butyl bicyclic systems, presumably because the bicyclic system enforces poor

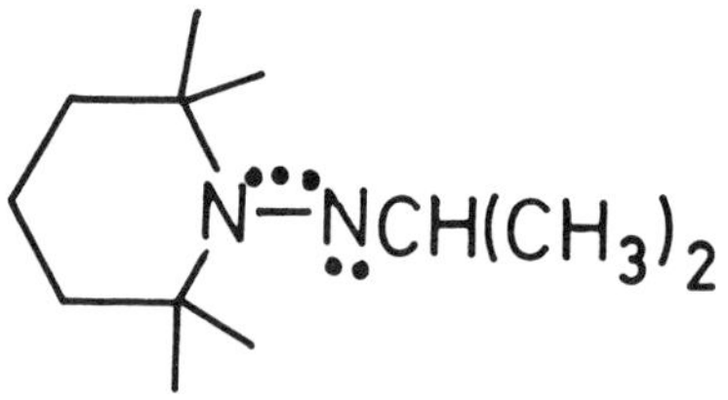

17 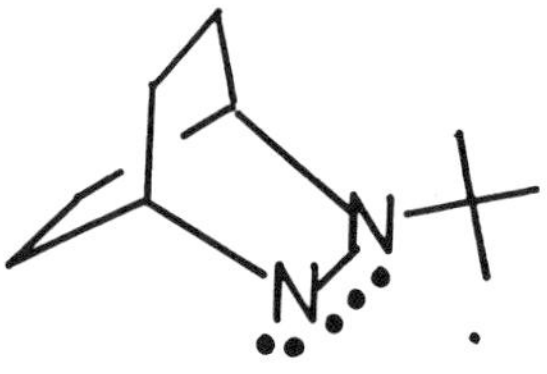   **18**

Scheme 3.

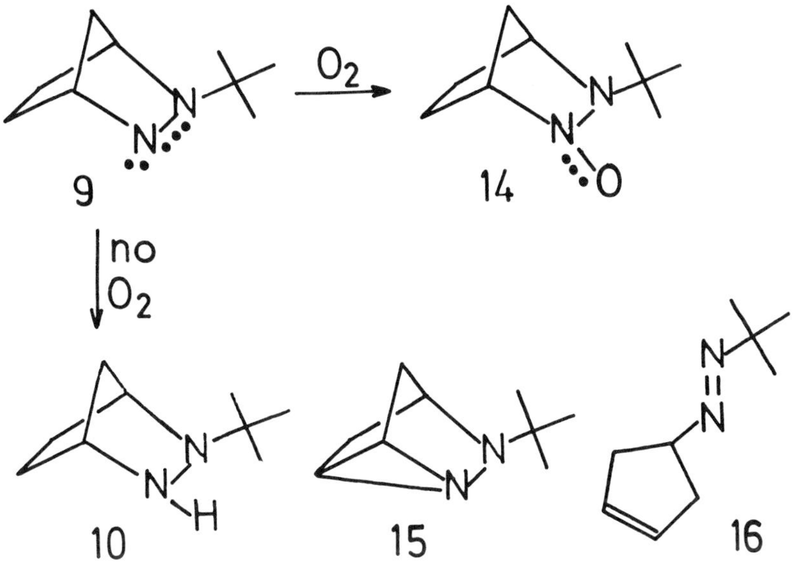

orbital allignment for loss of a tertiary radical.
Surprisingly, 18, the bicyclo[2.2.2]octyl analogue of
9, is far longer lived, having a lifetime in solution
of months instead of hours (26).

Air oxidation of 10 in cyclohexene gives remarkably high yields (>95% (28)) of a ca. 10:1 mixture of
15 and 16 (29). Obtaining the same oxidation products
in similar ratio in the autoxidation reaction as in the
self-decomposition of 9 suggests that 9 is the precursor of 15 and 16, although it is not clear why 9 is
not intercepted by oxygen under these conditions. In
acetonitrile, although 15 and 16 are still found in low
yield, the principal product is the diazenium salt 11,
with unknown counterion. Much remains to be learned
about this unusual autoxidation reaction (30).

References

1. For a review see, Lemal, D. M., in "Nitrenes", W.
   Lwowski, Ed., Interscience, 1970, pp. 345-403,
   New York.
2. For a good review of hydrazine conformational work
   through 1973, see Shvo, Y., in "The Chemistry of
   Hydrazo, Azo, and Azoxy Groups, 1975, Pt. 2.,
   pp. 1017-1095, St. Patai, Ed., J. Wiley and Sons,
   New York.
3. a) Hünig, S. (1964) Ann. 476, 32. b) Hünig, S.
   (1967) Pure Appl. Chem. 15, 109. c) Hünig S.,
   and Schilling, P. (1976) Ann., 1039, and previous
   papers in the series.
4. Adams, J. Q, and Thomas, J. R. (1963) J. Chem. Phys.
   39, 1904.
5. For a review, see Forrester, A. R., Hay, J.M., and
   Thomson, H. R. (1968) "Stable Free Radicals",
   Academic Press, New York, NY, pp. 137-166.
6. a) McBride, W. R., and Korse, H. W. ((1957) J. Am.
   Chem. Soc. 79, 572. b) Urry, W. H., Szecsi, P,
   Ikoku, C., Moore, D. M. (1964) ibid. 86, 2224.
   c) Cauquis, G., and Genies (1971) Tetrahedron Lett.
   4677. d) Urry, W. H., Gabriel, Z. L. F., Duggan,
   J. C., and Tseng, S. S. (1973) J. Am. Chem. Soc. 95,
   4338.
7. Cauquis, G., and Genies, M. (1970) Tetrahedron Lett.,
   2903, 3405.
8. a) Nelsen, S. F. (1966) J. Am. Chem. Soc. 88, 566.
   b) Brunning, W. H., Michejda, C. J., and Romans, D.
   (1967) Chem. Commun., 11.
9. a) Nelsen, S. F., and Hintz, P. J. (1972) J. Am.
   Chem. Soc. 94, 7108. b) Nelsen, S. F., Peacock, V.
   and Weisman, G. R. (1976) ibid. 98, 5269.
10. a) Nelsen, S. F., and Hintz, P. J. (1970) J. Am.
    Chem. Soc. 92, 6215. b) Nelsen, S. F., and Hintz,
    P. J. (1971) ibid. 93, 7104. c) Nelsen, S. F.,
    Weisman, G. R., Hintz, P. J., Olp, D., and Fahey,
    M. R. (1974) ibid. 96, 2916. d) Nelsen, S. F., and
    Echegoyen, L. (1975) ibid. 97, 4930.
11. Streitwieser, A., Jr. (1961) "Molecular Orbital
    Theory for Organic Chemists", John Wiley and Sons,
    Inc., New York, NY, pp. 173-199.
12. Miller, L. L., Nordblom, G. D., and Mayeda, E. A.
    (1972) J. Org. Chem. 37, 916.
13. a) Nelsen, S.F. and Buschek, J. M. (1973) J. Am.
    Chem. Soc. 95, 2011. b) Nelsen, S. F., Buschek, J.
    M., and Hintz, P. J. (1973) ibid. 95, 2013.
    c) Nelsen, S. F., and Buschek, J. M. (1974) ibid.
    95, 2392. d) Nelsen, S. F., and Buschek (1974)
    ibid. 96, 6982. e) Nelsen, S. F., and Buschek

(1974) ibid. 96, 6987.
14. a) Rademacher, P. (1973) Angew. Chem. 85, 410.
    b) Rademacher, P. (1974) Tetrahedron Lett., 83.
    c) Rademacher, P. (1975) Chem. Ber. 108, 1548.
    d) Rademacher, P., and Koopman, H. (1975) ibid.
    108, 1557.
15. Nelsen, S. F., and Weisman, G. R. (1976) J. Am.
    Chem. Soc. 98, 3281.
16. Kessel, C. R., unpublished work.
17. Lesko, R. M., and Turner, R. B. (1968) J. Am. Chem.
    Soc. 90, 6888.
18. Nelsen, S. F., Echegoyan, L., and Evans, D. H.
    (1975) J. Am. Chem. Soc. 97, 3530.
19. a) Nelsen, S. F., Echegoyan, L., Clennan, E. L.,
    Evans, D. H., and Corrigan, D. A. (1977) J. Am.
    Chem. Soc. 99, 1130. b) Evans, D. H., and Nelsen,
    S. F. (in press), in "Characterization of Solutes
    in Non-Aqueous Solvents", Mamantov, G., Ed.,
    Plenum Press, New York, NY.
20. Nelsen, S. F., and Kessel, C. R. (1977) J. Am. Chem
    Soc. 99, 2392.
21. Hollinsed, W. C., and Calabrese, J. C., unpublished
    work.
22. Nelsen, S. F., Hollinsed, W. C., and Calabrese, J.
    C. (1977) J. Am. Chem. Soc. 99, 4461.
23. a) Fantechi, R., and Helcke, G. A. (1972) J. Chem.
    Soc., Faraday Trans 2, 68, 924. b) Wood, D. E.,
    Wood, C. A., and Latham, W. A. (1972) J. Am. Chem.
    Soc. 94, 9278. c) West, R., and Bichlmeir, B.
    (1973) ibid. 95, 5897. d) Wiberg, N., Uhlenbrock,
    W., and Baumeister, W. (1974) J. Organomet. Chem.
    70, 259. e) Malatesta, V., and Ingold, K. U.
    (1973) ibid. 95, 6110. f) Malatesta, V., and
    Ingold, K. U. (1974) ibid. 96, 3949. g) Lunazzi,
    L., and Ingold, K. U. (1974) ibid. 96, 5558.
    h) Kaloa, R. A., Linazzi, L., Lindsay, D., and
    Ingold, K. U. (1975) ibid. 97, 6792. i) Malatesta,
    V., Lindsay, D., Horrwill, E. C., and Ingold, K. U.
    (1974) Can. J. Chem. 52, 864. j) Gravel, P. L.,
    and Pirkle, W. H., (1974) J. Am. Chem. Soc. 96,
    3335.
24. Depevre, R.-M., and Rassat, A. (1966) J. Am. Chem.
    Soc. 88, 3180.
25. Nelsen, S.F., and Landiss, R. T. II (1973) J. Am.
    Chem. Soc. 95, 5422.
26. a) Nelsen, S. F., and Landis, R. T. II (1973) J.
    Am. Chem. Soc. 95, 6454. b) Nelsen, S. F., and
    Landis, R. T. II (1974) ibid. 96, 1788.

27. a) Marnett, L. J., Smith, P., and Porter, N, A,
    (1973) Tetrahedron Lett. 108. b) Malatesta, V,,
    and Ingold, K. U. (1973) ibid., 3307, 3313,
28. Parmelee, W. P., unpublished results.
29. Nelsen, S. F., and Landis, R. T. II (1973) J, Am,
    Chem. Soc. 95, 2719.
30. It is a pleasure to thank my coworkers who have
    worked on the aspects of hydrazine oxidation dis-
    cussed here, P. J. Hintz, R. T. Landis ,II, J. M,
    Buschek, Dr. L. Echegoyan, G. R. Weisman, V, E,
    Peacock, W. C, Hollinsed, C. R. Kessel, and W, P,
    Parmelle. Prof. D.H. Evans has given us invalu-
    able assistance in the electrochemical aspects of
    this work.

RECEIVED December 23, 1977,

Aspects of Phosphoranyl Radical Chemistry

WESLEY G. BENTRUDE

Department of Chemistry, University of Utah, Salt Lake City, UT 84112

Phosphoranyl radicals ($\underset{\sim}{1}$) are tetracovalent species most typically formed by the oxidative addition reaction shown in

$$Z\cdot \ + \ \overset{..}{P}Z_3 \ \rightarrow \ Z\overset{.}{P}Z_3 \qquad \underset{\sim}{1} \qquad (1)$$

equation 1 ($\underset{\sim}{1}$). ESR evidence concerning the structure of such radicals is abundant and a very recent review is available ($\underset{\sim}{2}$). The majority are near-trigonal-bipyramidal (TBP) in structure

with the vacant position or phantom ligand equatorial, and often are represented by structure $\underset{\sim}{2}$. A number of theoretical calculations ($\underset{\sim}{3}$-$\underset{\sim}{7}$) are consistent with a somewhat distorted TBP structure and also with the results of anisotropic ESR investigations ($\underset{\sim}{8}$-$\underset{\sim}{12}$) which show a high degree of spin density residing on the apical ligands. Structure $\underset{\sim}{3}$ represents the HOMO calculated ($\underset{\sim}{3}$) for $\overset{.}{P}H_4$. Certain but not all phosphoranyl radicals phosphorus-substituted by aryl groups appear to be tetrahedral ($\underset{\sim}{13}$-$\underset{\sim}{17}$) with the odd electron in the π-system, $\underset{\sim}{4}$. A C_{3v} geometry is proposed for $Ph\overset{.}{P}Cl_3$ ($\underset{\sim}{18}$,$\underset{\sim}{19}$).

The superficial similarity of $\underset{\sim}{1}$ to pentacovalent phosphorus systems is apparent in their common TBP geometries and is emphasized by structure $\underset{\sim}{2}$. Structure $\underset{\sim}{2}$ is useful for stereochemical representations and also because it is easily written, though we acknowledge its inability to represent accurately electronic structure.

Phosphoranyl radicals ($\underset{\sim}{5}$) are postulated as probable intermediates in several types of radical processes involving trivalent phosphorus compounds ($\underset{\sim}{1}$) as demonstrated in the scheme below.

(α and β scission processes were first proposed by Walling ($\underline{1}$).)

Alkoxy radical addition followed by rapid β-scission (process a)
leads to overall oxidation. Alternatively, a rapid β scission reaction (path b) of $\underset{\sim}{5}$ yields the product of substitution. Addition
of X· (path c) followed by β scission is a third overall reaction,
a free radical Arbuzov (reaction 2) ($\underline{20}$).

$$X· + ROP(OEt)_2 \rightarrow \underset{\sim}{5} \rightarrow XP(O)(OEt)_2 + R· \qquad (2)$$

Both product and ESR studies show the ratio of β/α scission
(oxidation/substitution) to be dependent on changes in the stabilities of radicals R· and X·. Effects of structure changes on
the competition depicted by equation 3 are recorded in Table I
($\underline{21}$).

TABLE I. E_a and $\underline{A}$-Factor Effects on the Competition between
Oxidation and Substitution in the Reaction RS· + R'P(OEt)$_2$

RS·	R'	$E_\alpha - E_\beta^a$	A_α/A_β	k_β/k_α (60°C)
$\underline{i}$-PrS[b]	PhCH$_2$	0.36 ± 0.01	0.81	2.12
$\underline{t}$-BuS[c]	PhCH$_2$	0.48 ± 0.06	0.44	4.74
$\underline{p}$-Me C$_6$H$_4$CH$_2$S[c]	PhCH$_2$	9.9 ± 0.1	1.3	13.2
$\underline{i}$-PrS	$\underline{t}$-Bu	3.2 ± 0.2	5.4	22.9

[a]95% confidence limit   [b]Based on GLC measurements of $\underline{i}$-PrSP(S)
(OEt)$_2$ and PhCH$_2$P(S)(OEt)$_2$ [c]Based on toluene and PhCH$_2$P(S)(OEt)$_2$
[d]Based on i-PrSP(S)(OEt)$_2$ and t-BuP(S)(OEt)$_2$

The increase in β/α ratio in the series $\underline{i}$-PrS·, $\underline{t}$-BuS·,
$\underline{p}$-MeC$_6$H$_4$CH$_2$S· and corresponding increase in $E_\alpha-E_\beta$ are consistent
with the above idea. So is the effect of change of R' from PhCH$_2$
to $\underline{t}$-Bu. It is notable that these parameters vary with both RS·

and R'· change. This rules out possible exclusive control of β/α scission by configurational effects. E.g. $\underset{\sim}{6}$ might undergo only α

scission making the proportion of α scission dependent solely on the statistics of $\underset{\sim}{6}$ formation, which are likely to be unrelated to R'· stability. The importance of configurational effects in addition to radical stability factors is emphasized by the interpretation (22) given the decrease in overall rate of decomposition of $\underset{\sim}{7}$ observed with bulky but relatively stable R' such as t-butyl. A reversible permutational isomerization which puts bulky substituents where they interact sterically with each other prior to α scission was postulated. The kinetics of phosphoranyl radical decay via C-O β scission (reaction 4) also has been investigated by ESR. E_a values which reflect changes in R· stability are found (23).

The rate of α or β scission of a phosphoranyl radical intermediate also can greatly effect the overall reactivity of a given radical towards a particular trivalent phosphorus derivative (26). Alkoxy radicals react with trialkyl phosphites with k_1 (reaction 4) about 10^8 sec^{-1} mol^{-1} and E_a~2 kcal/mol (24). For R equal t-

$$RO· + P(OEt)_3 \xrightarrow{k_1} RO\overset{·}{P}(OEt)_3 \xrightarrow[-R·]{k_2} OP(OEt)_3 \qquad (4)$$

butyl, subsequent β-scission has E_a=8-10 kcal/mol. (24,25). Phenyl radical similarly adds very rapidly to phosphites (27). In these reactions, the first step is rate-determining. With Et· by contrast a free   radical Arbuzov process, reaction 5, does not occur (28) although ΔH for the reaction is favorable by

$$Et· + (EtO)_3P \underset{\leftarrow}{\rightarrow} Et\overset{·}{P}(OEt)_3 \xrightarrow{-Et·} EtP(O)(OEt)_2 \qquad (5)$$

>40 kcal/mol. ESR work shows (23) that Me· adds reversibly to trialkyl phosphites. However, if a very rapid β scission can follow Me· or Et· addition, then an overall reaction ensues (23, 28), reaction 6. Rapid α scission also can trap the initial

$$Et· + (EtO)_2POCH_2Ph \underset{\leftarrow}{\rightarrow} \underset{PhCH_2O}{\overset{Et}{\underset{/}{\overset{\diagdown \cdot}{P}}}}(OEt)_2 \rightarrow EtP(O)(OEt)_2 + PhCH_2· \qquad (6)$$

adduct (29), as in process 7 with R· various alkyls. The same sorts of reactions occur with dialkylamino radicals (28). Over-

$$Et\cdot \; + \; RP(OEt)_2 \; \underset{\leftarrow}{\rightarrow} \; \overset{\overset{\displaystyle R}{\underset{\displaystyle |}{\diagdown}}}{\underset{\underset{\displaystyle Et}{\diagup}}{P}}(OEt)_2 \; \rightarrow \; EtP(OEt)_2 \; + \; R\cdot \qquad (7)$$

all reactivity in these systems requires a <u>rapid second step</u>.
A third class of radicals are simply <u>too stable</u> and don't
add to phosphorus at all. The $PhCH_2\cdot$ of reaction 6 does not give
$PhCH_2P(O)(OEt)_2$. Isopropyl radical also is unreactive in processes 6 and 7 (<u>29</u>).

These three reactivity cases are illustrated in Figure 1.
The isoenergetic representation of the phosphoranyl radicals,
$R'OP(OR)XY$, is a minor oversimplification as they likely will
have stabilities which depend somewhat on substituent. $RO\cdot$ adds
irreversibly, but radical $X\cdot$ forms a relatively weak X–P bond in
a reversible addition which gives substitution or Arbuzov pro-
duct only if subsequent α or β scission is rapid (low $\Delta G^\ddagger$). The
species $Y\cdot$ falls in the third reactivity class, because the P–Y
bond is too weak to form at all.

The superficial structural similarity of phosphoranyl
radicals, 2, and truly pentacovalent phosphorus derivatives, 8, is
quite striking as noted earlier. Similar trends in substituent
apicophilicities are noted also (<u>30,31</u>). It is therefore of
interest to ascertain whether or not the analogous sort of Mode

$$\underset{\underset{\displaystyle 8}{\sim}}{} \quad \overset{5}{F}\!\!\!\diagdown \overset{\overset{\displaystyle F^1}{\underset{\displaystyle |}{}}}{\underset{\underset{\displaystyle F_3}{|}}{\underset{}{P}}}\!\!-\!F^2 \; \underset{(1435)}{\overset{M_1}{\rightleftharpoons}} \; \overset{3}{F}\!\!\!\diagdown \overset{\overset{\displaystyle F^5}{\underset{\displaystyle |}{}}}{\underset{\underset{\displaystyle F_4}{|}}{\underset{1}{\overset{}{P}}}}\!\!-\!F^2 \qquad (8)$$

one (M_1 – see Musher (<u>32</u>)) permutational process which is known
to be rapid for various PX_5 (where Z may be all the same or
different) is also operative with $\cdot PZ_4$. E.g., the process illustrated above for PF_5, has a rate too fast to allow kinetic study
by NMR. Calculations (<u>33</u>) place limits of 5 kcal/mol on its
activation energy.

Two methods for gaining information on the types of permutations available for $\cdot PZ_4$ are ESR studies and stereochemical investigations. In the former, effects of temperature on hyperfine
splittings, line shapes and patterns are observed. In the stereochemical approach, the overall stereochemistry of a free-radical
oxidation, substitution, or Arbuzov reaction is determined. One
then asks whether the operation of a given permutational mode on
the presumed phosphoranyl radical intermediate is allowed by the
stereochemical outcome. For example, if 9 were to undergo the

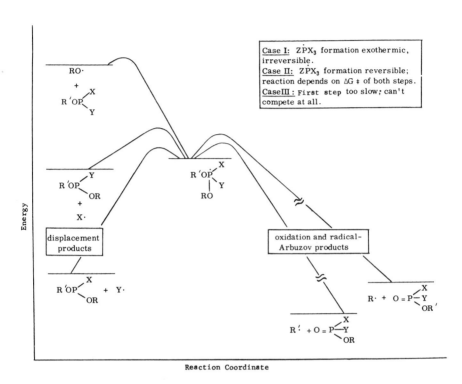

Case I: ŻPX₃ formation exothermic,
irreversible.
Case II: ŻPX₃ formation reversible;
reaction depends on ΔG ‡ of both steps.
CaseIII : First step too slow; can't
compete at all.

Figure 1.   Reactivities of radicals toward trivalent phosphorus derivatives

sort of M_1 permutation of substituents shown below prior to β scission, phosphine oxide of configuration <u>opposite</u> to that

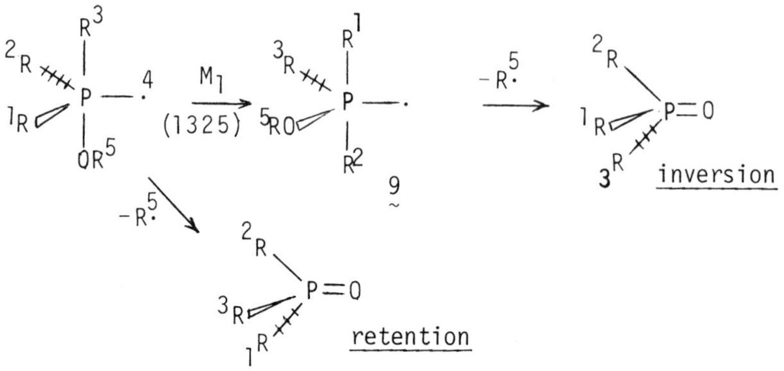

formed on direct β scission of $\underset{\sim}{9}$ would result.

Our stereochemical studies have involved the use of five- and six-membered ring derivatives of trivalent phosphorus, $\underset{\sim\sim}{10}$

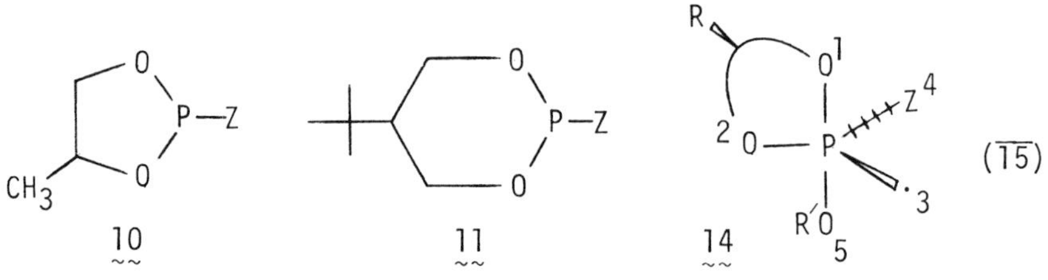

$$(\overline{15})$$

and $\underset{\sim\sim}{11}$, whose cis or trans geometries and conformations we have characterized thoroughly ($\underline{34},\underline{35}$). $\underline{t}$-Butoxy radicals, generated thermally or photochemically, were found to transfer oxygen to both five- and six-membered ring phosphites ($\underset{\sim\sim}{10}$ and $\underset{\sim\sim}{11}$ with Z = CH_3O) nearly stereospecifically with <u>retention</u> of configuration about phosphorus. Since these results are already in the litera-ture ($\underline{36}$), the data will not be reproduced here.

In discussing the implications of these results, it is help-ful to be able to refer to the various permutational isomers by the sorts of designations now commonly applied to pentacovalent phosphorus species ($\underline{37},\underline{38}$). In equation 9 we show reaction of a cis isomer, $\underset{\sim\sim}{12}$, (generalized to represent a ring of any size) to give the permutamer, $\underset{\sim\sim}{13}$. This isomer, in the formalism employed in discussion of pentacovalent phosphorus intermediates, results

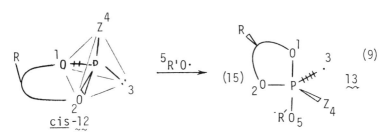

from facial attack opposite the ring oxygen numbered one. The
designation (15) shows which ligands are apical and indicates that
the groups 2, 3, and 4 are arrayed in clockwise order. Isomer
($\overline{15}$) from the same reaction of trans-12 is shown above (structure
14) for comparison.

Figure 2 is a topological representation ($\underline{37},\underline{38}$) which shows
the M_1 interconversions of all 18 permutamers (the vertices of
the hexasterane diagram) which can be formed from reactions of
R'O· with cis- and trans-12. The edges of the diagram connect
permutamers which can be interconverted by M_1 isomerizations.
The C-T plane (perpendicular to the page) separates initial
adducts which could be formed from attack of R'O· on either cis
(left side) or trans phosphite. The P-O plane separates permutamers whose β scission gives trans product from the cis phosphate
generating ones. The position of R'O in each permutamer is
shown by eq or ap. The retentive nature of the observed oxidations excludes 14, $\overline{24}$, and 35 as initially formed phosphoranyls
from cis-12 or $\overline{14}$, 24, $\overline{35}$ from trans-12. Clearly, no extensive
amount of M_1 isomerization of any of the remaining potential
initial adducts occurs, e.g. $\overline{25} \rightarrow 14$ or $13 \rightarrow \overline{24} \rightarrow 15$, as loss of
retentive oxidation stereospecificity would result. Since $\Delta G^{\ddagger}$
for β-scission of these intermediates with R' equal t-Bu is at
least 11 kcal/mol ($\underline{24},\underline{25}$), the M_1 permutations have overall
barrier greater than this.

Further simplification of this scheme is possible. Consider
reaction of cis-12. From ESR evidences it is clear that only
odd electron equatorial phosphoranyls have enough stability to
be observed ($\underline{2}$). Initial adducts with odd electron apical ($\overline{13}$,
23, 34), therefore, should be rapidly converted to more stable
ones: 24, $\overline{14}$, $\overline{45}$, $\overline{25}$, or 15. Moreover, preferential initial
introduction of R'O apical is also likely ($\overline{45}$, $\overline{25}$, 15 formation)
since apical bonds are expected to be longer and presumably
weaker (3,4). Ab initio calculations ($\underline{3}$) for the reaction H· +
$PH_3 \rightarrow$ ·PH_4 favor apical attack as well. (Such an assumption is
generally made in pentacovalent phosphorus chemistry. ($\underline{37}$)) We
can thus center our attention on $\overline{45}$, $\overline{25}$, and 15.

That intermediate $\overline{45}$ ($\underline{14}$) should give cis-phosphate (retention) is not surprising since inversion would require isomeriza-
tion via ($\overline{13}$) or (23) prior to β-scission. Such odd-electron-

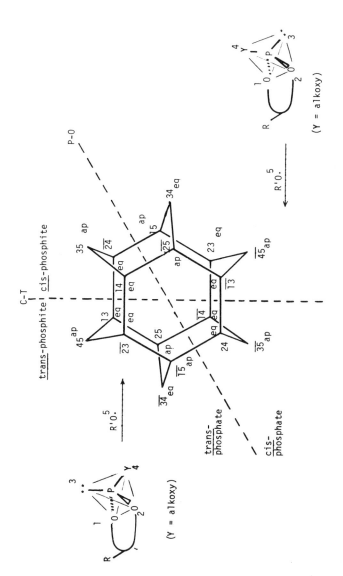

Figure 2. Formation and permutational isomerization of cyclic tetraalkoxy phosphoranyl radicals

apical intermediates may well be relatively high energy species.

However, the evidence is very strong (30) that in such reaction
systems those phosphoranyl radicals observed by ESR have five-
membered dioxa ring systems attached to phosphorus apical-
equatorial as in 15 (15) and its mirror form 25‡. This applies
to six-membered ring 1,3-diaza species too (39). The apparent
failure of the isomerization of radical 15 to 16 (24) (a thermo-
neutral process which keeps the odd electron equatorial) to compete with loss of R'· (R' = t-Bu), indicates that the cyclic
phosphoranyl radicals are not simple analogs of pentacovalent
phosphorus intermediates, at least where permutational properties
are concerned. Thus by contrast to the apparent high $\Delta G^{\ddagger}$ of
>11 kcal/mol assigned to process M_1 of 15 (15 → 24), that for the
isomerization of the corresponding pentaalkoxy compound, 17,
must be fewer than 10 kcal/mol. This follows since the equilibra-

---

‡The arguments which follow also apply to permutamers 24 and
14 which have ap-eq ring attachment and differ from 25 and 15 only
in that the R'O and CH_3O groups are interchanged. ESR work also
suggests that the rapid interconversions 25 ⇄ 24 and 15 ⇄ 14
occur. (See later discussion.)

tion of CH_3O groups in $\underset{\sim\sim}{18}$, which requires the CH_2 to become
apical (energetically unfavorable), has a $\Delta G^{\ddagger}$ value of only
10 kcal/mol $(\underline{40})$. (See earlier comments on PF_5 barriers.)
An objection to the above is the possibility that $\underset{\sim}{16}$ $(\overline{24})$
cannot undergo β scission since R'O has become equatorial. How-
ever cis/trans mixtures of $\underset{\sim}{19}$ $(\underline{36})$ gave the outcome expected from
direct β-scission of R'O equatorial as illustrated below for the
cis-19. Mode 1 isomerization of the intermediate to place R'O

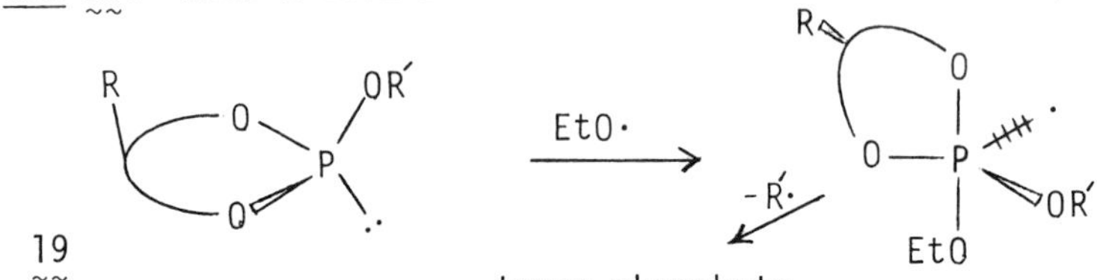

$\underset{\sim\sim}{19}$ trans-phosphate

apical as in $\underset{\sim}{15}$ would have inverted product stereochemistry.
Likewise, the stereochemistry of free-radical Arbuzov re-
action of $Me_2N\cdot$ with cis/trans 5- and 6-membered benzyl phosphites
failed to reveal any evidence of such an M_1 permutation $(\underset{\sim\sim}{20} \rightarrow \underset{\sim\sim}{21})$.

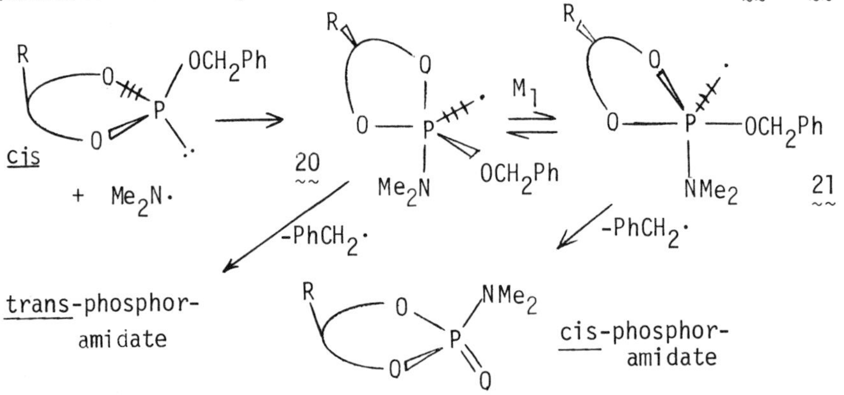

Product phosphoramidate trans/cis ratios corresponded closely to
cis/trans ratios of the starting phosphites $(\underline{20})$. Some driving
force for the permutation $\underset{\sim}{20} \rightarrow \underset{\sim}{21}$ could come from the slightly
greater apicophilicity of Me_2N over $PhCH_2O$ $(\underline{39})$. This is com-
pensated for, however, by the greater rate of β-scission giving
$PhCH_2\cdot$ than that for the $\underline{t}$-butoxy radical.
Free-radical substitution processes occur only rarely at
carbon but more often at heteroatoms. The stereochemistry of
such a process cannot normally be ascertained, however. Trivalent phosphorus derivatives are perhaps uniquely suitable for
such studies since phosphorus is sufficiently stable configura-
tionally, and free radical substitution at phosphorus occurs with
ease.
For reaction 10, we find $(\underline{28},\underline{29})$ that the series of substituents shown constitute a kind of substitution order in which

$$X \cdot \ + \ YP(OEt)_2 \ \rightarrow \ \overset{X}{\underset{Y}{\overset{\displaystyle |}{\underset{\displaystyle |}{P}}}} \cdot (OEt)_2 \ \overset{-Y \cdot}{\rightarrow} \ XP(OEt)_2 \qquad (10)$$

$$RO \cdot \ > \ \phi O \cdot \ > \ R_2N \cdot \ > \ R \cdot \ (X \ and \ Y)$$

any one may serve as X· to replace any of those (Y) to the right of it.

In Figure 3 is pictured the formation of intermediate 22 in a manner analogous to that proposed for the oxidation and Arbuzov reactions. Subsequent M_1 permutation places the leaving group in the apical position. Departure of Y· then is predicted to yield product with <u>retention</u> of configuration at phosphorus.

We previously published (41) the results of study of the stereochemistry of reaction 11 which occurs with inversion of

$$\text{11} \quad Z = \underline{t}\text{-Bu, PhCH}_2 \qquad (11)$$

phosphorus configuration. We have since also looked at the five-membered ring system, 10, (Z equal $\underline{t}$-Bu, PhCH$_2$) and again observed (42) inversion of configuration about phosphorus in the

10   23   24

product phosphoramidite (10, Z = Me$_2$N). With the 5-membered ring, structure 22 should be especially favored.

Most recently we have examined (43) the reaction of 23 (R = $\underline{t}$-Bu or Me) with R'O· to give phosphite 24. Our earlier substitution stereochemical studies are open to some ambiguity of interpretation in that the intermediate phosphoranyl radical is too unstable to be detected by ESR, so no circumstantial evidence is available supporting its presence or suggesting its geometry. Inversion is most easily explained then by intermediate 25 or its corresponding transition state. In fact, since Me$_2$N· additions are reversible, initial intermediate 22 could conceivably not lead to product at all but break down instead to reform Me$_2$N· and

reactant which then give product via 25. By contrast, RO·
additions to trivalent phosphorus are irreversible (44,45).

25   R'O   26

Furthermore, ESR results (39) show clearly the formation of
radicals of structure 26 from R'O· and an unsubstituted phosphoramidite analog of 23 (R equals H, Et_2N equals Me_2N).
     In Table II are found results of some of our studies of
reactions of EtO· and sec-BuO· with 23 (R equal Me and t-Bu). The
product ratios are not the thermodynamic equilibrium ones. The
conclusion is that here again substitution is with inversion about
phosphorus. As before, the stereochemical outcome is inconsistent
with an M_1 permutation of 22.

TABLE II. Substitution Stereochemistry

R	R'O·	cis / trans 23	cis/trans 23 consumed[a]	trans/cis 24[b] formed
CH_3	EtO·	37/63	40/60	39/61
CH_3	sec-BuO·	36/64	39/61	39/61
t-Bu	EtO·	19/81	25/75	24/76

[a]At 14-37% consumption of 23 [b]Yields 64-85%

     A further question remains. What other permutational modes
of the five (plus the identity) categorized by Musher (32) are
consistent with the observed stereochemistries? In a series of
definitive ESR investigations, B. P. Roberts and coworkers have
shown (30,39) that very rapid M_4 permutational isomerizations are
undergone by certain cyclic phosphoranyl radicals, 27. The socalled M_4 (exocyclic) process is more rapid and operative at
lower temperatures than is the M_4 (ring) permutation. M_4
(exocyclic) for 26 occurs (39) with rate constant about 10^7 sec^{-1}
at $-120°$. Although the exchange rates for M_4 (exocyclic) are
dependent on the relative apicophilicities (30), such processes
are surely rapid at room temperatures where our stereochemical
work was done. Shown in Figure 4 is the predicted effect of an

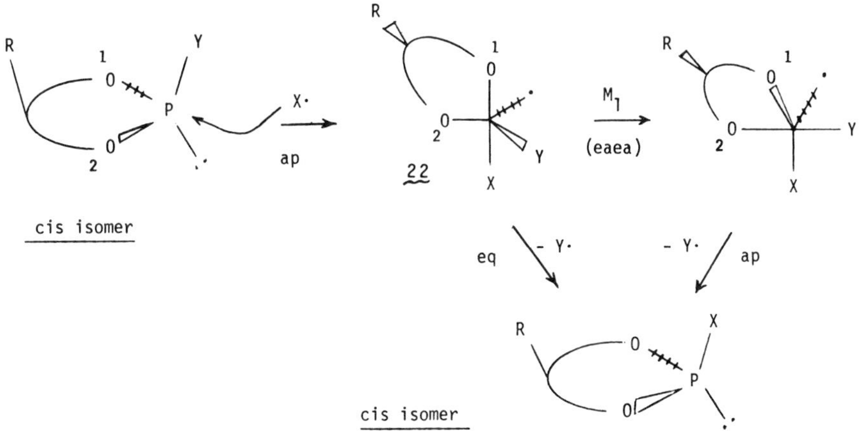

Figure 3. Potential effect of M₁ permutation on substitution stereochemistry

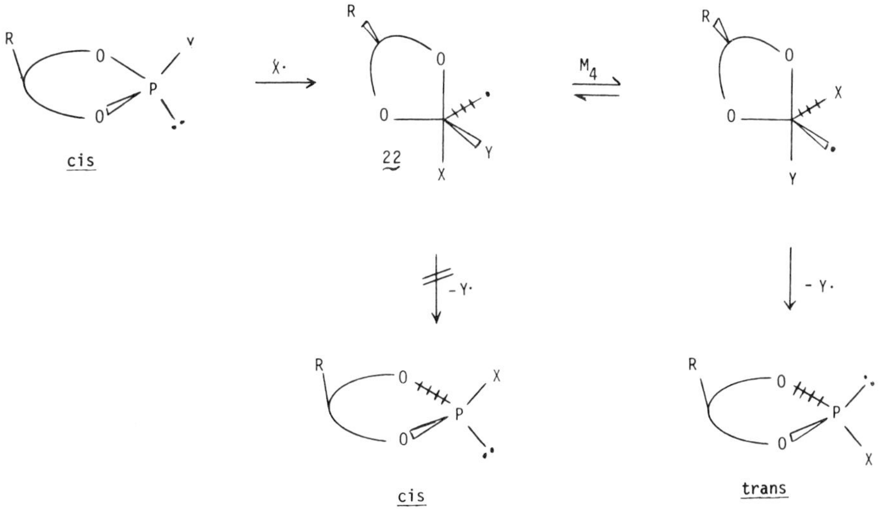

Figure 4. Potential effect of M₄ permutation on substitution stereochemistry

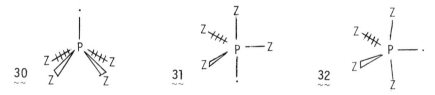

28   27   29

M_4 (exocyclic) isomerization on substitution stereochemistry.
This process moves the departing Y (Me_2N) into an apical
position where its leaving (presumably favored by microscopic
reversibility) gives product of <u>inverted</u> phosphorus configuration
as is observed experimentally. An M_4 (ring) permutation of the
type 27 → 28 would not alter substitution stereochemistry, but
neither would it place the Me_2N apical.

In Figure 5 are shown the effects of M_4 (exocyclic) isomerizations on the stereochemistries of oxidation and free-radical
Arbuzov as well as the substitution process discussed above. All
three are stereochemically <u>consistent with</u> rapid M_4 (exocyclic)
permutations although the results do not require that such processes be rapid, especially in the oxidation and Arbuzov reactions.
Mode 5 processes (socalled disrotatory Berry) also fit the stereochemistry but not the ESR results. Modes 2 and 3 are ruled out
by our stereochemical work. <u>It is to be emphasized that the
statements as to allowed permutation modes are made under the
assumption that odd electron apical and ring diequatorial
structures are excluded energetically.</u>

Theoretical calculations (3,5) predict that a square pyramidal, odd-electron-apical geometry (30) is much higher in energy

30   31   32

than the optimized distorted trigonal bipyramidal one (32).
Values for the energy increase include: 35 kcal/mol, ab initio,
Z = H (3); 39 kcal/mol, ab initio, Z = F (3); 25 kcal/mol,
CNDO/2, Z = F (5). If the physical mechanism for an M_1 process
with odd electron as equatorial pivot involves 30 as the barrier
geometry (a true Berry pseudorotation process), then the reluc-
tance of phosphoranyl radicals to undergo such M_1 isomerizations
(equation 12) is reasonable. By comparison, this energy dif--

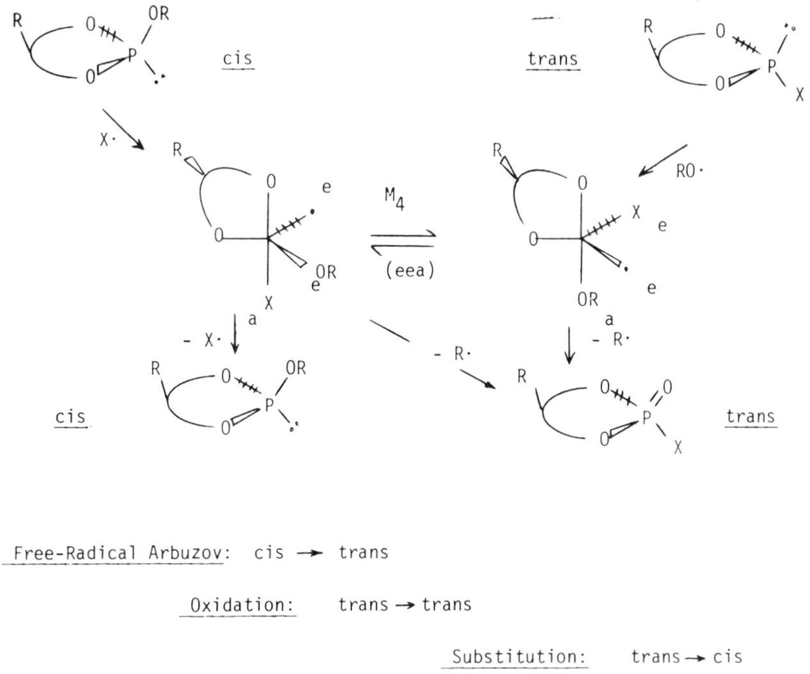
cis
trans
X·
RO·
M_4
(eea)
- X·
- R·
cis
trans
Free-Radical Arbuzov: cis → trans
Oxidation: trans → trans
Substitution: trans → cis

Figure 5. Stereochemical effects of M_4 isomerizations on stereochemistries of free radical oxidation, Arbuzov, and substitution reactions

$$(12)$$

ference for PZ₅ compounds is calculated to be <5 kcal/mol ($\underline{3},\underline{33}$). Clearly, theory predicts a Berry mechanism M₁ exchange of this type to be much more rapid with PZ₅ than with ·PZ₄. This is consistent with both the stereochemical and ESR findings.

Moreover, it is notable that the calculations predict ·PZ₄ structures with odd electron apical, $\underline{31}$, to be of lower energy than those with geometry $\underline{30}$, being 20 kcal/mol ($\underline{3}$) above that of the optimized geometry for ·PH₄ (ab initio) and 3.4 kcal/mol ($\underline{5}$) for ·PH₃OH (CNDO/2). Again it is apparent that a phosphoranyl radical is not simply a pentacovalent phosphorus species with the odd electron as electropositive phantom ligand. A further significance of this is that M₁ processes <u>via</u> intermediates or barrier geometries with the odd electron apical may be more rapid than those with odd electron as equatorial pivot. Such a structure ($\underline{33}$) could then represent an intermediate in a two-step M₁ x M₁ process which is permutationally equivalent to the M₄. A sequential M₂ x M₂ process exchanging substituents 3 and 5, then 4 and 5 of $\underline{27}$ will accomplish the same thing. As has been

noted ($\underline{32}$), once restraints as to permutamers which are energetically accessible are removed, it is not possible, except for M₁, to separate individual permutation modes (M₂ through M₅) from combinations of other modes which are permutationally equivalent.

Fluxional behavior has been noted as well by ESR for acyclic phosphoranyls: ROṖF₃ ($\underline{46}$); (EtO)₂Ṗ(NMe₂)₂ ($\underline{39}$); $\underline{t}$-BuOPH₃ ($\underline{47}$); and ROṖ(CH₃)₃ ($\underline{22}$). The mode of exchange has not been assigned experimentally.

In a paper delivered at this symposium J. A. Kampmier of
the University of Rochester presented data on radical substitutions at sulfide sulfur. The effects of stereochemical constraints on the products formed led to the conclusion that a
linear array of attacking radical, sulfur, and leaving radical
($X^{0}_{\cdot} - \cdot S - - - Y^{0}_{\cdot}$) is required. If a sulfuranyl radical ($R_3S\cdot$) is
present in such reactions, permutational isomerizations should
allow other groups to assume apical positions where they then
could depart. A high barrier to permutation may be operative in
such systems, as is found (48) by ESR for $(RO)_3S\cdot$. Evidence
that $(RO)_3S\cdot$ results from apical introduction of $RO\cdot$ was also
reported (48).

Acknowledgement. The author thanks the National Science
Foundation, the National Cancer Institute of the Public Health
Service, and the Petroleum Research Fund for grants supporting
work described here.

REFERENCES

1. This topic was recently reviewed: Bentrude, W. G., in "Free
    Radicals," vol. 2, Chapter 22, Kochi, J. K., Ed., Wiley-
    Interscience, New York, 1973.
2. For a review see: Schipper, P., Jansen, E. H. J. M., and
    Buck, H. M., Topics in Phos. Chem. (1977), 9, 407.
3. Howell, J. M. and Olsen, J. F., J. Am. Chem. Soc. (1976),
    98, 7119.
4. Hudson, A. and Treweek, R. F., Chem. Phys. Lett., (1976),
    39, 248
5. Gorlov, Y. F. and Penkovsky, V. V., Chem. Phys. Lett., (1975)
    35, 25.
6. Colussi, A. J., Morton, J. R., and Preston, K. F., J. Phys.
    Chem, (1975), 62, 2004
7. Hudson, A. and Witten, J. T., Chem. Phys. Lett., (1974), 29,
    113.
8. Gillbro, T. and Williams, F., J. Am. Chem. Soc., (1974), 96,
    5032.
9. Nishikida, K. and Williams, F., J. Am. Chem. Soc., (1975),
    97, 5462.
10. Kerr, C. M. L., Webster, K., and Williams, F., J. Phys. Chem.
    (1975), 79, 2650.
11. Hasegawa, A., Ohnishi, K., Sogabe, K., and Miura, M.,
    Molecular Phys., (1975), 30, 1367.
12. Nelson, D. J. and Symons, M. C. R., J. Chem. Soc., Dalton
    Trans, (1975), 1164.
13. Rothius, R., Luderer, R. K. J., and Buck, H. J., Recl. Trav.
    Chim. Pays. Bas, (1972), 91, 836.
14. Rothius, R., Fontfreide, J. J. H. M., and Buck, H. M., Recl.
    Trav. Chim. Pays - Bas, (1973), 92, 1308.

15. Rothius, R., Fontfreide, J. J. H. M., van Dijk, J. M. F., Recl. Trav. Chim. Pays - Bas, (1974), 93, 128.

16. Boekstein, G., Jansen, E. H. J. M., and Buck, H. M., J. Chem. Soc., Chem. Commun., (1974), 118.

17. Davies, A. G., Parrott, M. J., and Roberts, B. P., J. Chem. Soc., Chem. Commun., (1974), 973.

18. Berclaz, T., Geoffroy, M., and Lucken, E. A. C., Chem. Phys. Letters, (1975), 36, 677.

19. Symons, M. C. R., Chem. Phys. Letters, (1976), 40, 226.

20. Bentrude, W. G., Alley, W. D., Johnson, N. A., Murakami, M., Nishikida, K., and Tan, H.-W., J. Am. Chem. Soc., (1977), 99, 4383.

21. Bentrude, W. G. and Rogers, P. E., J. Am. Chem. Soc., (1976), 98, 1647.

22. Cooper, J. W. and Roberts, B. P, J. Chem. Soc., Perkin 2, (1976), 808.

23. Davies, A. G., Griller, D., and Roberts, B. P., J. Chem. Soc., Perkin 2, (1974), 2224.

24. Davies, A. G., Griller, D., and Roberts, B. P., J. Chem. Soc, Perkin 2, (1972), 993.

25. Watts, G. B., Griller, D., and Ingold, K. U., J. Am. Chem. Soc., (1972), 94, 8784.

26. Bentrude, W. G., Fu, J. J. L., and Rogers, P. E., J. Am. Chem. Soc., (1973), 95, 3625.

27. Fu, J. J. L., Bentrude, W. G., and Griffin, C. E., J. Am. Chem. Soc., (1972), 94, 7717.

28. Khan, W. A., unpublished results.

29. Hansen, E. R., Khan, W. A., and Rogers, P. E., unpublished results.

30. Cooper, J. W., Parrott, M. J., and Roberts, B. P., J. Chem. Soc., Perkin 2 (1977), 730.

31. Dennis, R. W., Elson, I. H., Roberts, B. P., and Dobbie, R. C., J. Chem. Soc., Perkin 2 (1977), 889.

32. Musher, J. I., J. Chem. Ed., (1974), 51, 94.

33. Strich, A., and Veillard, J., J. Am. Chem. Soc., (1973), 95, 5574.

34. Bentrude, W. G. and Tan, H. W., J. Am. Chem. Soc., (1976), 98, 1850.

35. Bentrude, W. G., Tan, H.-W., and Yee, K. C., J. Am. Chem. Soc., (1975), 97, 573.

36. Bentrude, W. G., Johnson, N. A., Rusek, Jr., P. E., Tan, H.-W., and Wielesek, R.A., J. Am. Chem. Soc., (1976), 98, 5348.

37. Mislow, K., Accounts Chem. Res., (1970), 3, 321.

38. Lauterbur, P. C. and Ramirez, F., J. Am. Chem. Soc., (1968), 90, 6722.

39. Dennis, R. W. and Roberts, B. P., J. Chem. Soc., Perkin 2, (1975), 140.

40. Gorenstein, D., J. Am. Chem. Soc., (1970), 92, 644.

41. Bentrude, W. G., Khan, W. A., Murakami, M, and Tan, H.-W.,

J. Am. Chem. Soc., (1974), 96, 5566.
42. Khan, W. A., Kosugi, Y., and Murakami, M., unpublished results.
43. Nakanishi, A., unpublished results.
44. Bentrude, W. G. and Wielesek, R. A., J. Am. Chem. Soc., (1969), 91, 2406.
45. Bentrude, W. G. and Min, T. B., J. Am. Chem. Soc., (1976), 98, 2918.
46. Elson, I. H., Parrott, M. J., and Roberts, B. P., J. Chem. Soc., Chem. Commun., (1975), 586.
47. Krusic, P. J. and Meakin, P., Chem. Phys. Letters, (1973), 18, 347.
48. Cooper, J. W., Roberts, B. P., J. Chem. Soc., Chem Commun., (1977) 228.

RECEIVED December 23, 1977.

Radical Ion Chemistry

Radical, Anion, Anion-Radical Reactions with Organic Halides

SHELTON BANK and JANET FROST BANK

Department of Chemistry, State University of New York at Albany, Albany, NY 12222

Interesting similarities in the behavior of the three reactive species, having one or two electrons, led us to study their reactions with organic halides from the point of view that comparisons of product and kinetic data could lead to correlations, and of equal importance the lack thereof, of the various mechanisms.

Aromatic radical anions have been studied in depth beginning in 1867 with Bertholet's discovery (1), and continuing with Schlenk and Bergmann's pioneering work (2) around the turn of the century. Although radical anions exhibit complex chemical behavior, their ability to function as reducing agents by electron transfer processes is reasonably well understood. In this regard, reduction of organic halides was selected as a model for election transfer reactions of radical anions.

The products of the reaction of radical anions, for example sodium naphthalene, with bromides and chlorides are the hydrocarbon, the olefin and alkylated dihydronaphthalene. Reaction with iodides gives hydrocarbon dimer in addition.The mechanism as shown is a result of numerous product and kinetic studies (3, 4, 5). Two steps that merit further investigation are the

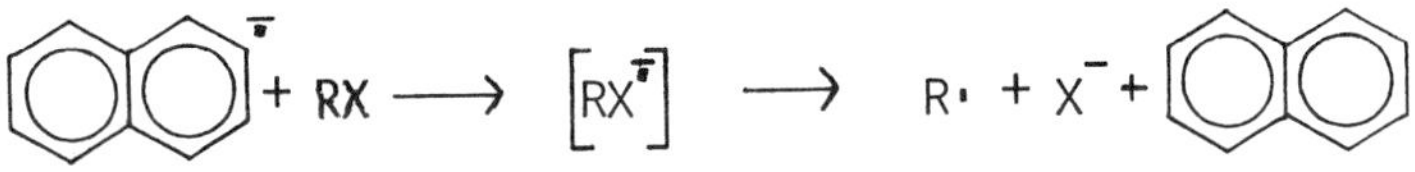

(1)

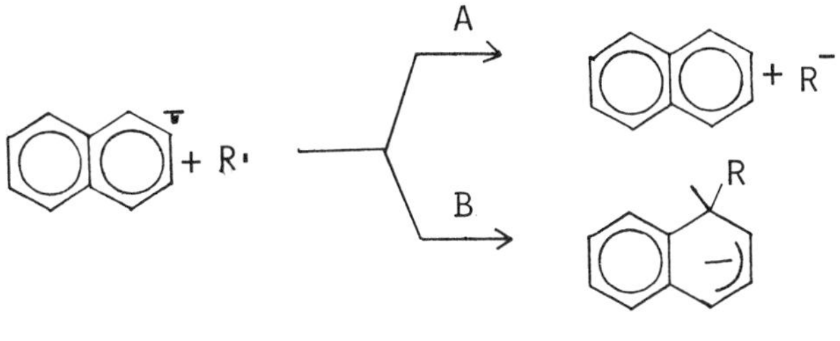

(2)

© 0-8412-0421-7/78/47-069-343$05.00/0

$$R^- \quad \overset{SH}{\underset{RX}{\underbrace{}}} \quad \begin{array}{l} RH \\ RR + R(-H) + RH + X^- \end{array} \qquad (3)$$

$$\text{(structure)} \quad \overset{SH}{\underset{RX}{\underbrace{}}} \quad \text{(structures)} + X^- \qquad (4)$$

initial electron transfer (eq. 1) and the coupling
reaction (eq.2).
Two possibilities for the first step are that the species,
$RX^{\mp}$, is a transition state for a single-step bond dissociation,
and that the species is an intermediate organic halide radical
anion. The first possibility is generally accepted, however,
extensive kinetic data (4C) for a series of organic radical
anions and halides indicate a change in the nature of the transition state in going from a highly exothermic to a less exothermic reaction. These changes could result from differences
in timing along the reaction path in a one-step bond cleavage,
or from a superposition of the two proposed mechanisms.
In an attempt to resolve this dilemma, the reaction rates
were measured for a series of organic bromides with sodium
anthracene (Table I) and correlated with two model systems. The
model for the first possibility, a one-step process with bond
dissociation, is the tri-n-butyltin hydride reaction with these
same halides (6). Correlation of the reaction rates would
indicate that the transition state for anthracene radical anion
reduction is similar to the transition state for radical formation by tri-n-butyltin radical. On the other hand, the model
for the second possibility, the organic halide reduction potential, is a measure of organic halide radical anion formation (7).

Table I. Second-Order Rate Constants for Reaction of Sodium
Anthracene with Alkyl Bromides in THF at 0°C.[a]

Alkyl Bromide	k, M^{-1}sec^{-1}	k(rel.)	Logk(rel.)
n-Butyl	242[b]	1.0	0.0
sec-Butyl	938[c]	3.9	0.59
tert-Butyl	2402[c]	9.9	1.0
Phenyl		0.0015[d]	-2.82
Benzyl		25.0[e]	1.40

(a) Measured with stopped-flow apparatus at 775 nm (b) The rate
was unaffected by the addition of tert -butanol in excess.
(c) tert- Butanol was added to prevent build-up of strongly
absorbing anion. (d) From ref 4c at 20°C. (e) Estimated from
competition experiments (8, 9).

The relationship in Figure 1 between the known radical
reaction and the radical anion reaction is monotonic and there-

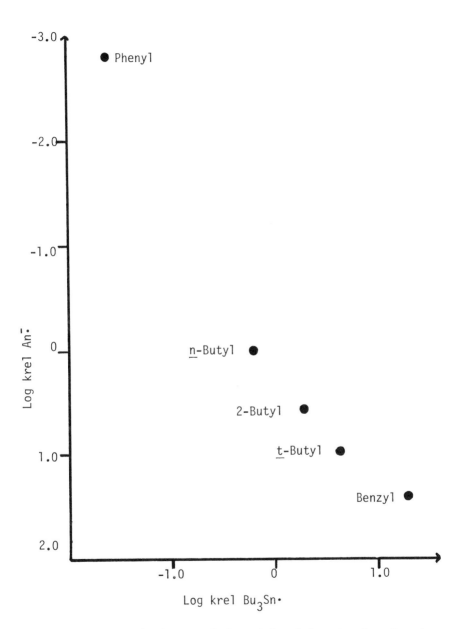

Figure 1.   The relationship between the logs of the relative rates of reaction of so-dium anthracene and tri-n-butyltin hydride with select bromides

fore similar reaction factors are likely to be involved. This
definitive correlation compared to the decided lack of correlation in Figure 2 indicates that for sodium anthracene the
formation of RX$\bar{\cdot}$ is not involved to any great extent, but rather
the transition state resembles that for a radical reaction.

In contrast, for reaction of organic halides with metallic
magnesium, Whitesides (7) found a poor correlation with the
rates of tri-n-butyltin hydride, and a reasonable correlation,
especially with primary bromides, with reduction potentials,
suggesting the formation of RX$\bar{\cdot}$. Sodium naphthalene reductions
also correlate with reduction potentials for primary halides
(4b). Moreover, reaction rates are faster in solvents favoring
loose ion pairs over tight, further evidence for an early transition state involving election transfer and little bond
dissociation. (4c)

Therefore, one can conclude that for the less exothermic
reduction by sodium anthracene,eq. 1 is a one-step radical
formation. For radical anions of higher reduction potential,
the possibility for the species, RX$\bar{\cdot}$, must still be considered.

Turning now to equation 2 in the mechanism, the genesis
of the alkylated aromatic raises the question of possible contributions from displacement reactions, and more generally the
question of possible stereochemical integrity. Steps 5 and 6
were considered

$$\text{(5)}$$

$$\text{(6)}$$

to contribute foremost to the mechanism (10) before experiments
by Sargent (5b) and Garst (3c) showed little dependence either
on the structure or on the halogen of the alkyl halide in the
ratio of alkylated to reduced products. Data from experiments
with chiral substrates (11) however, suggests that an SN_2 process
might contribute to some extent. An important prediction of such
a path is of course inversion of configuration at the reaction
site. This is moreover a stringent test for the timing of steps
1 and 2A, for if there is a stereochemical preference for step
2A and if reaction is faster than racemization of the radical,
then stereochemical integrity can be preserved.

Bridgehead tertiary halides were chosen as the experimental
model because rear-side attack is not possible, and therefore, if
displacement is involved, there would be a decrease in the amount
of alkylated products. The results in Table II are consistent

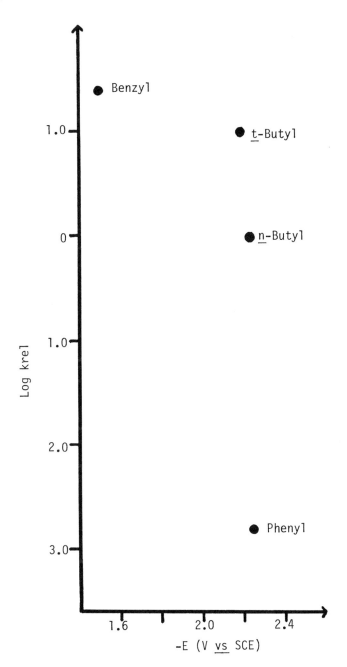

Figure 2. The relationship between the log of the relative rates of reaction of sodium anthracene with select bromides and their reduction potentials

Table II. Reaction of Bridgehead Bromides with Sodium
Naphthalene

Bromide	R-H%	R-Naph-H%.[c]
1-Adamantyl	41[b]	59[d]
1-Bicyclo[2.2.2.]octyl	42[b]	(58)
4-Bicyclo[2.2.2.]-1-azooctyl	50[b]	(50)

(a) In THF at 20°C.
(b) Analyses were by vpc on a U.C. W 98 column.
(c) Determined by difference from the yields of aliphatic
     products.
(d) Isolated by liquid chromatography and characterized by
     mass spectroscopy and proton n.m.r.

with yields from primary halides. (4b) They agree as well with
experiments by Sargent (5b) who found 61% alkylation from reaction with tert-pentyl iodide. In all cases it appears that the
alkylation reaction is insensitive to stereochemistry which in
turn is inconsistent with an SN_2 mechanism. The insensitivity
further indicates that the radical, which is clearly implicated,
has sufficient time to equilibrate before reaction. These
results additionally provide estimated rate limits for reactions
bearing on stereochemical integrity. From competition experiments, the rate constant for the coupling step was determined to
be ~1x10^9 M^{-1} sec^{-1} (12). Accordingly, loss of stereochemical
integrity must be faster to account for the commonality of
results with the acyclic and bridgehead bromides. A parallel
scheme for electron transfer processes with anions (eq. 8 and 9)
is suggested and this precludes any stereochemical integrity
unless there is some special feature about the coupling step.

The tenfold increase in rate (Table I) in going from a
primary to a tertiary bromide is in direct contrast to the
expected structural effect for a displacement reaction. This
supports further the conclusion that alkylation products arise
neither by an SN_2-like process, nor by any process with strong
stereochemical requirements.

A potential, common surface for the chemistry of anions and
radicals involves organometallic compounds and organic halides.
In addition to the usual two-election displacement (eq. 7), a
stepwise scheme (eq. 8 and 9) can explain the products of reaction. The electron transfer step (eq 8) has been implicated by

$$An^- + RX \longrightarrow An-R + X^- \qquad (7)$$
$$An^- + RX \longrightarrow An^{\cdot} + R^{\cdot} + X^- \qquad (8)$$
$$An^{\cdot} + R^{\cdot} \longrightarrow An-R \qquad (9)$$

the observation of radicals using ESR (13) and CIDNP (14) techniques. These observations prompted a search for the quantitative contributions of radicals and their role in the transition
state for anion reactions.

To provide the quantitative data we have focused attention on the kinetics of the reaction of select anions with halides. This approach concentrates on the rate determining step and the factors that affect it. Two kinds of anionic species have been investigated, 9-alkyl-10-lithio-9,10-dihydroanthracenyl anions and lithio di-and tri-pehnylmethyl anions.

The stereochemistry of the alkylanthracenyl anion reactions with halides has in fact fascinating results. In the main, reaction with primary halides leads to a predominance of the cis product, while reaction of the secondary halides leads to the trans (15). We have used kinetic measurements to sort out the factors influencing the transition states for cis and trans products.

Table III. Second-Order Rate Constants for Reaction of 9-Alkyl-10-Lithio-9,10-Dihydroanthracene with Alkyl Bromides[a]

Bromide	9-Alkyl	k(corr.)[b]	cis/trans	k(cis)	k(trans)
n-Hexyl	H	2673			
	Et	1473	76/24	1124	346
	iPr	574	59/41	337	237
	tBu	312	10/90	31	280
i-Propyl	H	62			
	Et	40	25/75	10	30
	iPr	37	13/87	5	32
	tBu	36	2/98	0.7	35

(a) Measured with stopped-flow apparatus described previously, in THF at 20° at 400 nm. (b) Absolute rate constants were corrected for the elimination reaction (<15%).

The rate and product date in Table III indicate that, for both primary and secondary halides, the transition state leading to cis product is a marked function of the size of the substituent in the 9-position. In direct contrast, the transition state leading to the trans product is insensitive to the substituent.

The geometry of the anion is not known with certainty and in fact might be a mixture of conformers (16), nevertheless, the geometry of the transition state leading to the cis product is most likely an axial-axial orientation shown as A. This orientation

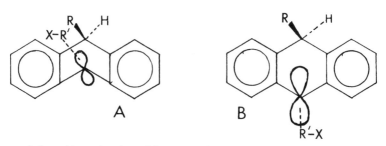

accommodates the steric effects and the stereochemical preferences and allows anion stabilization by π orbital overlap. In the case of <u>trans</u>, whereas there are several possible conformers leading to the product, it is clear that the 9-substituent has little affect on the reaction rate. A reasonable model for the transition state is depicted in B. This absence of a steric effect suggests alternative mechanisms, perhaps involving electron transfer. We delay consideration of this point until a later section.

We consider now the relationships between anions and radical anions by comparing the corresponding reaction rates with organic halides reported in Table IV.

Table IV. Absolute and Relative Rate Constants for Anion and Radical Anion Reactions with Primary and Secondary Halides [a]

Halide/ Anion	$\underline{n}$-$C_6H_{13}Cl$	$\underline{n}$-$C_6H_{13}Br$	$\underline{n}$-$C_6H_{13}I$	$\underline{sec}$-C_3H_7Br	$\underline{sec}$-C_3H_7I
Anth$\cdot^-$	1.3×10^{-1}	6.6×10^2	4.4×10^4	2.6×10^3	
AnthH	10	2.7×10^3	2.2×10^4	6.2×10	3.4×10^3
Napth$\cdot^-$	4.0×10^2	1.5×10^5	4.4×10^7		
NaphH	8.4				
Φ_2CH^-	10	2.7×10^3	5.0×10^4	1.9×10^2	5.0×10^3
Φ_3C^-		3.1×10	2.0×10^3	1.9	(5.0×10^2) [b]
$\frac{Anth\cdot^-}{AnthH}$	0.021	0.24	2.0	41.5	
$\frac{\Phi_2CH^-}{\Phi_3C^-}$		86.0	25	103	(10)

(a) In lit/mole/sec. Measured with stopped-flow apparatus.
(b) Uncorrected for elimination reaction.

Of initial interest is the comparative reactivity of the anthracenyl radical anion and the anion with a primary bromide. The anion is four times more reactive than the radical anion. This reactivity order means, therefore, that dialkylation is faster than initial alkylation of anthracene radical anion by primary bromide (k4b>kl).

The opposite is true for naphthalene where the radical anion is faster than the anion in reaction with primary chloride. The difference arises from disparate reactivity of the respective radical anions and shows the danger in extrapolating from one radical anion-anion pair to another. Another point of interest is the much greater reactivity of naphthalene radical anion vs. anthracene, which suggests alternate mechanisms for the electron transfer step, again raising the possibility for the existence of RX⁻ for the naphthalene system.

The comparative reactivity order for Anth⁻ vs. Anth⁻ with chlorides and bromides is contrary to what is expected on the basis of their respective reduction potentials ($\underline{17}$). This rate ratio changes by an order of magnitude in going from chlorides to bromides to iodides. For the iodides the rates of the two reactions are comparable but this surprising similarity may be fortuitous and may not in fact mean a similarity in mechanisms. Further discussion of this point appears in a subsequent section.

Focusing last on the rate ratios of di- and triphenylmethyl anions, we note that in contrast to the reduction potential order $\left(E_{\frac{1}{2}}\Phi_3C^- = 1.3v(\underline{18})>E_{\frac{1}{2}}\Phi_2CH^- = 1.1v\ (\underline{19})\right)$ but in concert with the basicity order ($\overline{pka\ \Phi_2CH^-} = 33.1 > pka\ \overline{\Phi_3C^-} = 31.5$) ($\underline{20}$), the diphenylmethyl anion is more reactive. This rate difference of about two powers of ten is ~3kcal/mole in $\Delta G\dagger$ for primary and secondary bromides and decreases to 1.7 kcal/mole in $\Delta G\dagger$ for primary and secondary iodides. It appears as if factors affecting basicity contribute more than factors affecting reduction potential for both primary and secondary systems.

The proposed comparisons of radical, radical anion and anion reactions are now described by two reactivity grids for organic halide reductions. Table V records patterns for a wide variety of reagents. There are definite trends, however for the chlorides, Figure 3 shows significant overlapping of reactivities of the three species. In particular, at log reactivity ratio of -2.7, there is an overlapping of all three reagents. The kinetic data as such is not likely to provide clear distinctions of mechanistic paths.

On the other hand reactivity ratios for the iodides offer a useful diagnostic for mechanistic distinctions. This is based on two considerations. First, there is no overlap of the three species. The full spread of SN_2 values for both dipolar aprotic solvents as well as protic solvents is quite distinct from the radical values, which in turn are distinct from radical anion valves. Second, the differences are sufficiently large to be useful. At the midpoints of the radical anion, radical and SN_2 ranges, the iodides would be 282, 68 and 6 times more reactive than the corresponding bromides. These values should lend confidence to predictions.

The second reactivity grid describes the structural effects

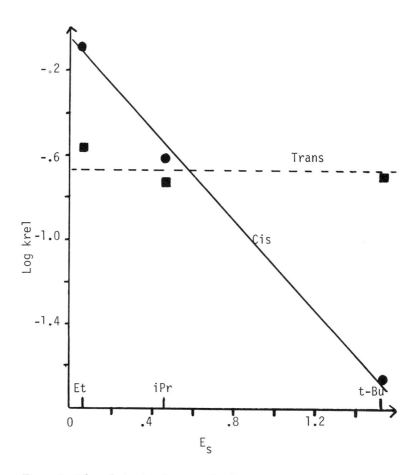

Figure 3. The relationship between the log of relative rates of reaction of 9-alkyl-10-lithio-9,10-dihydroanthracenyl anions with hexyl bromide, and the Taft steric constants for the alkyl groups

of the halide. The rate ratio of primary to secondary halides in Table VI fall into a pattern such that radical and radical anion reactivities are similar and at midpoint the secondary halide is three times more reactive than the primary. Displacement reactivity ratios are distinct and the primary is 10-15 times more reactive than the secondary.

Table VI. The Ratios of Reaction Rates for Primary vs. Secondary Halides with Various Reducing Reagents

Reagent	Rate 1°/2°	Ref.
$Bu_3Sn^\bullet H$ + RBr	0.33	(6)
$Cr(en)_2II$ + RBr	0.11	(21)
$Phenyl^\bullet$ + RI	0.57	(27)
$An^{\bullet -}$ + RBr	0.26	(4e)
SN_2 average	15.5	(24, 25)
Φ_2CH^- + RBR	14.2	(26)
Φ_2CH^- + RI	10.1	(26)
9-RAnH$^-$ + RBr	9.0 (trans)	(26)

We now attempt to utilize these diagnostics for our study of organometallic reagents and the reaction scheme in equations 7, 8 and 9. In table V, the reactivity ratios for chlorides is, as expected in the overlapping region of radical, radical anion and SN_2 mechanisms. The iodide ratio is distinct, however, and although the reactivities are high and near those for radicals, the patterns for structural changes correspond to displacement reactions. In this regard, for reaction of alkyl dihydroanthracenyl anions to give trans product, the similarities of the rate ratios with those of other anions mitigates against the mechanistic change.

The predominate mechanistic pathway for the reaction of these anions and halides is consistent with a two-electron process (eq. 7), however this kind of kinetic evidence cannot exclude contributions of ~10% from an alternative scheme and clearly any contributions below 1% are undetectable. To assess small contributions to any certainty will probably require structural changes that augment those contributions.

In conclusion, do the similarities and differences in the reactivity patterns for these reactive species with organic halides indicate a basic relationship or are these similarities fortuitous and does nature provide distinct pathways of comparable energy. Of the three mechanistic possibilities considered for radical anions, electron transfer, radical-like and nucleophic substitution, electron transfer is indicated for highly exothermic reactions, and a radical-like process is likely for the less

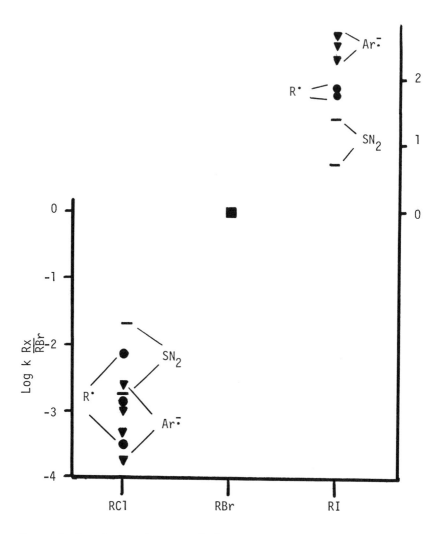

Figure 4.   Correlations of anion, radical, and radical anion reactions with organic halides and the halide effect

Table V. The Log of the Ratio of Second-Order Rate Constants for Chlorides and Iodides vs. Bromides with Various Reducing Reagents.

Reagent	RCl	RI	Reference
Bu_3SnH	-2.82	1.80	(6)
$Cr(en)_2$ (II)	-2.15	1.85	(21)
Py$^{\cdot}$ + Benzyl	-3.47		(22)
Naph$^{\bar{\cdot}}$	-2.58	2.50	(4d)
Pyrene$^{\bar{\cdot}}$	-2.96	2.27	(4d)
Anth$^{\bar{\cdot}}$	-3.71	2.57	(4d)
Py$^{\cdot}$ + NO_2-Benzyl	-3.36		(23)
SN_2 Typical	-1.7	0.7	(24)
ΦS^- in EtOH	-2.1	0.54	(25)
EtO$^-$ in EtOH	-2.62	0.28	(25)
N_3^- in DMF	-2.66	0.83	(25)
N_3^- in MeOH	-2.13	1.05	(25)
SCN$^-$ in DMF	-2.67	0.84	(25)
SCN$^-$ in MeOH	-2.31	0.48	(25)
Φ_2CH^- in THF	-2.44	1.35	(26)
HAnH$^-$ in THF	-2.43	0.91	(26)

exothermic. Except for the fact that energetics of reaction
sometimes are comparable for radical anion and displacement
reactions (9), there is no compelling evidence that mechanistic
similarities apply. In this regard a better concept of radical
anions is as radicals with high reduction potentials rather than
as anions. Moreover, the relatively weak basicity and reactivity
towards proton donors does not fit well with connotations of
anions (28).

For the case of anions, radical-like or pathways involving
single electron transfer are not in evidence from the work
described here. The products and reactivity patterns are expressed adequately by two-electron processes. A critical
difference between the one and two-electron process is the contribution of bond making to the latter. Apparently this provides
a substantial energy lowering that makes this path dominate in
the systems studied in spite of the fact that electron transfer
is possible. Conceivably this situation will be reversed for
these kinds of anions when bond making is rendered less important,
or when bond breaking more important, as for example _via_ steric
effects.

Acknowledgments. We are appreciative of the assistance of co-
workers and collaborators whose names appear in the references.
The financial assistance of NATO Grant No. RG 1069 is gratefully
acknowledged.

Literature Cited

(1). Berthelot, M., Ann. Chem. (Paris) (1867), 12, 155.
(2). Schlenk, W. and Bergmann, E., Annalen, (1928), 464, 1, and
      references cited therein.
(3). (a). Garst, J. F., Accounts of Chemical Research (1971),
          4, 400.
     (b). Garst, J. P., Barbas, J. T., and Barton, F. E.,
          J. Am. Chem. Soc., (1968), 90, 7159.
     (c). Garst, J. F., Roberts, R. D., and Abels, B. N.,
          J. Am. Chem. Soc., (1975), 97, 4926 and references
          cited therein.
(4). (a). Sargent, G. D., Cron, J. N. and Bank, S., J. Am. Chem.
          Soc., (1966), 88, 5363.
     (b). Bank, S. and Bank, J. F., Tetrahedron Lethers, (1969),
          5433.
     (c). Bank, S. and Juckett, D. A., J. Am. Chem. Soc.,
          (1975), 97, 567.
     (d). Bank, S. and Juckett, D. A., J. Am. Chem. Soc.,(1976),
          98, 7742.
(5). (a). Sargent, G. D., and Browne, M. W., J. Am Chem. Soc.,
          (1967), 89, 2788.
     (b). Sargent, G. D., and Lux, G. A., J. Am. Chem. Soc.,

(1968), 90, 7160.
(c). Sargent, G. D., Tetrahedron Letters, (1971), 3279 and references therein.
(6). (a). Kuivila, H. G., Accounts of Chemical Research (1968), 1, 299.
(b). Kuivila, H. G. Menapace, L. W., and Warner, C. R., J. Am. Chem. Soc., (1962), 84, 3854.
(c). Menapace, L. W. and Kuivila, H. G., J. Am. Chem. Soc., (1964), 86, 3047.
(7). Rogers, R. J., Mitchell, H. L., Fujiuara, V., and Whitesides, G. M., J. Org. Chem., (1974), 39, 857.
(8). (a). Lee, Y., Ph.D. Thesis, SUNY at Albany, Sept., 1974.
(b). Lee, Y. and Closson, W. D., Tetrahedron Letters, (1974), 1395.
(9). Bank, S., and Thomas, S. P., J. Org. Chem., (1977), 42, 2858.
(10). Lipkin, D., Divis, G. J., and Jordan, R. W., 155th National Meetings of the American Chemical Society, San Francisco, California, Preprints, Div. Petrol. Chem. (1968), 13, D61.
(11). Mazoleyrat, J. P., and Welvart, Z., Chem. Comm., (1972), 547.
(12). Garst, J. F., and Barton, F. E., J. Am. Chem. Soc., (1974), 96, 523.
(13). (a). Russell, G. A. and Lamson, D. W., J. Am. Chem. Soc., (1969), 91, 3967.
(b). Fischer, H., J. Phys. Chem., (1969), 73, 3834.
(14). Lawler, R. G., in Lepley, A. R., and Closs, G. L. "Chemically Induced Magnetic Polarization", Wiley, New York, 1973.
(15). (a). Daney, M., Lapoyade, R., Mary, M., and Bouas-Laurent, H., J. Organomental Chem., (1976), 92, 267.
(b). Harvey, P. G., and Cho, H., J. Am. Chem. Soc., (1974), 96, 2434.
(c). Zieger, H. E. and Gelbaum, L. T., J. Org. Chem., (1972), 37, 1012.
(d). Panek, E. J., J. Am. Chem. Soc., (1974), 96, 7959.
(16). Fu, P. P., Harvey, R. G., Paschal, J. W., and Rabideau, R.W., J. Am. Chem. Soc., (1975), 97, 1145.
(17). Electron transfer from the anthracene radical anion to The dihydroanthracenyl radical proceeds readily and completely. Bank, S. and Bockrath, B., J. Am. Chem. Soc., (1972), 94, 6076, and references cited therein.
(18). Breslow, R., and Mazur, S., J. Am. Chem. Soc., (1973), 95, 584.
(19). House, H. O. and Weeks, P.D., J. Am. Chem. Soc., (1975), 97, 2785.
(20). Lowry, T. H. and Richardson, K. S., "Mechanism and Theory in Organic Chemistry", pp. 146-147, Harper and Row, New York, 1976, and references cited therein.

(21). Kochi, J. K. and Powers, J. W., J. Am. Chem. Soc., (1970), 92, 137.

(22). Mohammad, M. and Kosower, E. M., J. Am. Chem. Soc., (1971), 93, 2709.

(23). Mohammad, M. and Kosower, E. M., J. Am. Chem. Soc., (1971), 93, 2713.

(24). Streitwieser, A., "Solvolytic Displacement Reactions", pp. 30-31, Mc Graw-Hill, New York, 1962, and references cited therein.

(25). Lowry, T. H. and Richardson, K. S., Reference 20, p. 193.

(26). Bank, S., Bank, J., Daney, M., Labrande, B., and Bouas-Laurent, H., J. Org. Chem., (1977), 42, in press.

(27). Danen, W. C., Tipton, T. J. and Saunders, D. G., J. Am. Chem. Soc., (1971) 93, 5186.

(28). (a). Bank, S., and Bockrath, B., J. Am. Chem. Soc., (1971), 93, 430.

     (b). Bockrath, B. and Dorfman, L. M., J. Am. Chem. Soc., (1974), 96, 5708.

RECEIVED December 23, 1977.

Electrophilic Reactions of Aromatic Cation Radicals

HENRY J. SHINE

Department of Chemistry, Texas Tech University, Lubbock, TX 79409

A list of reactive intermediates in organic reactions would include the following: carbonium ions, carbanions, free radicals, carbenes (and nitrenes, etc.), arynes, anion radicals, and cation radicals. Among these the chemistry of all but the ion radicals has been well established and documented. Involvement of carbonium ions and free radicals in organic reactions was deduced long before these intermediates were detected directly (i.e., spectroscopically). In contrast, with the advent of esr spectroscopy a large number of ion radicals became identified and characterized before much interest was shown in their chemistry (1). This situation has changed during the last several years, particularly with respect to cation radicals. A number of reactions of cation radicals have been discovered and explored by both chemical and electrochemical techniques.

A cation radical is formed by the removal of an electron from a neutral molecule (eq. 1). The species so formed is at the same

$$M - e^- \longrightarrow M^{\cdot+} \qquad (1)$$

time a cation and a radical (the remaining unpaired electron). Its chemistry therefore is of two kinds - both cationic and radical in nature.

The one-electron oxidation of eq. 1 may be achieved with chemical oxidants (e.g., sulfuric and perchloric acids, polyvalent metal ions) and anodic oxidation. The former methods are very useful for preparative-scale chemistry, while the latter is more versatile for exploring the mechanisms of cation-radical reactions (2, 3). Physical methods for gas-phase (pulse-radiolysis, electron-impact) and solution or glass-phase work (photoionization) can also be used.

It is understandable that electron-rich molecules, i.e., aromatics, heteroaromatics and alkenes, are among the most well explored in present cation radical chemistry, since such molecules are relatively easily oxidized. Some aromatics and heteroaromatics, furthermore, give cation radicals which are so stabilized by electron delocalization that they can be isolated as crystalline salts. It is, for the most part, cation radicals of this sort (from compounds 1-11) with which we have been working and whose chemistry will be described.

All of the cation radicals of 1-11 have been isolated as solid perchlorate salts. Compounds 1 (4) and 2 (5) are best oxidized by perchloric acid in anhydrous solvents. The cation radicals of 3 (6), 5 (7), 7, 8 (8), 10 (9), and 11 (10) can be made by oxidation with iodine-silver perchlorate. The crystalline cation radical perchlorates can be separated out or used in mixture with the solid silver iodide. The cation radical of 4 (7), as well as that of 3(6) is obtained by an interesting disproportionation reaction of the parent compound and its 5-oxide in perchloric acid (eq. 2). The reaction probably involves the dication and the

$$\underset{\overset{\parallel}{O}}{S} + \underset{S}{\diagup} + HClO_4 \longrightarrow 2 \ \overset{\cdot +}{\underset{S}{\diagup}} \ , \ ClO_4^- + H_2O \qquad (2)$$

$$\underset{\overset{\parallel}{O}}{S} + 2H^+ \rightleftharpoons \overset{++}{\underset{S}{\diagup}} + H_2O \qquad (3)$$

$$\overset{++}{\underset{S}{\diagup}} + \underset{S}{\diagup} \longrightarrow 2 \ \overset{\cdot +}{\underset{S}{\diagup}} \qquad (4)$$

parent (eqs. 3,4), but this has never been established. Although we have prepared the perchlorate of 6 chemically [using potassium dichromate in perchloric acid (11)], the more reliable method has been anodic oxidation in ethyl acetate solution (12). The cation radical of 10 is also obtainable anodically (13), but the method we prefer is electron exchange with 6·+ (eq. 5) (14). We ran into

$$10 + 6 \ \overset{\cdot +}{}\ ClO_4^- \longrightarrow 10 \ \overset{\cdot +}{}\ ClO_4^- + 6 \qquad (5)$$

so many problems with the dimerization of 7·+ that we abandoned its preparation in favor of 8·+ (8). In contrast, the preparation of 9·+ is so easy that oxidation by iodine alone gives 9·+I⁻ (15). We have also prepared a number of crystalline cation radical tetrafluoroborates by reaction of the parent substance with NOBF₄ (16), but are not satisfied yet that they can entirely replace isolated perchlorates for chemical studies.

Chemical studies amount mostly to reactions with nucleophiles; that is, the known chemistry is mostly of the cationic nature of cation radicals. Most of the reactions, furthermore, have the stoichiometry of eq. 6, where ArH and Nu⁻ are used in

$$2ArH^{\cdot +} + Nu^- \longrightarrow ArNu + ArH + H^+ \qquad (6)$$

general terms. The stoichiometry shows that a molecule of parent
compound is obtained for each molecule of substitution product,
and it is in this way, of course, that unpaired electrons eventually become paired.

A number of cases are now known in which the details of the
reaction shown in eq. 6 have been worked out, principally by
Blount (17, 18) and Parker (19, 20). The particular reactions
will be described, but, in general terms, most of the reactions
now documented occur by what Blount (17) and others term the
half-regeneration mechanism (eq. 7-9). The rate-determining

$$ArH^{\cdot +} + Nu^- \rightleftharpoons (ArH-Nu)^{\cdot} \qquad (7)$$

$$(ArH-Nu)^{\cdot} + ArH^{\cdot +} \rightleftharpoons (ArH-Nu)^+ + ArH \qquad (8)$$

$$(ArH-Nu)^+ + B^- \longrightarrow ArNu + HB \qquad (9)$$

step in this sequence may be either in eq. 8 or eq. 7, depending
on the cation radical, nucleophile, and medium.

Reactions With Water

Polynuclear aromatics, e.g., perylene (9), and dibenzodioxin
(12, 21) cation radicals are converted into quinones. The reaction must involve quite a number of steps since the stoichiometry
is complex, eq. 10. Organosulfur cation radicals (except $3^{\cdot +}$) are

$$6ArH^{\cdot +} + 2H_2O \longrightarrow 5ArH + O=Ar=O + 6H^+ \qquad (10)$$

converted into sulfoxides, with the stoichiometry of eq. 11, where
$R_2S^{\cdot +}$ represents the cation radicals of, for example, 1 and 2.
Reaction of $1^{\cdot +}$ with water has been studied in detail and has been

$$2R_2S^{\cdot +} + H_2O \longrightarrow R_2S=O + R_2S + 2H^+ \qquad (11)$$

shown by Blount to be a half-regeneration reaction, which is in
fact third order in water (eq. 12) (17). This is in contrast to

$$-d(1^{\cdot +})/dt = k_{total}[1^{\cdot +}]^2[H_2O]^3/[H_3O^+] \qquad (12)$$

the proposal for a disproportionation mechanism, made when this
reaction was first discovered (4).

Reactions With Ammonia and Amines

The organosulfur cation radicals $1^{\cdot +}$, $2^{\cdot +}$, $4^{\cdot +}$, and $5^{\cdot +}$ react
with ammonia and alkyl- and dialkylamines at their sulfur atoms.
In principle, reaction with ammonia should give the sulfilimine
salt, as in eq. 13. In practice, we have achieved this only with

$$2 \; {}^{\cdot +}_{\diagdown S \diagup} \;, \; ClO_4^- + 2NH_3 \longrightarrow \; \overset{\overset{+}{N}H_2}{\underset{ClO_4^-}{\underset{||}{\diagdown S \diagup}}} \; + \; \diagdown S \diagup \; + \; NH_4ClO_4 \qquad (13)$$

$2^{\cdot +}$, the phenoxathiin sulfilimine being obtained if a rapid stream of ammonia is bubbled into a solution of $2^{\cdot +}$ (5). If the flow of ammonia is slow another product is obtained (12b). It is this type of product (12a, c, d) which is obtained from reaction of ammonia with $1^{\cdot +}$, $4^{\cdot +}$, and $5^{\cdot +}$, (7, 22). The products 12 arise apparently from formation and continued reaction of the sulfilimine.

Reaction of primary and secondary alkylamines with cation radicals has received limited attention. Sioda (23) reported that when DPA$^{\cdot +}$ reacted with ammonia and some primary alkylamines DPA was formed in half the amount of the DPA$^{\cdot +}$ used; that is, as if reaction had occurred according to eq. 6. However, the substitution product, if formed, was not sought. The only complete reactions we are aware of are our own with organosulfur cation radicals (7, 24-26). These react with primary amines to give N-alkylsulfilimine salts (eq. 14). Reaction with dialkylamines

$$2 \; {}^{+\cdot}_{\underset{ClO_4^-}{\diagdown S \diagup}} \;, \; + \; 2RNH_2 \longrightarrow \; \overset{\overset{+}{H}NR}{\underset{ClO_4^-}{\underset{||}{\diagdown S \diagup}}} \;, \; + \; \diagdown S \diagup \; + \; RNH_3ClO_4 \qquad (14)$$

gives N,N-dialkylaminosulfonium salts (eq. 15). Over 60 examples

$$2 \; {}^{+\cdot}_{\underset{ClO_4^-}{\diagdown S \diagup}} \;, \; + \; 2R_2NH \longrightarrow \; \overset{\overset{+}{N}R_2}{\underset{ClO_4^-}{\underset{||}{\diagdown S \diagup}}} \;, \; + \; \diagdown S \diagup \; + \; R_2NH_2ClO_4 \qquad (15)$$

these two reactions have been carried out. Tertiary alkylamines do not react in this way. Among the few reactions we have carried out we observed only electron transfer (eq. 16) (7). We do not yet know what has happened to the amine; that is, if $R_3N^{\cdot +}$ was

$$\overset{+\cdot}{\diagdown S \diagup} \; + \; R_3N \longrightarrow \; \diagdown S \diagup \; + \; ? \qquad (16)$$

$$R = Me, \; Et$$

formed and underwent other reactions, such as dealkylation. The failure of trimethyl- and triethylamine to react is somewhat surprising in view of the reactivity of pyridine (see below) and the reaction of triethylamine with DPA$^{\cdot +}$ to give 13 (27). On the other hand the oxidation potentials of trialkylamines are in the

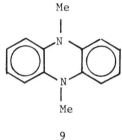

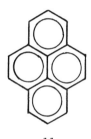

1, X = S (Thianthrene)	6, X = O (Dibenzodioxin)
2, X = O (Phenoxathiin)	7, X = NH (Phenoxazine)
3, X = NH (Phenothiazine)	8, X = NPh
4, X = NMe	
5, X = NPh	

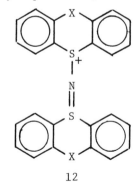

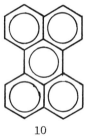

9
(5,10-Dimethyl-5,10-
dihydrophenazine)

10
(Perylene)

11
(Pyrene)

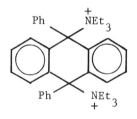

12

13

12a, X = S; 12c, X = NMe
12b, X = O; 12d, X = NPh

range 0.8–1.0 V (vs. SCE) as compared with 2.1 V for pyridine, so electron transfer from the tertiary amines is a lot easier than from pyridine.

A side reaction that occurs between primary alkylamines and the organosulfur cation radicals is the formation of 12 (particularly 12a, c, d). We do not yet know how this occurs.

The reaction of pyridine (and picolines and lutidines) with cation radicals is one of the earliest to be established and documented. In fact, it is referred to as the pyridination reaction. Reaction occurs at the pyridine's nitrogen atom. That is, electrophilic aromatic substitution (see below) does not occur, the pyridine ring being too unreactive for this. The pyridine behaves as a nucleophile and leads, for example, with DPA$^{\cdot+}$ to 14, and with 10$^{\cdot+}$ to 15. The reaction with DPA$^{\cdot+}$ has been particularly well studied, and has been shown to be a half-regeneration reaction (28).

Reaction of the organosulfur cation radicals gives ring-pyridinated products, not S-pyridinated (e.g., 16 not 17). We supplied an explanation for this and its difference from the ammonia and alkylamine reactions [and the water reaction (6)]. It is that at some stage of product formation a dication intermediate is required (18). In order for the intermediate to lead to a

$$\underset{18}{\overset{\overset{\displaystyle +}{\underset{\displaystyle \|}{HNR_2}}}{\underset{\displaystyle S}{\|_+}}} \quad \xrightarrow{\text{B}^-} \quad \overset{\overset{\displaystyle +}{\underset{\displaystyle \|}{NR_2}}}{\underset{\displaystyle S}{\|}} \quad + \text{ HB} \tag{17}$$

stable product proton loss must be possible, and this cannot occur from the dication 17. This idea has been validated by Blount, who has shown, kinetically, however, that 17d (X = NC$_6$H$_5$) is formed but is not stable to reaction with pyridine (eq. 18) (18). That

$$17d + Py \longrightarrow 16d + PyH^+ \tag{18}$$

is, the pyridination reaction in that case is second order in pyridine. Blount has also shown kinetically that 17a (X = S) is formed and is unstable to reaction with water (17). But in this case reaction occurs again at sulfur (eq. 19), the sulfoxide being formed.

$$17a + H_2O \longrightarrow \overset{\overset{\displaystyle O}{\underset{\displaystyle \|}{}}}{\underset{\displaystyle S}{}} + PyH^+ + H^+ \tag{19}$$

Reaction of arylamines with cation radicals usually leads to electron transfer (eq. 20). The coupling of arylamines leading,

$$ArH^{\cdot+} + ArNH_2 \longrightarrow ArH + ArNH_2^{\cdot+} \tag{20}$$

for example, to benzidines, is a cation radical reaction explored
a great deal electrochemically, but is outside the scope of our
discussion.

Reaction With Ketones

One would not ordinarily regard ketones as nucleophiles. An
interesting reaction occurs, though, between dialkyl- and alkyl
aryl ketones and organosulfur cation radicals. The products are
β-ketoalkylsulfonium salts, the stoichiometry of whose formation
follows the usual pattern (eq. 21). A variety of ketones has
been used (29-31), such as acetone, butanone, methyl isopropyl

$$2 \quad \overset{\cdot+}{\diagup S \diagdown} \quad + \quad R_2CHCOR' \quad \longrightarrow \quad \diagup S \diagdown \quad + \quad \overset{R_2CCOR'}{\underset{\diagup S \diagdown}{|}} \quad + \quad H^+ \quad (21)$$

ketone, indanone-1, tetralone-1, and some cycloalkanones. The
mechanism of reaction has not been investigated, but it has been
suggested that reaction occurs between the cation radical and the
enolic form of the ketone (31). Those β-ketosulfonium salts with
an α-H are readily converted into sulfur ylides by treatment with
base. In some cases the ylide is so readily formed that it is
obtained directly from the cation radical reaction. The β-ketosulfonium salts have also been found very useful for preparing
α-substituted ketones by nucleophilic displacement (e.g., eq. 22).

$$Nu^- \quad + \quad \overset{CH_2COR}{\underset{\diagup S \diagdown}{|}} \quad \longrightarrow \quad NuCH_2COR \quad + \quad \diagup S \diagdown \quad (22)$$

Reactions With Alkenes and Alkynes

In contrast with the great wealth of information on both
cationic and free-radical additions to alkenes and alkynes, not
too much is known about analogous cation radical reactions. The
most commonly known among these are additions of aminium radicals.
The overall reaction is that of, say, an N-chloramine or N-nitros-
amine in acid solution, brought about photochemically or by reaction with ferrous ion (eq. 23) (2). A chain reaction occurs in
which $R_2NH^{\cdot+}$ is a participant.

$$R_2\overset{+}{N}HX \quad + \quad \overset{\diagdown}{\underset{\diagup}{C}}=\overset{\diagup}{\underset{\diagdown}{C}} \quad \longrightarrow \quad R_2\overset{+}{N}H-\overset{|}{\underset{|}{C}}-\overset{|}{\underset{|}{C}}-X \quad (23)$$

We have discovered an entirely different addition of organosulfur cation radicals, resulting in the formation of alkane
disulfonium salts (eq. 24) and vinyl disulfonium salts (eq. 25)
(32).

$$2 \overset{\cdot+}{\underset{\diagdown}{S}} + \overset{\diagup}{\underset{\diagup}{C}}=\overset{\diagdown}{\underset{\diagdown}{C}} \longrightarrow \quad -\overset{\overset{\displaystyle \overset{+}{S}}{\displaystyle |}}{\underset{\displaystyle \underset{+}{\underset{\displaystyle S}{|}}}{C}}\!\!-\!\!-\!\!\overset{|}{\underset{|}{C}}- \qquad (24)$$

$$2 \overset{\cdot+}{\underset{\diagdown}{S}} + -C\equiv C- \longrightarrow \quad \overset{\diagdown}{\underset{-S^+}{C}}=\overset{\overset{\displaystyle S^+}{\diagup}}{\underset{\diagdown}{C}} \qquad (25)$$

Here, again, we have used only the cation radicals 1 and 2. The phenothiazine cation radicals 4 and 5 react only very slowly. Among alkenes and alkynes reacting successfully are ethene, propene, 2-butene, cycloalkenes, acetylene, propyne, diphenylacetylene. The products are written in eq. 24 and 25 as trans adducts but we have not yet studied the stereochemistry of addition. Whether or not thiiranium and thiirenium ion radical intermediates are involved is an interesting conjecture yet to be explored.

These additions must be radical in nature in that an adduct is presumably formed first and then pairs with another cation radical (e.g., eq. 26). At the same time, however, the unsaturated compound is inactivated by electron withdrawing groups (e.g., as in propargyl chloride), so that the transition state to addition must also have much cationic character.

$$-\overset{|}{\underset{\underset{+}{\underset{\diagdown S}{|}}}{C}}-\overset{|}{\underset{|}{C}}\cdot + \overset{+\cdot}{\underset{\diagdown}{S}} \longrightarrow -\overset{|}{\underset{\underset{+}{\underset{\diagdown S}{|}}}{C}}-\overset{\overset{\displaystyle \overset{+}{S}}{\displaystyle |}}{\underset{|}{C}}- \qquad (26)$$

The adducts themselves are interesting classes of compounds in which the potential for competition between elimination and substitution reactions is being studied.

Reactions With Aromatics

Cation radicals react as electrophiles with aromatics. The scope of this type of substitution is not too well explored, though. Amination of aromatics with dialkylaminium radicals has been known for some time (2). In recent years the arylation of organosulfur cation radicals was discovered in our own laboratories (eq. 27) (33, 34). The reaction is limited in that the

$$2 \overset{\cdot+}{\underset{\diagdown}{S}} + C_6H_5X \longrightarrow \overset{\overset{\displaystyle C_6H_4X(\underline{p})}{\displaystyle |}}{\underset{\diagdown}{S^+}} + \overset{}{\underset{\diagdown}{S}} + H^+ \qquad (27)$$

substituent X must be a good electron donor (OMe, OH, Me, NHAc), and that reaction occurs, apparently, only at the para-position. It was thought initially that the reaction was preceded by dis-proportionation of the cation radical, the electrophile then being the dication (33), but this was shown by Parker to be in-correct (19, 20). In the case of anisole and 1·+ the reaction follows the half-regeneration steps (eqs. 7-9) (19). In the case of phenol, though, reaction follows these steps in acidic but not neutral solutions. In the latter, the reaction is first-order in 1·+, the initial complex undergoing rate-determining deprotona-tion followed by faster oxidation by the second molecule of cation radical (eq. 28, 29) (20). Not many other examples of this type arylation are known. Anthracene and 9-phenylanthracene

$$(1-ArOH)^{\cdot+} \longrightarrow (1-ArO)^{\cdot} + H^+ \tag{28}$$

$$(1-ArO)^{\cdot} + 1^{\cdot+} \longrightarrow ArOH + 1 \tag{29}$$

cation radicals are arylated at the 9,10- and 10-positions respectively (35). Even benzene and chlorobenzene can be used, whereas their reactions with 1·+ are too slow to be useful.

We have more recently discovered a more versatile method of arylating heterocyclic organosulfur cation radicals, using organometallics R_2Hg and R_2Zn (eq. 30) (14). The reaction is even wider in scope in that dialkylmetals can be used too, which,

$$2 \quad \overset{\cdot+}{\swarrow^S\searrow} + R_2M \longrightarrow \overset{}{\swarrow^S\searrow} + \overset{\overset{R}{|}_+}{\swarrow^S\searrow} + RM^+ \tag{30}$$

of course, is not possible as an analogue of direct arylation (eq. 27). The advantage of the newer method is that substitution at any position (o-, m-, p-) in the aryl ring can be achieved exclusively provided the right Ar_2M is used (e.g., di-o-tolylmer-cury, di-m-chlorophenylmercury). Furthermore, moderately deacti-vated aryl rings can be introduced (phenyl and chlorophenyl) which was not possible in the older method. Apparently, the polariza-tion of the aryl-metal bond is sufficient to make reaction with the organosulfur cation radical feasible. Strongly deactivated rings, as in di-m-nitrophenyl- and di-perfluorophenylmercury can-not be used, though.

Although the cation radicals 1·+ and 2·+ react with R_2Hg and R_2Zn, 4·+ and 5·+ react only with the more active R_2Zn (R = Et, C_6H_5). Limited experiments, also, with 1·+ and BuLi and Grignard reagents resulted in electron transfer.

These organometallic reactions are of further interest in that they may proceed with retention of configuration, as in other reactions of organomercurials (36). That is, optically active groups from R_2*Hg would then be introduced on sulfur. Furthermore, groups such as neopentyl and camphenyl may be attachable to sulfur with the use of the appropriate R_2Hg (37), and this may turn out to be the only way of making such attachments.

But, we do not know yet how these organomercury reactions do occur. The good yields and the absence of by-products in most of the reactions suggest that the alkyl and aryl groups are not free of the organometal before attachment to sulfur occurs. Electron transfer from R_2Hg to the cation radical prior to alkyl- or aryl group transfer may occur in analogy with other reactions of R_2Hg (38), although the high oxidation potential (1.8 V vs. Ag/Ag^+) of diphenylmercury (39) is not too persuasive of this possibility. The fact that we carry out the reactions in air without apparent interference by oxygen (cf. 38) suggests that free alkyl radicals are not involved. These reactions and their stereochemistry are under study.

Reactions With Inorganic Anions

The reactions we refer to here are with halide (F^-, Cl^-, Br^-, and I^-), cyanide, nitrite and nitrate ions. We have carried out a number of reactions with some of the cation radicals among $1^{·+}$ - $10^{·+}$. The mechanisms of the reaction in some cases need to be re-examined.

The simplest reaction to be expected with halide ions is nucleophilic halogenation (eq. 31). This seemingly simple

$$2 \text{ ArH}^{·+} + X^- \longrightarrow \text{ArX} + \text{ArH} + H^+ \tag{31}$$

reaction has complexities, however, that need untangling. One of these complexities is electron exchange (eq. 32). This reaction

$$\text{ArH}^{·+} + X^- \rightleftharpoons \text{ArH} + 1/2 \text{ X}_2 \tag{32}$$

occurs in all cases of our experience (except that of $9^{·+}$) with iodide ion, and is used, in fact, in the iodometric assay of our cation-radical salts. The electron exchange is reversible, and in the case of iodine the reverse reaction is often used to prepare cation radicals. In that case a silver salt ($AgClO_4$, $AgBF_4$) is a co-reactant used to carry the equilibrium to the left. We do not, ourselves, know of cases in which iodide ion reacts as in eq. 31. The cation radical $9^{·+}$ is stable toward iodide ion.

The possibility of electron exchange with chloride and bromide ion is a problem. In several cases exchange is thought to occur and to be followed by electrophilic halogenation (eq. 33). The final result is no different from that of eq. 31 and

$$\text{ArH} + X_2 \longrightarrow \text{ArX} + HX \tag{33}$$

in most cases a proper distinction has not been made between the two processes.

Reaction of chloride ion with $1^{\cdot+}$ resulted in part in electron exchange since some chlorine was removed by nitrogen flow and was assayed (9%). Thianthrene (47%), thianthrene 5-oxide (18, 33%) and a very small amount (0.3%) of 2-chlorothianthrene (19) were also obtained (4). Since molecular chlorine slowly chlorinates thianthrene and since thianthrene 5-oxide can be prepared by oxidation of thianthrene with chlorine in wet solvents, the cation radical results are thought to "fit together" (eqs. 34-37). Nevertheless a proper kinetic and mechanistic study needs to be made.

$$1^{\cdot+} + Cl^- \longrightarrow 1 + 1/2\ Cl_2 \tag{34}$$

$$1 + Cl_2 \longrightarrow 1(Cl_2) \tag{35}$$

$$1(Cl_2) \longrightarrow 19 + HCl \tag{36}$$

$$1(Cl_2) + H_2O \longrightarrow 18 + 2HCl \tag{37}$$

Reaction of $1^{\cdot+}$ with bromide ion has not been studied. Reaction of $3^{\cdot+}$ with chloride and bromide ion gave 3 (about 45%), the 3-halogenophenothiazine (20, about 35%) and some 3,7-dihalogeno-phenothiazine (21, about 5%) (6). The formation of 21 was particularly perplexing, and led to the feeling that electron exchange (eq. 32) had occurred and electrophilic halogenation was responsible for the formation of 20 and 21. We do not know though that we can rule out successive half-regeneration reactions (eq. 38-42) since it is probable that the relative oxidation potentials of 3 and its halogenoderivatives would not entirely inhibit the successive steps.

$$3^{\cdot+} + X^- \longrightarrow (3-X)^{\cdot} \tag{38}$$

$$(3-X)^{\cdot} + 3^{\cdot+} \longrightarrow 3 + 20 + H^+ \tag{39}$$

$$20 + 3^{\cdot+} \longrightarrow 20^{\cdot+} + 3 \tag{40}$$

$$20^{\cdot+} + X^- \longrightarrow (20-X)^{\cdot} \tag{41}$$

$$(20-X)^{\cdot} + 3^{\cdot+} \longrightarrow 21 + 3 + H^+ \tag{42}$$

In some of our earlier work it was reported that electron exchange also occurred between $10^{\cdot+}$ and chloride and bromide ion, leading only to the formation of 10 (9). We now think that this is not entirely correct. We now know that the chloride ion reaction leads to 3-chloro- (22) and a dichloroperylene (23), presumed to be the 3,9-isomer (40). Whether or not their formation

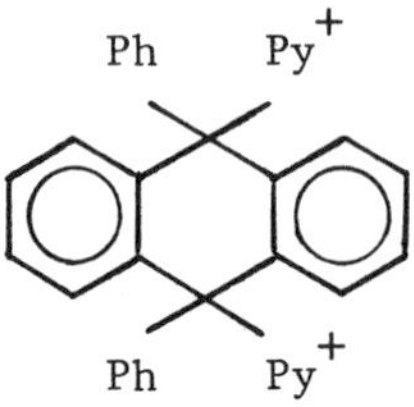

14

15

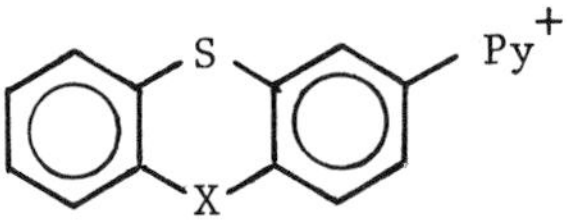

16

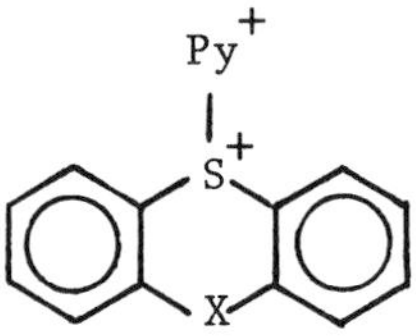

17

16a, 17a: X = S;   16c, 17c: X = NMe
16b, 17b: X = 0;   16d, 17d: X = NPh

Py^+ =

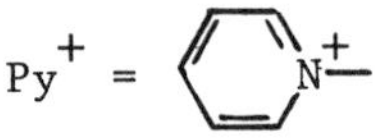

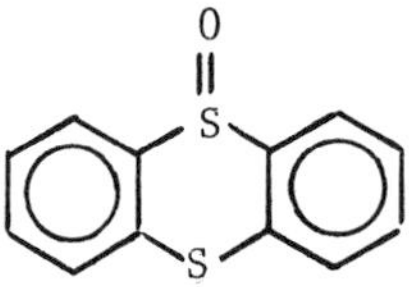

18 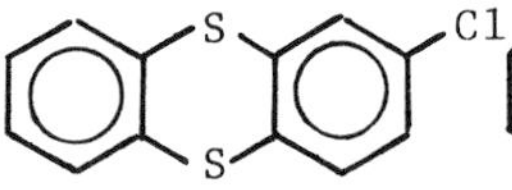   19 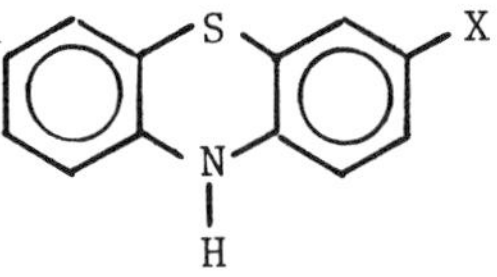   20

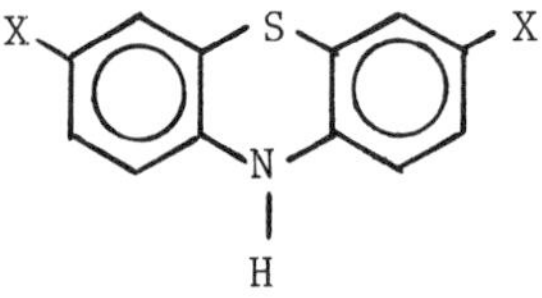

21    22 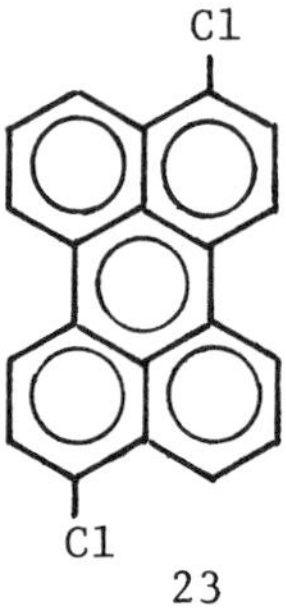   23

is preceded by electron exchange is still not settled but we feel
that it probably is not. The bromide ion reaction also leads to
the halogenoperylenes, but here preliminary electron exchange
still appears to be likely. Reaction of chlorine with perylene
also rapidly gives 22 and 23 (40). In fact the preparation of 22
appears, from the literature, never to have been achieved earlier.
The great difficulty with the perylene work is in separating
perylene from the monohalogeno and, to a lesser extent, the
dihalogenoperylenes. Separation of the chloro compounds has now
been achieved by multiple-plate TLC and column chromatography,
and the products have been identified by mass spectrometry.

Fluoride-ion reactions are the most perplexing of all. A
fairly large number of anodic fluorinations have been reported in
the literature (41). These are anodic oxidations of aromatics
and alkenes at potentials well below that of fluoride ion. Yet,
examples of fluorination of isolated cation radicals are, so far,
very rare. Reaction of fluoride ion with $3^{\cdot+}$ gave 3 (38%), its
3,10'-dimer (13%) and the well-known green dimer cation (6). In
leading to the dimer, fluoride ion has behaved as a base. For
some years it appeared that fluoride ion was too poor a nucleophile
to react as in eq. 31, but we believe now that this is not
correct. Mass spectrometry has shown that a monofluoro-N-phenyl-
phenoxazine dimer is obtained from $8^{\cdot+}$ (8). Most recently mass
spectrometry has also shown that a small amount of fluoroperylene
is formed from $10^{\cdot+}$, in contrast with our earlier report (9).
The major problem encountered here is in separating fluoroperylene
and perylene. This has not been possible at all by TLC (40).
It is evident now, therefore, that fluoride ion is indeed a poor
nucleophile but reaction does occur to a small extent. Some
cation radicals are inert even in anodic fluorination reactions,
e.g., that of tetraphenylethylene (42). However, why many other
anodic fluorinations are successful and "chemical" reactions are
not is yet to be answered. Possibly, the answer lies in the
second oxidation step (eq. 43) which may be easy at an anode but
difficult by, say, $ArH^{\cdot+}$ in "chemical" reactions.

$$(ArHF)^{\cdot} \; -e^{-} \longrightarrow (ArHF)^{+} \longrightarrow ArF + H^{+} \qquad (43)$$

The summation appears to be that halide ion reactions need
much further study. The only reaction which has been studied
kinetically is that of $DPA^{\cdot+}$ with chloride ion, a reaction which
follows the half-regeneration pathway, leading eventually to
dichloro-DPA (43). In contrast, incidentally, $DPA^{\cdot+}$ and bromide
ion undergo exclusive electron exchange (23).

Our experiences with nitrite-ion reactions show that both
nitration and oxygen-atom transfer can occur. Thus, $6^{\cdot+}$ and $10^{\cdot+}$
give 2-nitrodibenzodioxin (12) and 3-nitroperylene (10) respectively according to eq. 6. The cation radical of zinc tetra-
phenylporphyrin is similarly nitrated at one of the pyrrolic
carbon atoms (44). The perylene reaction is so facile that it

can be carried out in situ, that is by shaking a solution of
perylene with silver nitrite and iodine, but the perylene must be
in large excess, say five-fold, over the iodine, to avoid poly-
nitration, suggesting that in this case nitrogen dioxide may be
the nitrating agent (10, 14). Pyrene can be nitrated similarly
(10). The cation radical $9^{\cdot+}$ is converted entirely into mono-
nitro-10 in what appears to be a two-stage reaction. We think
that the first stage follows eq. 6, and then the 9 formed in that
stage is slowly nitrated by nitrous acid (15). Reaction of $3^{\cdot+}$
gives 3-nitrophenothiazine also according to eq. 6 (6), and this
is in accord with nitration of 6 with ferric chloride-nitrite ion
(45). In contrast, $1^{\cdot+}$ and $2^{\cdot+}$ are converted entirely into the
corresponding 5-oxides (6, 46). The last reactions have been
interpreted (6) as in eq. 44, but we do not really know if,

$$\overset{\cdot+}{\diagup S \diagdown} \; + \; NO_2^{-} \; \longrightarrow \; \overset{ONO}{\underset{\cdot}{\diagup S \diagdown}} \; \longrightarrow \; \overset{O}{\underset{\|}{\diagup S \diagdown}} \; + \; NO \qquad (44)$$

perhaps, electron exchange does not occur first, that is, that if
reaction is not between, say, 1 and NO_2. The oxidation potential
of nitrite ion in acetonitrile is 0.96 V (vs. SCE), so that electron exchange with nitrite in cation radical reactions is a distinct possibility, as has been noted by Dolphin (47). This
applies not only to reactions of $1^{\cdot+}$ and $2^{\cdot+}$ but to the other
cation radical reactions too. In that they are all "clean"
reactions it would appear that if electron exchange does occur
in these reactions the NO_2 must react immediately and completely
with the redox partner. A more complicated reaction has been
noted with $5^{\cdot+}$ and nitrite ion, the products being not only
10-phenylphenothiazine 5-oxide but nitro- and dinitro-10-phenyl-
phenothiazine 5-oxides, too (48).

Reactions with nitrate ion are similar to those of nitrite
ion. That is, $6^{\cdot+}$ gave 2-nitrodibenzodioxin (12) and $1^{\cdot+}$ gave
thianthrene 5-oxide (6). The last reaction was interpreted
analogously to eq. 44, but the mechanism of the ring nitration is
entirely unknown. Anodic nitrations which appear to be cation
radical-nitrate ion reactions can be found in the literature, and
their mechanisms are also unsolved.

Not much is known about the last of our inorganic ion reactions, that of cyanide ion. A number of anodic "cyanations" and
cyanation-methoxylation reactions are in the literature. These,
anodic oxidations in solutions (methanol usually) of cyanide ion,
are reactions of cyanide with cation radicals. Yet, the only
successful "chemical" reactions, we are aware of are with $10^{\cdot+}$
which gave small yields of 1- and 3-perylene nitrile (13), and
with zinc octaethylporphyrin cation radical, which gives 68%
of meso-cyanooctaethylporphyrin (49).

Acknowledgements

Support for research in cation-radical chemistry was received
from the Robert A. Welch Foundation (Grant D-028), the National
Science Foundation (Grant CHE 75-02794), and Texas Tech University
Institute for Research.

Abstract

Aromatic and heteroaromatic cation radicals are readily prepared by one-electron oxidation of the parent compound. A number
of these cation radicals are sufficiently stable to be isolated
as solid salts, usually the perchlorates. These cation radicals
react with a variety of nucleophiles, such as water, halide ions,
nitrite, and cyanide ions. Reactions with purely aromatic cation
radicals lead to ring-substituted aromatics. Reactions of aromatic organosulfur cation radicals occur more often at the sulfur
atom, and lead to sulfonium salts of various kinds. Examples of
the last type are reactions with aromatics (e.g., anisole),
dialkyl- and diarylmercurials, primary and secondary alkylamines,
and dialkyl and alkylaryl ketones. Addition of organosulfur
cation radicals to alkenes and alkynes is also described. It is
also possible that a cation radical and nucleophile undergo electron exchange instead of or prior to the substitution reaction.
This possibility is particularly strong in reactions with halide
ions and nitrite ion, and in these cases subsequent product-
forming reactions may, in fact, be those of electrophilic substitution. Discussion of reactions of this kind is given.

Literature Cited

1. Kaiser, E. T., and Kevan, L. (Ed.), "Radical Ions." Interscience, New York, N. Y., 1968.
2. Bard, A. J., Ledwith, A., and Shine, H. J., in "Advances in
   Physical Organic Chemistry," V. Gold and D. Bethel (Ed.),
   Vol. 13, pp. 155-278, 1976.
3. Eberson, L., and Nyberg, K., in "Advances in Physical Organic
   Chemistry," V. Gold and D. Bethel (Ed.), Vol. 12, pp. 1-129,
   1976.
4. Murata, Y., and Shine, H. J., J. Org. Chem. (1969), 34, 3368.
5. Mani, S. R., and Shine, H. J., J. Org. Chem. (1975), 40,
   2756.
6. Shine, H. J., Silber, J. J., Bussey, R. J., and Okuyama, T.,
   J. Org. Chem. (1972), 37, 2691.
7. Bandlish, B. K., Padilla, A. G., and Shine, H. J., J. Org.
   Chem. (1975), 40, 2590.
8. Wu, S.-M., unpublished work.
9. Ristagno, C. V., and Shine, H. J., J. Org. Chem. (1971), 36,
   4050.

10. Ristagno, C. V., and Shine, H. J., J. Am. Chem. Soc. (1971), 93, 1811.
11. Padilla, A. G., unpublished work.
12. Shine, H. J., and Shade, L. R., J. Heterocycl. Chem. (1974), 11, 139.
13. Shine, H. J., and Ristagno, C. V., J. Org. Chem. (1972), 37, 3424.
14. Bandlish, B. K., unpublished work.
15. Henderson, G. N., and Pendarvis, R. O., unpublished work.
16. Bandlish, B. K., and Shine H. J., J. Org. Chem. (1977), 42, 561.
17. Evans, J. F., and Blount, H. N., J. Org. Chem. (1977), 42, 976.
18. Evans, J. F., Lenhard, J. R., and Blount, H. N., J. Org. Chem. (1977), 42, 983.
19. Svanholm, U., Hammerich, O., and Parker, V. D., J. Am. Chem. Soc. (1975), 97, 101.
20. Svanholm, U., and Parker, V. D., J. Am. Chem. Soc. (1976), 98, 997.
21. Cauquis, G., and Maurey-Mey, M., Bull Soc. Chim. France (1972), 3588.
22. Shine, H. J., and Silber, J. J., J. Am. Chem. Soc. (1972), 94, 1026.
23. Sioda, R. E., J. Phys. Chem. (1968), 72, 2322.
24. Shine, H. J., and Kim, K., Tetrahedron Lett. (1974), 99.
25. Shine, H. J., and Kim, K., J. Org. Chem. (1974), 39, 2537.
26. Bandlish, B. K., Mani, S. R., and Shine, H. J., J. Org. Chem. (1977), 42, 1538.
27. Blount, H. N., J. Electroanal. Chem. (1973), 42, 271.
28. Shang, D. T., and Blount, H. N., J. Electroanal. Chem. (1974), 54, 305.
29. Kim, K., and Shine, H. J., Tetrahedron Lett. (1974), 4413.
30. Kim, K., Mani, S. R., and Shine, H. J., J. Org. Chem. (1975), 40, 3857.
31. Padilla, A. G., Bandlish, B. K., and Shine, H. J., J. Org. Chem. (1977), 42, 1833.
32. Bandlish, B. K., Mani, S. R., and Shine, H. J., Abstracts 172nd Meeting of the American Chemical Society, San Francisco, 1976, ORGN 094.
33. Silber, J. J., and Shine, H. J., J. Org. Chem. (1971), 36, 2923.
34. Kim, K., Hull, V. J., and Shine, H. J., J. Org. Chem. (1974), 39, 2534.
35. Svanholm, U., and Parker, V. D., J. Am. Chem. Soc. (1976), 98, 26942.
36. Jensen, F. R., and Rickborn, B., "Electrophilic Substitution of Organomercurials," McGraw-Hill, New York, N. Y., 1968, pp. 64, 178.

37. Makarova, L. G., and Nesmeyanov, A. N., "The Organic Compounds of Mercury," Vol. 4 in "Methods of Elemento-Organic Chemistry," A. N. Nesmeyanov and K. A. Kocheshkov (Ed.), North-Holland, Amsterdam, 1967.
38. Chen., J. Y., Gardner, H. C., and Kochi, J. K., J. Am. Chem. Soc. (1976), 98, 6150.
39. Maironovskii, S. G., Russ. Chem. Rev. (1976), 45, 298.
40. Bandlish, B. K., and Stephenson, M. T., unpublished work.
41. Rozhkov, I. N., Russ. Chem. Rev. (1976), 45, 615.
42. Rozhkov, I. N., Aliev, I. Ya., and Knunyants, I. L., Izvest. Akad. Nauk. SSSR, Ser. Khim. (1976), 1418.
43. Evans, J. F., and Blount, H. N., J. Org. Chem. (1976), 41 516.
44. Padilla, A. G., Wu, S.-M., and Shine, H. J., J. Chem. Soc. Chem. Commun. (1976), 236.
45. Danecke, J., and Wanzlick, H.-W., Justus Liebigs Ann. Chem. (1970), 740, 52.
46. Mani, S. R., unpublished work.
47. Johnson, E. C., and Dolphin, D., Tetrahedron Lett. (1976), 2197.
48. Bandlish, B. K., Kim, K., and Shine, H. J., J. Heterocycl. Chem. (1977), 14, 209.
49. Evans, B., and Smith, K., Tetrahedron Lett. (1977), 3079.

RECEIVED December 23, 1977.

23

Use of the Semidione Spin Probe to Study Molecular Rearrangements (1)

GLEN A. RUSSELL, K. SCHMIDT, C. TANGER, E. GOETTERT, M. YAMASHITA, Y. KOSUGI, J. SIDDENS, and G. SENATORE

Department of Chemistry, Iowa State University, Ames, IA 50011

Radical ions derived formally by the one electron reduction of α-diones are termed 1,2-semidiones.

$$R-\overset{\text{O}}{\underset{\text{O}}{\overset{\|}{C}}}-\overset{\|}{\underset{\text{O}}{C}}-R + e^- \rightarrow R-\overset{\cdot}{\underset{\text{O}}{C}} \doteq \overset{\cdot}{\underset{\text{O}}{C}}-R \rightleftarrows R-\overset{\overset{\text{O}^-}{\|}}{C} \doteq \overset{\cdot}{\underset{\cdot\text{O}}{C}}-R \qquad (1)$$

Spin density is about equally distributed over the four atoms in the π-system, an estimate of the carbonyl carbon spin density being 0.35 for cis-dimethylsemidione and 0.28 for cyclohexane-1,2-semidione (2). This leads to considerable double bond character between the carbonyl carbon atoms of a cis-semidione in the range of 30%. Use of the valence bond structure 1 emphasizes

1

this double bond character and classifies the 1,2-semidione as an olefin derivative. This paramagnetic olefin derivative can be observed by esr spectroscopy during molecular reorganizations which may intimately involve the carbonyl carbon atoms. Since the paramagnetic center may be involved directly in the reaction the semidione group is not really a spin label for a double bond, when the carbonyl carbon atoms are directly involved in the rearrangement reaction.

Semidiones exist in equilibrium with the α-dione and the enediol dianion, equation 2. Low temperatures and ion-pairing favor the diamagnetic components of

$$2 \; RC(O\cdot)=C(O^-)R \; \rightleftarrows \; RCOCOR + RC(O^-)=C(O^-)R \qquad (2)$$

equation 2. Acyclic semidiones exist as a thermodynamic mixture of the cis and trans-isomers shown in
equation 1, at least in a static system of DMSO plus
alkali metal cation. The equilibrium is dramatically
affected by ionic association; for example with
lithium as the gegenion only the cis(lithium chelated)-
ion pair can be observed in DMSO whereas with cesium
as the gegenion only the trans(free ion)-semidione
is observed (3). With sodium or potassium mixtures
of the cis and trans-species are observed. Perturbation of equilibrium (2), for example by UV irradiation does not affect the observed cis/trans-ratio.
This is not because the cis and trans-semidione are
readily interconverted by rotation about the partial
double bond. Instead the further equilibrium of
Scheme 1 can serve to establish a thermodynamic ratio
of the two isomers.

Scheme 1

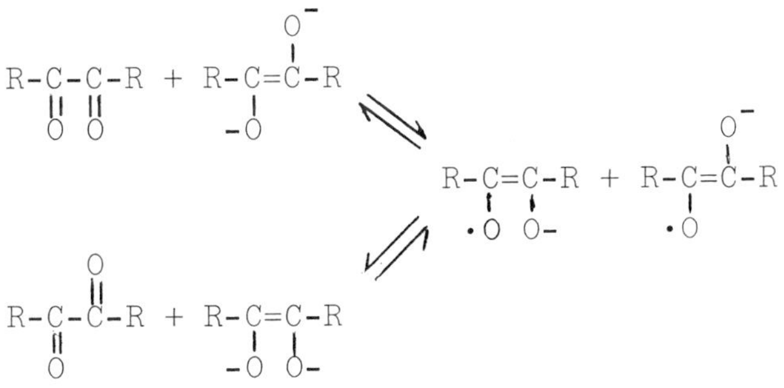

The semidione group more nearly fulfills the
definition of a spin label when the conformational
equilibria of the cycloalkane-1,2-semidiones are considered. Thus, by esr spectroscopy the $\Delta H^{\ddagger}$ for ring
flip of this cyclohexene derivative has been measured
as 4 kcal/mole which corresponds to a magnetic co-
alescence temperature of -85° (4). This value can
be compared with a $\Delta F^{\ddagger}$ of 5.3 kcal/mole observed for

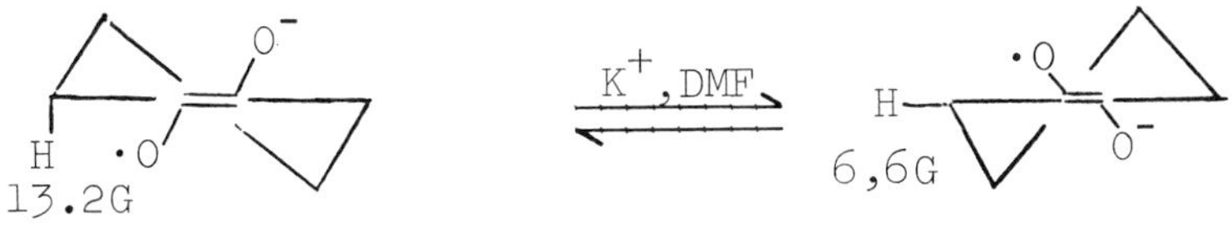

ring inversion of cis-3,3,4,5,6,6-d_8-cyclohexene in
bromotrifluoromethane by p.m.r. (coalesence temperature = -165°) (5). It appears that in this case the
semidione is a reasonable spin label for the olefin.
    At 4-t-butyl group freezes the cyclohexane semi-
dione into conformation 2 up to at least 90°. On the
other hand for a 4-methyl substituent both 2 and 3 are

populated with 2 being preferred by an enthalapy difference of 1.4 kcal/mole ([3]/(2] = 0.13 at 40°) (4).
    Cycloheptane-1,2-semidione is a rigid species
(τ >10^{-7} sec.) up to at least 90°. From an analysis of
the esr hyperfine interactions it can be concluded that
the single populated conformation is the staggered conformation 4 (6).

a^H(G)	C_3	C_4	C_5
equatorial	0.28	2.05	0.54
axial	6.60	0.28	<0.05

Cycloheptane semidione illustrates a rich long range
hyperfine splitting due to π-σ delocalization. This
interaction is maximized when the carbon-hydrogen bond
involved forms one leg of a coplanar zigzag arrangement of bonds and the carbonyl carbon p_z-orbital. Two
further examples of this long range interaction are
given in structures 5 and 6.

(0.20)
↘ (7.61)
(6.5)H H(0.4)   (3.45)H H H H(1.05)   (7.78)H H(1.37)
2.3 H   H(2.88)
   H   O·   (6.34)H   O·
        H H   H
       H   2.3 O⁻   H (0.98) O⁻   H H   O⁻
<0.1   (0.10)
                                              (0.47,0.15)

5, aᴴ in G   6a, aᴴ in G   6b, aᴴ in G
  (ref. 7)   (ref. 6)   (ref. 6)

Valence Isomerization of Semidiones

The first case of molecular rearrangement in an aliphatic semidione was discovered by chance in an investigation of the long-range splitting in bicyclo-[3.1.0]hexan-2,3-semidion, Scheme 2 (8).

Scheme 2

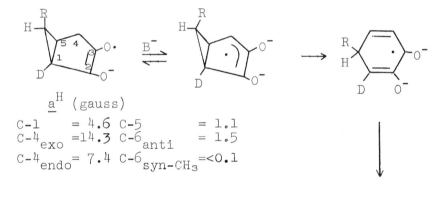

aᴴ (gauss)

C-1   = 4.6   C-5   = 1.1
C-4$_{exo}$ =14.3   C-6$_{anti}$ = 1.5
C-4$_{endo}$= 7.4   C-6$_{syn-CH_3}$=<0.1

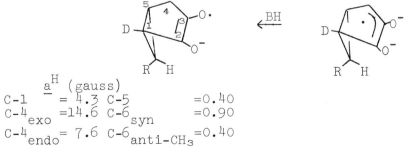

aᴴ (gauss)
C-1   = 4.3   C-5   =0.40
C-4$_{exo}$ =14.6   C-6$_{syn}$ =0.90
C-4$_{endo}$= 7.6   C-6$_{anti-CH_3}$=0.40

   The isomerization of Scheme 2 occurs more rapidly
the higher the concentration and the stronger the base.
This, as well as hydrogen-deuterium exchange at the
$C\text{-}4_{exo}$ position implicate the radical dianion. The
reaction also requires a steric driving force at 25°.
The isomerization is not observed when R=D and is
faster when $R=C_2H_5$ or CH_2OCH_3 than when $R=CH_3$. With
$R=CH_3$ the reaction goes to completion, presumably because the isomer with the $C\text{-}6_{anti}$ methyl group is
thermodynamically more stable because of non-bonded
interactions. These observations have led us to search
for analogous bicyclic-monocyclic equilibria in systems
analogous to the classical cycloheptatriene-norcara-
diene and cyclooctatetraene-bicyclo[4.2.0]octatriene
systems.

   Paramagentic Analogues to the Cycloheptatriene-
Norcaradiene Equilibrium. The 1,4-semidione (7) prepared as shown in Scheme 3 unquestionably exists in
the bicyclic structure (9). This is easily ascertained

Scheme 3

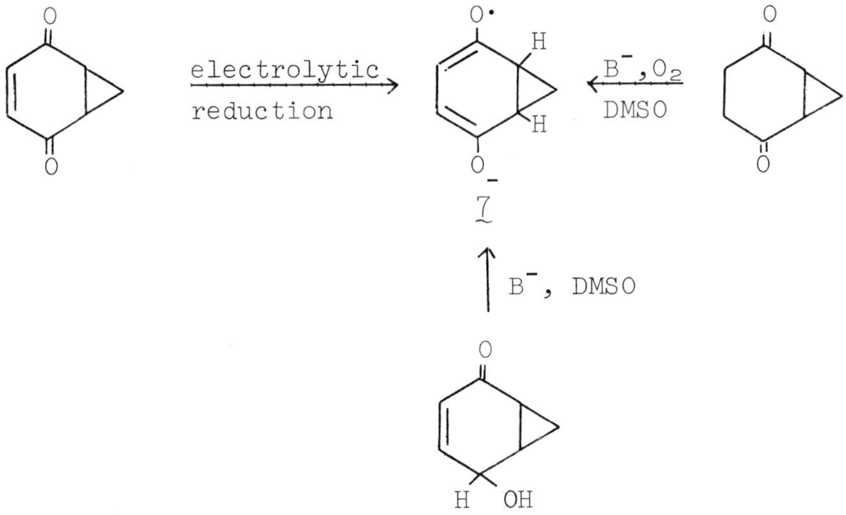

by an examination of 7 and its analogue with methyl
groups at the bridgehead positions. For the monocyclic valence isomer, equation 3 , one would expect
$a_{1,5}^H$ to be approximately equal in magnitude to

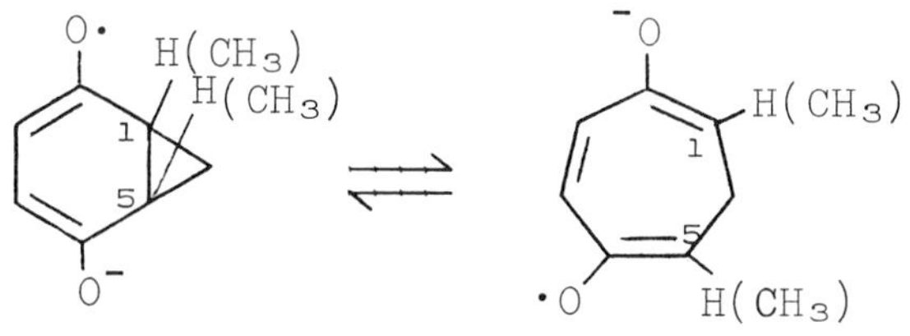

$$(3)$$

$\underset{\sim}{7} \qquad \underset{\sim}{8}$

$a_{CH_3}^H(1,5)$ since as is well known, (e.g., ethyl radical)
that $|Q_{CH}^H| \cong |Q_{CCH_3}^H|$. Instead the following values of
the hsc are observed, as ecpected for the bicyclic
structure.

Chart 1. Hyperfine Splitting Constants (gauss) for
Bicyclo[4.1.0]hept-3-ene-2,5-semidiones(<u>10</u>).

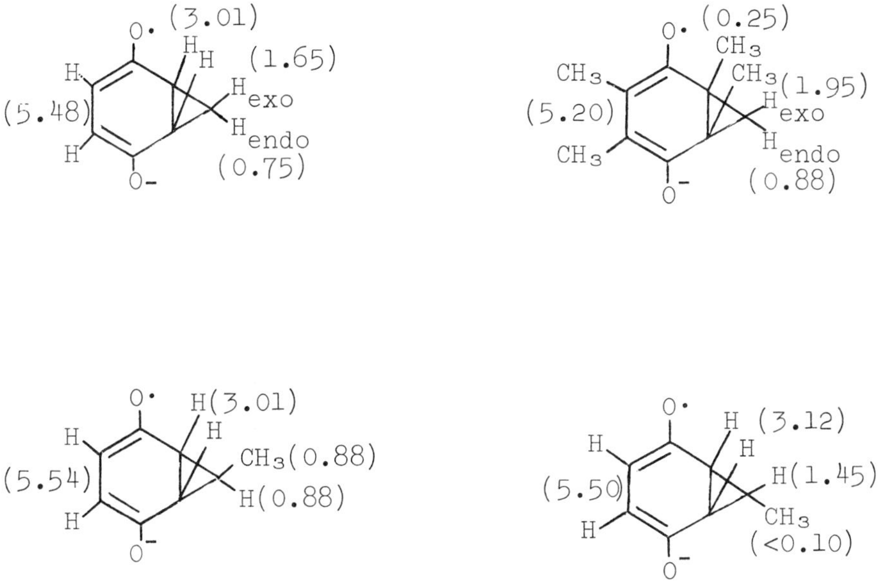

We next examined the question of 8 being thermally
accessable to allow a syn-anti isomerization in 7a,b
as observed in the bicyclo[3.1.0]hexane system. It
was expected that 7b would be more stable than 7a

because of non-bonded interactions. The isomeric
enedione precursors to 7a and 7b were synthesized and
did not isomerize. Upon electrolytic reduction each
dione initially formed its own 1,4-semidione which
could be observed in the absence of the isomeric semidione. The equilibrium of equation 4 is either
excluded or else must occur very slowly. However,
upon extensive reduction either of the 1,4-enediones
gave the same ratio of 7a/7b = 4 parts/96 parts. The
simpliest explanation is that upon extensive reduction
the dianions have an appreciable concentration.
Although equilibrium (4) must occur slowly, the
analogous equilibrium (5) apparently occurs quite
readily at 25°. The observed 96/4 ratio of 7b/7a

probably reflects not only the thermodynamic stability
of these isomers but may also reflect the various disproportion equilibrium constants similar to those of
equations 2 and Scheme 1. The latter do not seem to
be too important because the 96/4 ratio of semidiones
did not vary significantly with the extent of reduction once an appreciable fraction of the starting
dione had been reduced.
      The formation of the 2,5-semidiones in the bicy-
clo[4.2.0]octane system has also been investigated.
The two isomeric saturated diones shown in Scheme 4
both gave the same semidione upon base-catalyzed oxidation. Perhaps again valence isomerization at the
dianion stage has occurred although in this case a
base-catalyzed stepwise epimerization of the bridgehead hydrogen atoms in the semidione may be responsible for the isomerization assumed to occur for the

syn,syn-7,8-dimethyl isomer. Hydrogen-deuterium
exchange at the bridgehead position occurs readily
under the reaction conditions.

Scheme 4

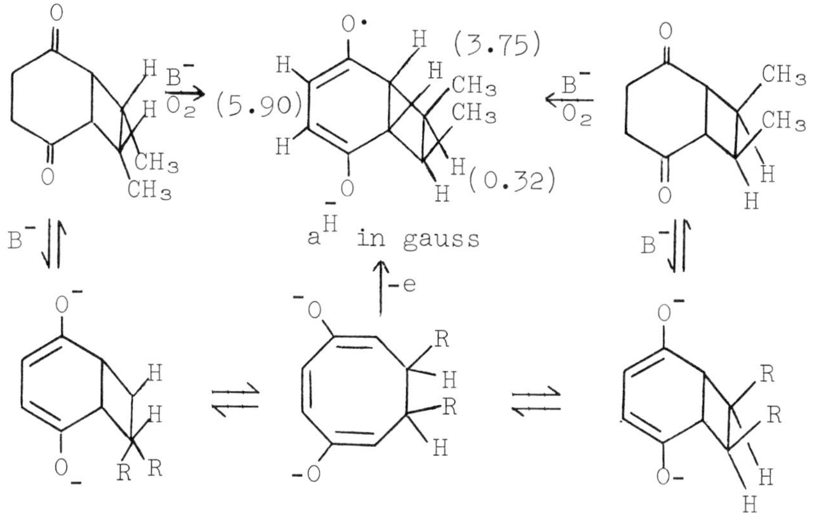

1,4-Semidiones Derived from (CH)8 Hydrocarbons.
We have examined the possible valence isomerization
shown in equation 6. Semidione 9 is stable and easily

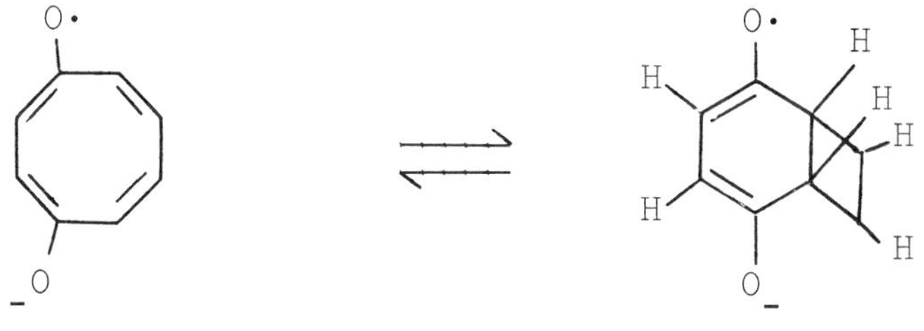

$$(6)$$

prepared as shown in Scheme 5.

Scheme 5

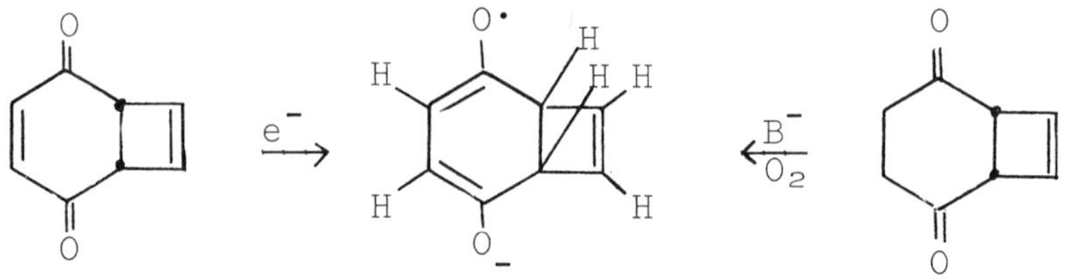

Attempts to reduce the known cycloocta-1,3,6-triene-5,8-dione (11) electrochemically failed to produce any interpretable esr signals. The reduction polarographically involved a clean 2-electron wave (-1.0 v. relative to s.c.e.) apparently the result of 8 being more easily reduced than the parent dione. Base catalyzed oxidation of octa-1,3-diene-5,8-dione produced semidione 9 and a new species assigned structure 10 (equation 7).

(7)

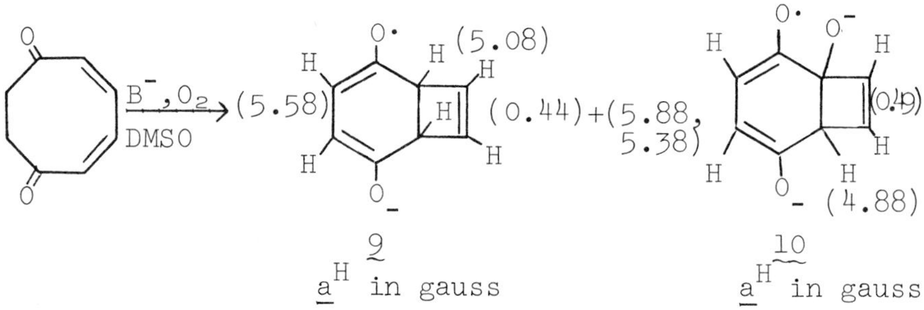

The source of 9 is not certain because the monocyclic diene-dione was prepared by thermolysis of the bicyclic precursor and traces of this precursor could be responsible for the observation of 9. The formation of 10 requires valence isomerization between the monocyclic and bicyclic structure since semidione 10 is not formed from oxidation of 9 or its bicyclic precursor. Scheme 6 presents a possible rationalization for the formation of 10.

Scheme 6

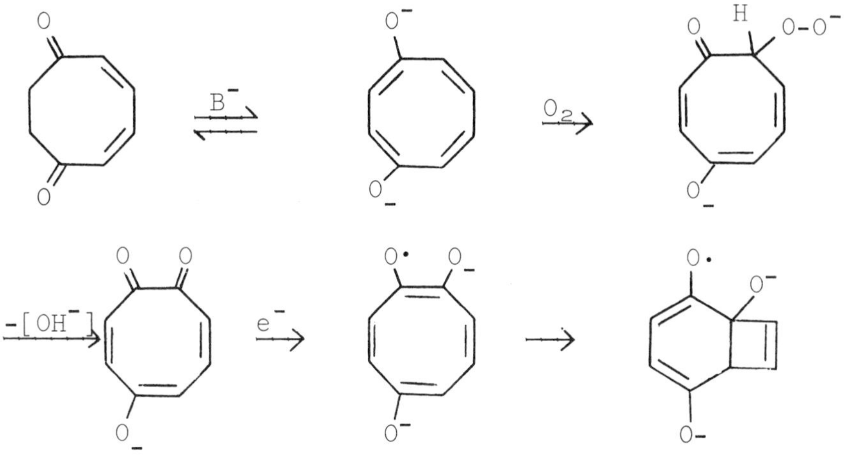

10

1,2-Semidiones Derived from $(CH)_6$ Hydrocarbons.
The possibility of reaction 8 was investigated by the
electrolytic reduction of the stable dione 11 (12).

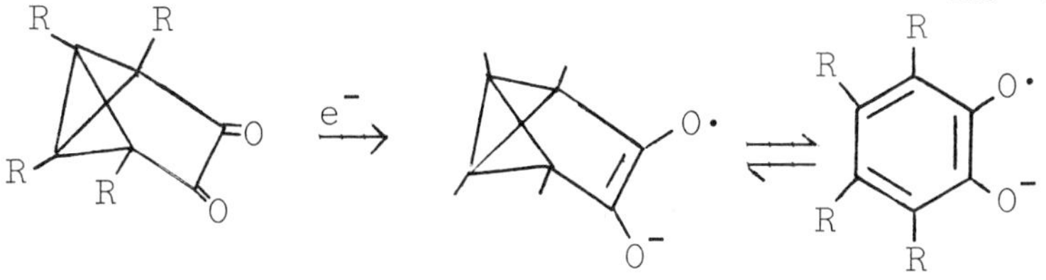

$$R=CH_3 \tag{8}$$

11   12   13

Dione 11 when reduced at -40° gave a broad singlet
absorption which might be 12. The paramagnetic species
was not stable and could be observed only when elec-
trolysis was occurring. At 25° electrolytic reduction
of 11 produced only 13. Apparently the expected
valence isomerization of this system occurs readily
at 25° but it is impossible to state whether rearrangement involves 12 or the analogous dianion.

1,2-Semidiones Derived from $(CH)_8$ Hydrocarbons.
A number of attempts have failed to produce the 1,2-
semidione derived from cyclooctetraene (14), equation 9. Treatment of the appropriate α-benzoyloxy

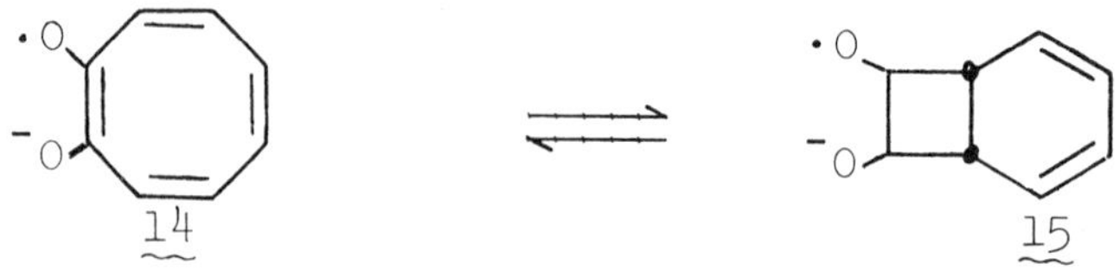

$$(9)$$

$$\underset{\widetilde{14}}{}\qquad\qquad\underset{\widetilde{15}}{}$$

ketone with base/DMSO in a flow system has up to now
failed to produce the semidione $\underset{\sim}{14}$, Scheme 7.

Scheme 7

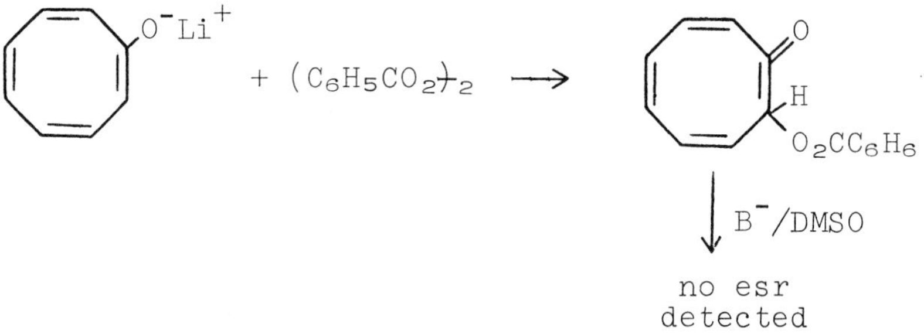

$$+ (C_6H_5CO_2)_2 \longrightarrow$$

no esr
detected

Two derivatives of $\underset{\sim}{15}$ have been produced via the
acyloin condensation of diesters. Unfortunately the
acyloin condensation works very poorly for the diesters
in Chart 2 and even in the presence of trimethylsilyl
chloride pure acyloin condensation products could not
be isolated. The spectra of semidiones $\underset{\sim}{15a}$ and $\underset{\sim}{15b}$

Chart 2.   Acyloin Dondensations of cis-5,6-carboethoxy-
           cyclohexa-1,3-dienes.

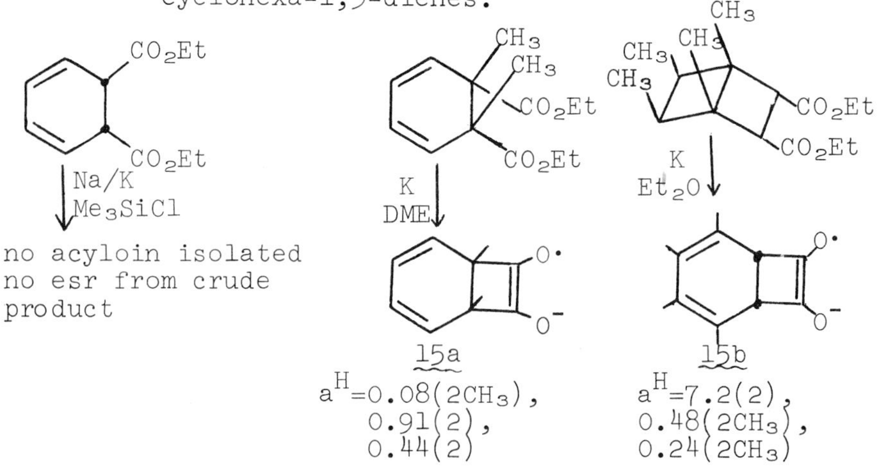

no acyloin isolated
no esr from crude
product

$$a^H=0.08(2CH_3),$$
$$0.91(2),$$
$$0.44(2)$$

$$a^H=7.2(2),$$
$$0.48(2CH_3),$$
$$0.24(2CH_3)$$

are shown in Figure 1. The large difference in hfsc for α-H and α-methyl require that these groups occupy bridgehead positions which is possible only in the bicyclic structure. The bicyclo[2.2.2]octadienesemidione (16) is perfectly stable and shows no rearrangement to structure 15 (equation 10).

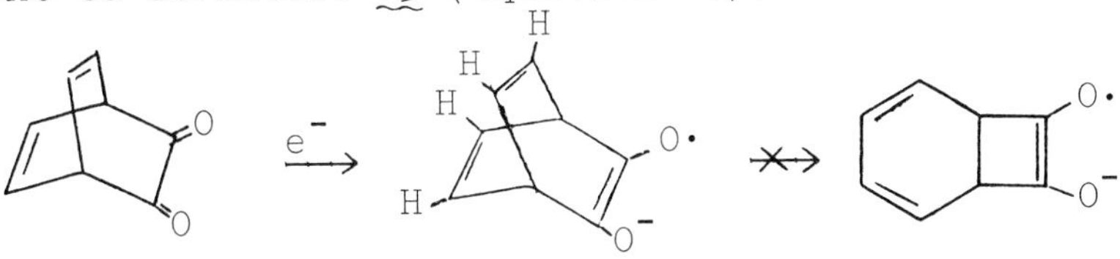
O
O
e⁻
H
H
H
H
O·
O⁻
O·
O⁻

(10)

a^H=0.61G(4)

16

The hfsc for hydrogen atoms in a large number of cyclobutane semidiones are in the range 10-14 gauss (11,13). The value of 7.2 gauss assigned to structure 15b is definitely too low. This suggests that perhaps for 15b (but presumably not for 15a) a rapid equilibrium may exist between the bicyclic and monocyclic structures (i.e., equation 9).

1,2-Semidiones Derived from $(CH)_{10}$ Hydrocarbons. Treatment of 17 or 18 with oxygen in basic DMSO produce the extremely stable semiquinone 19 (12-14). With more limited amounts of oxygen 20 can be detected as the precursor to 19 (Scheme 8). It seems quite

Scheme 8

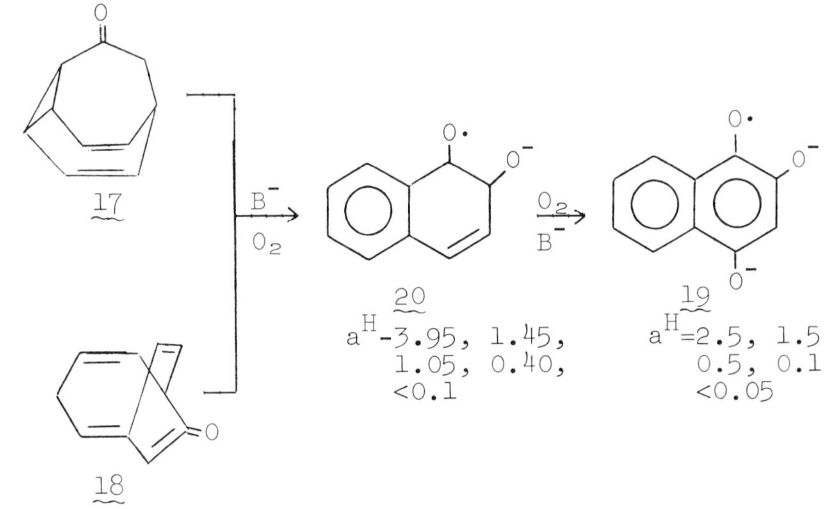
O
17
B⁻
O₂
O·
O⁻
O₂
B⁻
O·
O⁻
O⁻
20
aH-3.95, 1.45,
1.05, 0.40,
<0.1
19
aH=2.5, 1.5
0.5, 0.1
<0.05
O
18

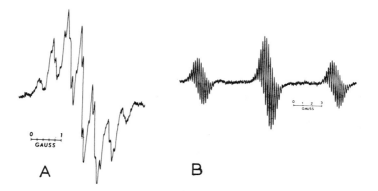

Figure 1.   ESR spectra of bicyclo[4.2.0]octa-2,4-diene-7,8-semidiones
in dimethyl sulfoxide solution at 25°C: (A) spectrum assigned to 15(a)
(1,6-dimethyl); (B) spectrum assigned to 15(b) (2,3,4,5-tetramethyl).

reasonable that the observed rearrangement-oxidation
occurs via some oxidation state of the 9,10-dihydro-
naphthalene derivative 21.

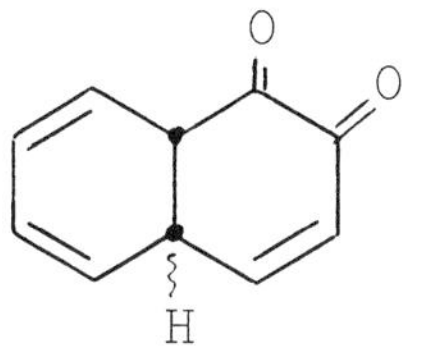

21

Working with more limited amounts of oxygen it becomes
apparent that there are additional stable paramagnetic
intermediates preceeding the formation of 20. The
semidione of 18 can be detected but it is very sensitive to traces of oxygen. When prepared from the
acetoxy ketone (Scheme 9), 22 is the only paramagnetic

Scheme 9

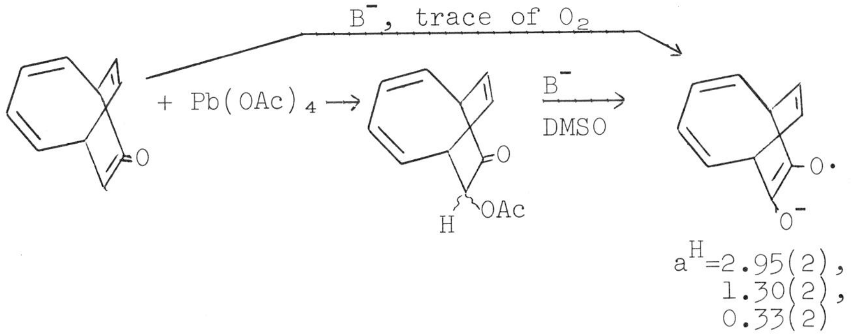

$$a^H=2.95(2),$$
$$1.30(2),$$
$$0.33(2)$$

22

species detected. Furthermore 22 is quite stable in
DMSO in the absence of oxygen for extended periods of
time. Treatment of 17 with traces of oxygen at 0° in
DMSO form an isomeric semidione which is also quite
stable but is converted by additional oxygen into 19.
The hfsc for this species suggest structures 23 or
possibly a rapidly equilibration mixture of 23 and 24.
There is no evidence for any equilibration between
this species and 22 although such an interconversion

Scheme 10

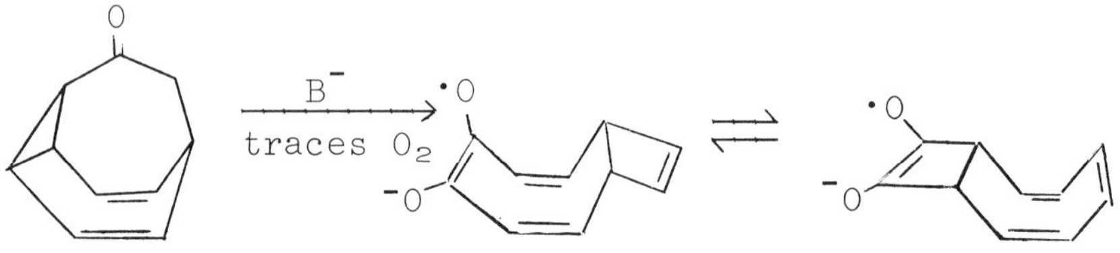

$$a^H=5.46(2),0.86(2),0.27(2),0.13(2)$$

could be readily achieved by a sigmatropic shift
(equation 10). Both 22 and 23 are destroyed when

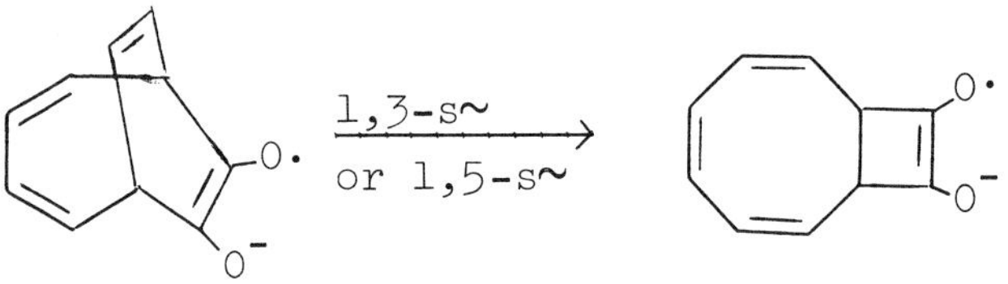

(10)

treated with oxygen to yield cleanly 19; intermediate
degrees of oxidation give mixtures of 19 and 22 or 23.
Molecular oxygen will remove the electron from a semidione to yield the free dione. Can we in this instance be observing the molecular rearrangement of the
diones? Scheme 11 presents a working hypothesis which
seems to fit all of the observed facts.

Scheme 11

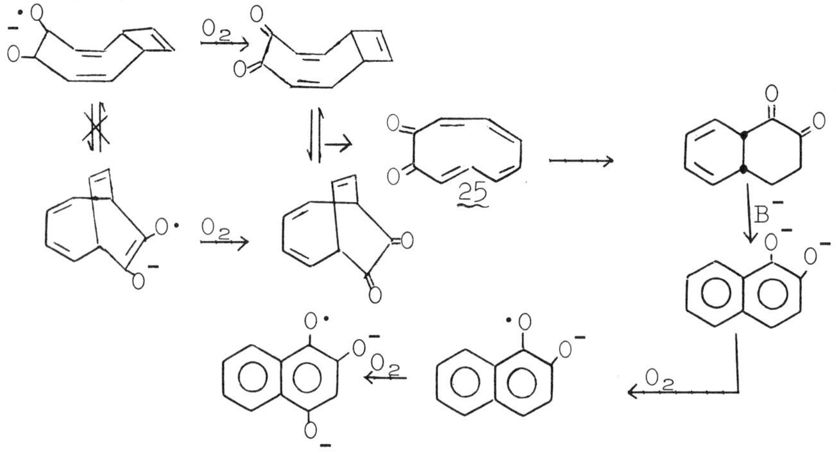

UV-Photolysis of semidione 23-24 forms still another apparently isomeric radical anion which also is destroyed by oxygen with the formation of 19. The photochemically formed radical anion possesses a low symmetry, perhaps 8 chemically different hydrogen atoms (equation 11). One reasonable suggestion for

$$23 \xrightarrow{h\nu} [\text{semidione}] \xrightarrow{O_2} 19 \qquad (11)$$

$$a^H = 1.52, 0.76, 0.43$$
$$0.34, 0.17(4)$$

the structure of this semidione is the monocyclic cyclodecatetraenesemidione, $25 \cdot ^-$.

Perhaps the most important conclusion to be reached from this study of the isomeric semidione derived from $(CH)_{10}$ hydrocarbons, is that in this case the radical anions are themselves thermally stable at $25°$ in respect to valence isomerization. On the other hand apparently the diones rapidly equilibrate. It begins to appear that when the following three oxidation states are considered, that one or the other of

$$4n, \; 4n + 2 \qquad\qquad 4n + 1$$

dione or dianion will possess the proper number of π or σ electrons for a low energy transition state of the Hückel-type ($4n + 2$) or of the Möbius-type ($4n$). The radical anion, on the other hand, never possesses the proper number of electrons to completely stabilize either a Hückel or Möbius-transition state. Finally, the overal course of the oxidation of ketones 17 and 18 at $25°$ appears to resemble the high temperature pyrolysis or the photolysis of the corresponding olefins which proceeds to naphthalene via the 9,10-dihydronaphthalenes (equation 12) (15).

$$(12)$$

1,3-Sigmatropic Rearrangements of Semidiones

A number of pairs of semidiones radical anions, which could be interconverted by a 1,3-sigmatropic rearrangement, are recognized to not undergo thermal isomerization at 25°. Some of the pairs of semidiones are listed in Chart 3. Nevertheless, the formation of

Chart 3. Non-equilibrating Pairs of Semidiones

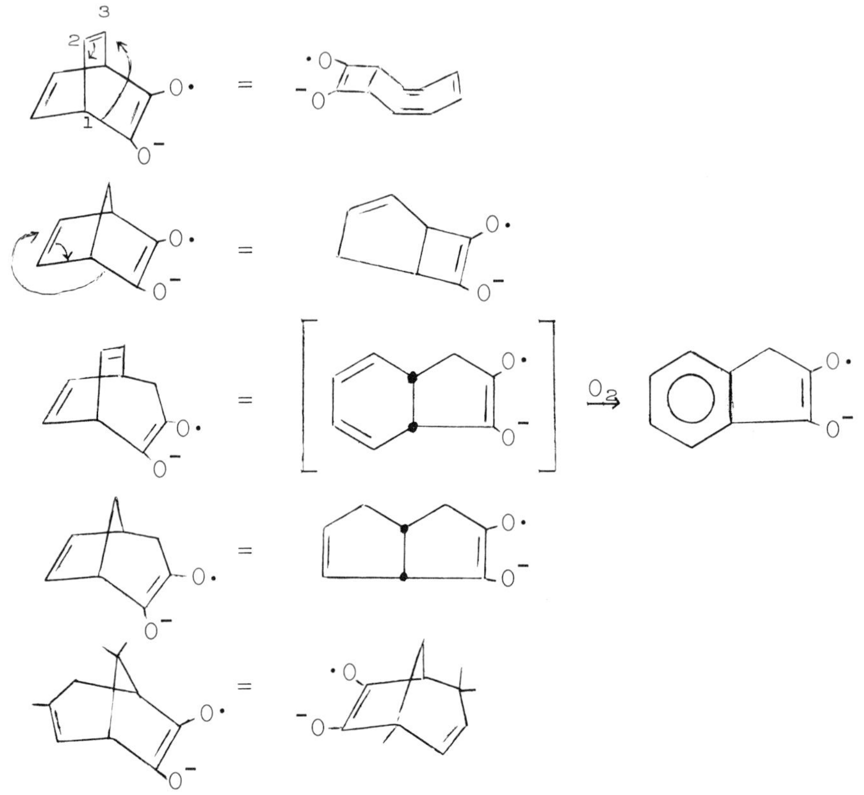

semidiones from α-acetoxy ketones plus base or from a methylene ketone plus base and oxygen has in two different cases led to the products of a 1,3-rearrangement.

When exo- or endo-hydroxynorbornene-2-one is treated with base in DMSO the first esr signal that can be detected (~0.1sec. after mixing) is that of the rearranged semidione, bicyclo[3.2.0]hept-2-ene-6,7-semidione (16). On the other hand when norbornene-2,3-dione is reduced electrolytically, or 3-alkoxy-norbornene-2-ones are oxidized in basic solution the unrearranged semidione is observed, Scheme 12. In both

Scheme 12

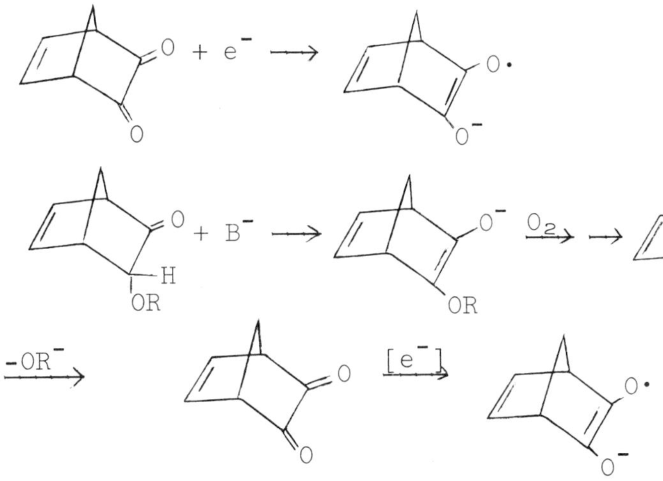

of these cases the unrearranged semidione can be formed
without going through the enediol dianion. On the
other hand, starting from the α-hydroxy ketone the
enediol dianion is first formed and it is apparently
the dianion which undergoes molecular rearrangement,
Scheme 13.

Scheme 13

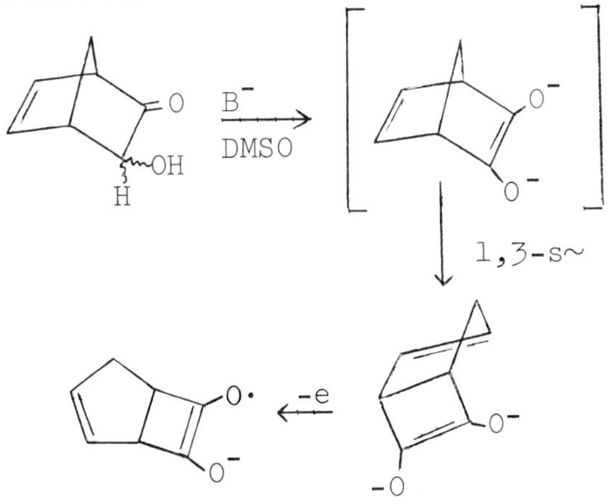

The rearrangement of the dianion is apparently reversible.
Thus, the three monomethyl derivatives shown in
Scheme 14 all produce roughly the same ratio of the
isomeric semidiones 26 and 27. A second case where

Scheme 14

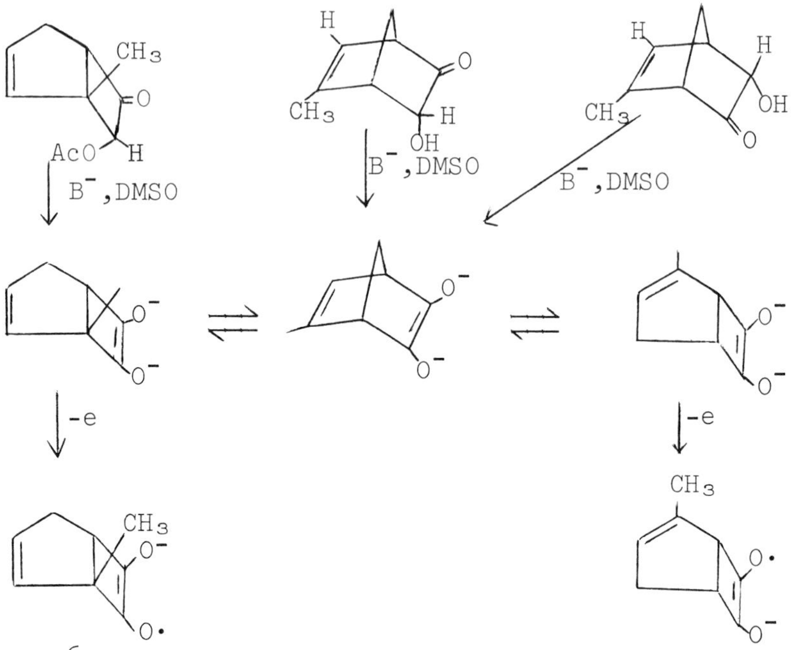

26   27

this type of rearrangement has been noted is in the oxidation of l-carboethoxy-2,8,8-trimethylbicyclo-[3.2.1]oct-2-en-6-one (27).

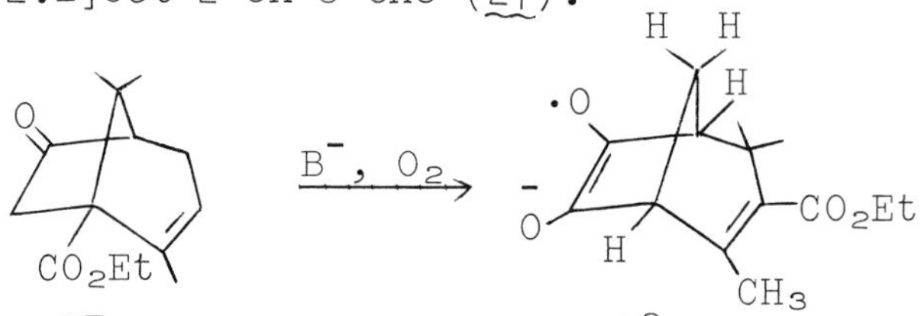

27   28

The structure of the semidione 28 is assigned as the basis of the hfsc which cannot be rationalized with an unrearranged structure. On the other hand, 29 and 30 yield the unrearranged semidiones. The assignment of

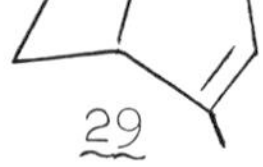

29 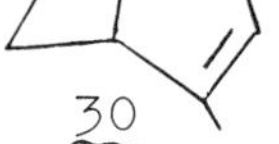   30

hfsc in the [3.2.1]-system are shown in Chart 4.

Chart 4.   Bicyclo[3.2.1]hept-2-ene-6,7-semidiones, a^H
           in gauss

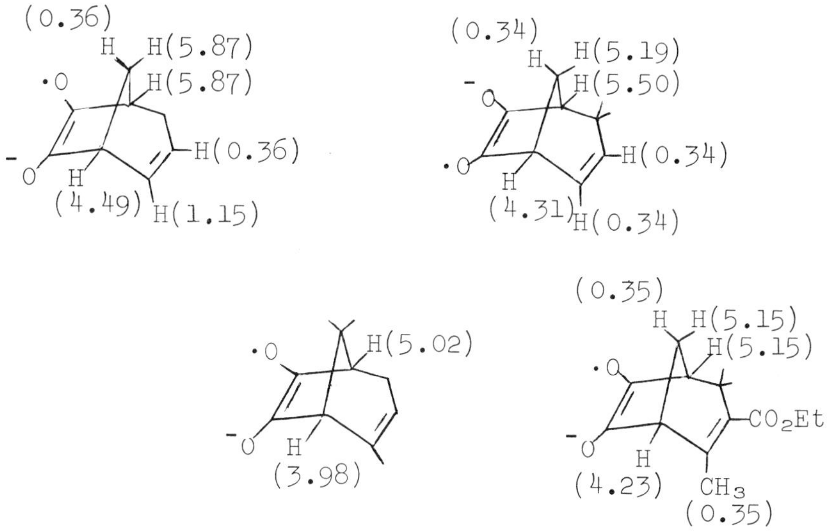

The presence of the carboethoxy group in 27 facilitates
the rearrangement. This is quite reasonable if the
rearrangement proceeds via the enediol dianion.
Scheme 15 illustrates how this could occur in a stepwise manner. A stepwise rearrangement may also be

Scheme 15 $E = EtO_2C-$

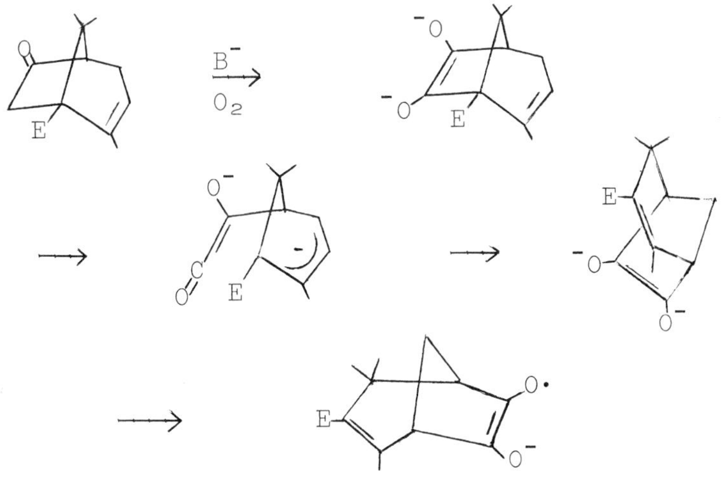

involved in the rearrangement of the dianion of the
enediol in the bicyclo[2.2.1]heptene system. Scheme 16
illustrates how this process may occur.

Scheme 16

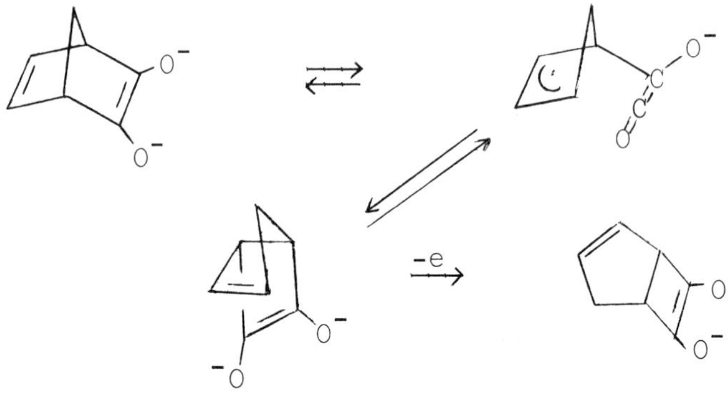

Another possible rationalization involves the
formation of a transition state for a concerted rearrangement having a sextet of electrons. The concerted 1,3-migration would involve transition state 31.
The argument can be advanced that a pair of electrons

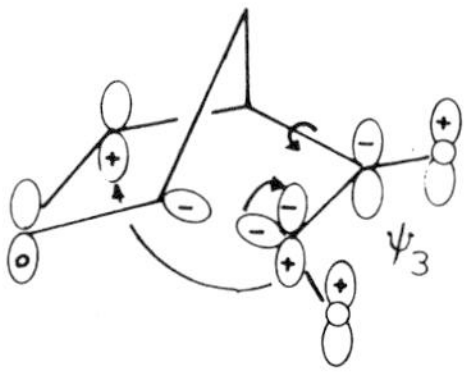

31

in the carbonyl p_z-orbital will be more nearly available in the dianion than in the semidione since ρ_c in
the dianion will be approximately twice ρ_c in the
radical anion (approximately 0.38 and 0.76, respectively).

Photochemical 1,3-sigmatropic rearrangements
occur readily for many of the ketones related to the
semidiones shown in Chart 3. Observation of thermal
rearrangements for the radical anions would in a way
be an example of photochemistry without light. However it appears that these reactions require more
than just having a single electron in the π^*-orbital
of the carbonyl group as in the semidione. Photochemical rearrangements of the thermally stable

semidiones are, of course, also possible. Thus the
facile transformation of 32⇌33 has been observed.

32

a^H=0.91,0.40(3),
    0.14(2) gauss

33

a^H=8.20,0.40(3),0.21,
    0.14 gauss

Conclusions

Open shell species, such as the semidiones, are
not more prone to undergo pericyclic reactions than
their closed shell analogs. Indeed, it appears as if
one or the other of the diamagnetic analogues (dione
or dianion) undergo rearrangement more readily than
the radical anion. In general this seems to be con-
nected with the diamagnetic molecules possessing
either the $4n$ or $4n+2$ π plus σ electrons required for a
low energy Möbius or Hückel transition state. On the
other hand the dianion or diketone may simply posess
higher energy contents (because of electrostatic considerations), thus leading to a lower energy barrier
for a thermodynamically favorable rearrangement of the
bicyclic skeleton.

Abstract

Bicyclic—monocyclic valence isomerization of a
radical dianion in the bicyclo[3.1.0]hexanesemidione
system has been demonstrated. Symmetrical 1,4-
semidiones formally derived from cycloheptatriene and
cyclooctatetraene perfer to exist in the bicyclic
(4.1.0 and 4.2.0) structures. Bicyclic—monocyclic
valence isomerization in the bicyclo[4.1.0]hept-3-ene-
2,5-dione system occurs more readily for the dianion
than for the radical anion. Several radical anions
derived from the $(CH)_{8-10}$ annulenes are reported. In
the case of the 1,2-oxygenated derivatives of $(CH)_{10}$
the dianions or radical anions are stable, but the
diones undergo valence isomerization and under oxida-
time conditions are converted to 4-hydroxynaphthalene-
1,2-semiquinone. Enediol dianions in the bicyclo
[2.2.1]hepta-2,5-dione and 1-carboalkoxybicyclo[3.2.1]
octa-2,6-diene systems have been observed to undergo

thermal 1,3-signatropic rearrangements under conditions
where the semidiones are stable. Photochemical 1,3-
sigmatropic rearrangement has been observed for bicyclo
[3.2.0]hept-7-ene-2,3-semidiones.

Acknowledgement

We gratefully thank Dr. E. Weissberger for a gift
of ketones 29 and 30 and Dr. W. Erman for ketone 27.

References Cited

(1)   Aliphatic Semidiones. XXXII. This work was sup-
      ported by grants from the National Science Found-
      ation and the Petroleum Research Fund.
(2)   Russell, G. A., Strom, E. T., Talaty, E. R.,
      Weiner, S. A., J. Am. Chem. Soc. (1966), 88, 1998.
(3)   Russell, G. A., Laurson, D. F., Malkus, H. L.,
      Stephens, R. D., Underwood, G. R., Takano, T.,
      Malatesta, V., J. Am. Chem. Soc. (1974), 96, 5830.
(4)   Russell, G. A., Underwood, G. R., Lini, D. C.,
      J. Am. Chem. Soc. (1967), 89, 6636.
(5)   Anet, F. A. L., Haq, N. Z., J. Am. Chem. Soc.
      (1965), 87, 3147.
(6)   Russell, G. A., Keske, R. G., Holland, G.,
      Mattox, J., Givens, R. S., Stanley, K., J. Am.
      Chem. Soc. (1975), 97, 1892.
(7)   Russell, G. A., Holland, G. W., Chang, K.-Y.,
      Keske, R. G., Mattox, J., Chung, C. S. C.,
      Stanley, K., Schmitt, K., Blankespoor, R.,
      Hosugi, Y., J. Am. Chem. Soc. (1974), 96, 7237.
(8)   Russell, G. A., McDonnell, J. J., Whittle, P. R.,
      Givens, R. S., Keske, R. G., J. Am. Chem. Soc.
      (1971), 93, 1452.
(9)   Russell, G. A., Ku, T., Lokensgard, J., J. Am.
      Chem. Soc. (1970), 92, 3833.
(10) Russell, G. A., Dodd, J. R., Ku, T., Tanger, C.,
      Chung, C. S. C, J. Am. Chem. Soc. (1974), 96,
      7255.
(11) Oda, M., Kayama, Y., Miyazaki, H., Kitahara, Y.,
      Angew Chem. Int. Ed. (1975), 14, 418.
(12) Heldeweg, R. F., Hogeveen, H., Tet. Lettr. (1975),
      1517.
(13) Russell, G. A., Whittle, P. R., Keske, P. G.,
      Holland, G., Aubuchon, C., J. Am. Chem. Soc.
      (1972), 94, 1693.
(14) Russell, G. A., Blankespoor, R. L., Trahanovsky,
      K. D., Chung, C. S. C., Whittle, P. R., Mattox,
      J., Myers, C. L., Penny, R., Ku, T., Kosugi, Y.,
      Givens, R. S., J. Am. Chem. Soc., (1975), 97, 1906.

(15) Maier, G., "Valenzisomerisierungen", 133-137,
      Verlag Chemie GmbH, Weinheim, 1972.
(16) Russell, G. A., Schmitt, K. S., Mattox, J., J.
      Am. Chem. Soc. (1975), 97, 1882.
(17) Russell, G. A., Whittle, P. E., Chung, C. S. C.,
      Kosugi, Y., Schmitt, K., Goettert, E., J. Am.
      Chem. (1974), 96, 7053.

RECEIVED December 23, 1977.

24

Aromatic Anion Radicals as Bronsted Bases

Correlation of Protonation Rates with Singlet Energies of the Precursor Aromatic Hydrocarbon*

HAIM LEVANON, P. NETA, and A. M. TROZZOLO

Radiation Laboratory and Department of Chemistry, University of Notre Dame, Notre Dame, IN 46556

One of the aims of physical organic chemistry is to derive generalizations regarding structure-reactivity relationships which can be related to the fundamental electronic properties of a molecule. In this paper are described experiments and interpretations of these experiments which seem to indicate that there is a fundamental correlation between the singlet energy of an aromatic hydrocarbon and the rate of protonation of its anion radical.

Anion radicals of aromatic hydrocarbons are known to undergo protonation in protic media (1-5). The spectrophotometric pulse radiolysis technique has been used to demonstrate this process in alcoholic solutions of several aromatic compounds (2). In such experiments, the anion radicals are produced by the reaction of the hydrocarbon with the solvated electron

$$Ar + e^-_{solv} \xrightarrow{k_1} Ar^{\cdot-} \tag{1}$$

Protonation can take place by reaction of the anion radical with the solvent molecule

$$Ar^{\cdot-} + ROH \xrightarrow{k_2} ArH^{\cdot} + RO^- \tag{2}$$

The additional protonation process is the reaction of $Ar^{\cdot-}$ with ROH_2^+ produced in the pulse

$$Ar^{\cdot-} + ROH_2^+ \rightarrow ArH^{\cdot} + ROH \tag{3}$$

which is negligible at low dose rates. Dorfman and coworkers (2) found that the rates of reaction 2 were strongly dependent on the acidity of the solvent, e.g., i-PrOH<EtOH<MeOH. In a particular solvent, k_2

varied over several orders of magnitude for the various compounds. Even the rate constants for protonation of two isomeric forms of the same compound, i.e., cis- and trans-stilbene, were found to be different (3).

In a previous study, the rates of reaction of sodium naphthalenide and anthracenide with water in THF as solvent were determined by a stopped-flow technique (4) and found to be faster for the former anion radical. The authors indicated, however, that this behavior seemed to be in disagreement with predictions of reactivities from molecular orbital calculations.

Rates of protonation of aromatic anion radicals also have been estimated from polarographic measurements by observation of changes in the curve shape upon protonation (5). These experiments were carried out in DMSO as solvent with various phenols as proton donors. This method, however, is applicable only within a limited range of protonation rates and is less reliable than the direct observation of the kinetic process. Furthermore, correlation of the protonation rates (5) with calculated electron densities does not appear to be satisfactory.

In an attempt to determine the factors which control the rate of protonation, we examined the correlation of these rates with some basic properties of the molecules, such as electron affinity, ionization potential, and singlet energy, which are known to be interrelated for alternant aromatic hydrocarbons.

Experimental Results

Protonation rate constants of various aromatic anion radicals in 2-propanol were measured by the pulse radiolysis technique as described previously for stilbene (3). The aromatic compound was dissolved in the alcohol, usually in the range of 10^{-5} to 10^{-3} M. Certain compounds, such as tetracene and the dibenzanthracenes, which were difficult to dissolve directly in the 2-propanol, were first dissolved in a small amount of dry THF and then diluted by the alcohol. The amount of THF in the irradiated solution was always less than 5%. The solutions were deoxygenated by prolonged bubbling with ultrapure nitrogen or argon and were kept in the dark. Irradiation was carried out at room temperature using 9 MeV electrons from an ARCO LP-7 linear accelerator. The pulse duration was usually 5 ns, although longer duration pulses were sometimes used to achieve higher doses.

Table I

Kinetics of Formation and Protonation of Aromatic Anion Radicals

No.	Aromatic Hydrocarbon	$\lambda(nm)^a$	$k_5(M^{-1}s^{-1})^b$	$k_2(s^{-1})^b$	$\Delta E_{s_1 \leftarrow s_0}(cm^{-1})^c$	$I(ev)^d$
	Alternant					
1	cis-Stilbene[e]	496	$<10^7$	6.4×10^5	31500	
2	Naphthalene[f]	810		5.7×10^5	32200	8.15
3	Phenanthrene[f]	1000		3.4×10^5	28900	7.86
4	Triphenylene[g]	410	$<10^8$	3.0×10^5	29900	7.89
5	trans-Stilbene[e]	486	$<10^7$	7.3×10^4	30400	7.60
6	Anthracene[f]	720		4.7×10^4	26700	7.47
7	Chrysene[h]	460–500	$<10^8$	2.0×10^4	27700	7.60
8	Pyrene[i]	492	1.1×10^8	1.0×10^4	26900	7.41
9	1,2,3,4-Dibenzanthracene[i]	790	4.0×10^8	3.2×10^3	26700	7.44
10	Benzo(a)pyrene[i]	583	7.0×10^8	2.5×10^3	24700	7.12
11	Tetracene[j]	810	1.2×10^9	4×10^2	21200	7.04
12	1,2,5,6-Dibenzanthracene[i]	790	1.5×10^8	2×10^2	25300	7.38
13	Perylene[i]	576	1.4×10^9	20	23000	7.06
	Non-Alternant					
14	Acenaphthylene[k]	387	1.4×10^9	4×10^2	21500	8.22
15	Fluoranthene[j]	445	7.2×10^8	~7	25300	7.95
16	Azulene[i]	425	1.5×10^9	1.5	14200	7.43

Footnotes to Table I

a. wavelenths at which most of the kinetic measurements were done, usually it is one of the main peaks of the anion radical absorption.
b. The rate constants are accurate to $^+_-15\%$ except for the very low values which may have a larger experimental error.
c. taken from "Photophysics of Aromatic Molecules" by Birks, J.B., Wiley-Interscience, London (1970) p. 70, except for the values of acenaphthylene from Heilbronner, E., Weber, J.P., Michl, J. and Zahradnik, R., Theoret. Chim. Acta (Berl.), (1966) 6, 141.
d. Ionization potentials taken mostly from Boschi, R, Clar, E. and Schmidt, W., J. Chem. Phys. (1974) 60, 4406, and from Birk's book (footnote b, p. 457).
e. from ref. 3.
f. from Arai, S., Tremba, E.L., Brandon, J.R. and Dorfman, L.M., Can. J. Chem. (1967) 45, 1119, see also ref. 2.
g. obtained from K&K Laboratories.
h. purified by zone refining.
i. obtained from Aldrich Chemical Co., highest purity available.
j. obtained from Eastman Organic Chemicals
k. purified by sublimation.

The experiments consisted of observation of the
time profile of the formation and decay of the optical
absorption of the various anion radicals. For all
cases the transient spectrum was examined to verify
the position of the maxima. These peaks were found to
be in agreement with the literature values for the
corresponding anion radicals in MTHF glasses (6). The
wavelengths in which the kinetic measurements were
performed with each compound are given in Table I.
In some cases, such as chrysene and triphenylene, the
transient absorption did not decay down to zero at all
wavelengths. The spectra were then recorded before
and after decay in order to verify that the observed
decay is due to the dissappearance of the anion
radical. The remaining absorption was sometimes
assignable to the solvent radical (7). In most cases,
however, the kinetic measurements were carried out at
wavelengths above 450 nm where neither the solvent
radical nor the product of protonation show any
considerable absorption.

The irradiation of dilute alcoholic solutions
produces mainly solvated electrons and solvent radicals with no direct action on the solute. The solvent
radicals may be initially positive ions $\dot{R}OH^+$ which
will convert very rapidly into carbon centered radicals. For example, in 2-propanol the main radicals
will be $(CH_3)_2\dot{C}OH$. A small amount of acid, ROH_2^+,
is also produced.

The reaction of all polynuclear aromatic compounds with the solvated electron (reaction 1) is very
rapid (2,3), $>10^9 M^{-1}s^{-1}$, and in most cases it is
diffusion controlled. Under the experimental conditions used, this process was complete long before the
protonation took place.

The anion radicals can protonate by reaction with
the solvent molecule (reaction 2) and by the small
amount of acid produced by the radiation pulse (reaction 3). The latter reaction was minimized by use
of very low doses, producing only $1-2 \times 10^{-6}$ M of
ROH_2^+. However, when the rate of reaction 2 becomes
lower than $\sim 10^4 s^{-1}$ the contribution of reaction 3
cannot be neglected. It was, therefore, necessary to
overcome this complication by adding a small amount of
base. For this purpose, sodium metal was dissolved in
2-propanol prior to the experiment to produce a solution of $(CH_3)_2CHO^-Na^+$ in the required concentration.
Since the concentration of this base was only in the
mM range it was not expected to affect the rate of
reaction 2. Experiments have shown that the addition

of base has no effect on the rates of protonation
when they are higher than $10^4 s^{-1}$. A small decrease
in rate was noticed upon addition of base when
k_2 was in the range of 10^4-$10^3 s^{-1}$, as expected.
The use of base has another effect on this
system. The radical from 2-propanol, $(CH_3)_2\dot{C}OH$, dis-
sociates into $(CH_3)_2\dot{C}O^-$ with a mid-point at $7 \times 10^{-4} M$
of $(CH_3)_2 CHO^- Na^+$ (8).

$$(CH_3)_2\dot{C}OH + (CH_3)_2 CHO^- \rightleftharpoons (CH_3)_2\dot{C}O^- + (CH_3)_2 CHOH \quad (4)$$

The basic form of this radical is known to be a
stronger reducing agent than the neutral form (9).
Thus, it was found to reduce the better electron ac-
ceptors among the compounds used in the present study.
In such cases, the initial rapid formation of $\dot{A}r^-$
by reaction 1 is followed by a slower increase in
absorption due to reaction 5

$$Ar + (CH_3)_2\dot{C}O^- \rightarrow \dot{A}r^- + (CH_3)_2 CO \quad (5)$$

Rate constants for this electron transfer process were
determined from measurements of the rate of formation
at various concentrations of Ar and the results are
given in Table I along with the rates of protonation.

Discussion

The reactivities of aromatic compounds have been
correlated with various physical parameters, such as
ionization potentials, triplet energies, free valence,
and energy levels of lowest unoccupied orbitals (10,11).
It was noted that simplified HMO calculations were
sufficient to account for many experimental observations. However, the reactivities of aromatic anion
radicals toward proton donors have not been successfully correlated with any physical parameter. Such a
correlation may shed light on the mechanism and allow
prediction of unknown rates.
In the early work on aromatic anion radicals,
Paul, Lipkin, and Weissman (1) correlated the reactivities of these radicals with electron affinities
derived from their spectra. Bank and Bockrath (4)
compared the rates of protonation of naphthalene and
anthracene anion radicals and concluded that the
higher rate for naphthalene was in contradiction to
their prediction from molecular orbital localization
energy calculations. Fry and Schuettenberg (5) correlated the rates of protonation of various aromatic

anion radicals with the calculated electron densities
at the position of highest electron density in each of
these radicals. This correlation also appears unsatisfactory. It seems that parameters derived from
simple MO calculations are not adequate to explain the
reactivity of anion radicals.

In even alternant aromatic hydrocarbons, the
difference in energy between the lowest unoccupied
molecular orbital (LUMO) and the highest occupied one
(HOMO) is essentially equal to the singlet energy
difference $\Delta E_{S_1 \leftarrow S_0}$. The unpaired electron in the
anion radical occupies the LUMO. Upon protonation of
this radical an odd alternant structure is formed in
which the unpaired electron occupies the non-bonding
orbital. The exothermicity upon protonation is,
therefore, $\sim \frac{1}{2}\Delta E_{S_1 \leftarrow S_0}$. If this energy change affects
the rate of protonation one would expect a correlation
between these rates and $\Delta E_{S_1 \leftarrow S_0}$. Fig. 1 shows that a
plot of log k_2 vs $\Delta E_{S_1 \leftarrow S_0}$ gives a reasonable linear
dependence. Deviations from linearity appear to be
relatively larger when the rates of protonation are
low, probably because the lower rates are more susceptible to experimental complications, particularly
impurity effects. It should be noted that the protonation rates determined by Fry and Schuettenberg
(5) also give a reasonable straight line when we plot
them against $\Delta E_{S_1 \leftarrow S_0}$. As expected for alternant hydrocarbons, linearity is also obtained when log k_2 is
plotted versus the ionization potential I (plot not
shown, see values in Table I). In principle, k_2 can
be correlated also with electron affinity (12) or
with the polarographic reduction half-wave potential,
but the literature values for these parameters are
widely scattered. The molecular dimensions seem to
have little effect on k_2.

Non-alternant hydrocarbons do not appear to fit
easily into the linear relation with the singlet separation (Fig. 1). Also, they do not fit on the line
of log k_2 versus ionization potential. For example,
the ionization potential of azulene is similar to that
of anthracene but the rate of protonation of the former is four orders of magnitude slower (see Table I),
which is in line with the large difference between
their respective $\Delta E_{S_1 \leftarrow S_0}$ values. The same considerations hold also for fluoranthene and acenaphthylene.
Because of sparse experimental data the apparent fit
of these latter compounds with the line in Fig. 1
should be treated cautiously. These findings indicate

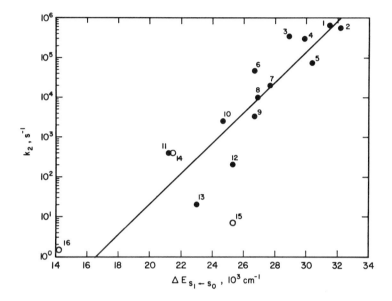

Figure 1. Correlation of protonation rate constants (k_2) with singlet–singlet separation ($\Delta E_{s_1 \leftarrow s_0}$). The compounds are identified by the numbers as given in Table I.

the relation between ionization potential and singlet-
singlet separation does not hold for non-alternant
hydrocarbons as discussed in detail by Michl and
Thulstrup (13) for the case of azulene.
The present study shows that the rate of protonation of an aromatic anion radical can be correlated with the energy level (LUMO) of the unpaired
electron. This finding suggests, on the basis of
Hammond's postulate (14), that the transition state
in the protonation process resembles the anion radical
more than it resembles the product (neutral radical).
The correlation described by Fig. 1 does not hold
for biphenyl and the three terphenyls whose rates of
protonation (2) deviate from the line, most probably
because of non-planarity.
The rates of reduction of the aromatic compounds
(Table I), appear to increase as $\Delta E_{S_1 \leftarrow S_0}$ and I decrease, in contrast with the behavior of k_2. This
trend is to be expected since k_5 reflects the
electron affinity which decreases when I and $\Delta E_{S_1 \leftarrow S_0}$
increase. The values of k_5 also can be related
to the polarographic half-wave potentials, $E_{\frac{1}{2}}$, of the
aromatic hydrocarbons and the general trend is, of
course, a decrease in k_5 when $E_{\frac{1}{2}}$ is more negative.

*The research described herein was supported by the
Division of Basic Energy Sciences of the Department of
Energy. This is Document No. NDRL-1839 from the Notre
Dame Radiation Laboratory.

Literature Cited

1.   Paul, D.E., Lipkin, D. and Weissman, S.I., J.
     Am. Chem. Soc. (1956) 78, 116.
2.   See review by Dorfman, L.M., Accounts Chem. Res.
     (1970) 3, 224, and references therein.
3.   Levanon, H. and Neta, P., Chem. Phys. Lett.
     (1977) 48, 345.
4.   Bank, S. and Bockrath, B., J. Am. Chem. Soc.
     (1972) 94, 6076.
5.   Fry, A.J. and Schuettenberg, A., J. Org. Chem.
     (1974) 39, 2452.
6.   Shida, T. and Iwata, S., J. Am. Chem. Soc. (1973)
     95, 3473.
7.   Simic, M., Neta, P. and Hayon, E., J. Phys. Chem.
     (1969) 73, 3794.
8.   Neta, P. and Levanon, H., J. Phys. Chem. (1977)
     81, 0000.

9. see e.g., Neta, P., Adv. Phys. Org. Chem. (1976)
    12, 223.
10. Streitwieser, A., "Molecular Orbital Theory for
    Organic Chemists," Wiley, New York (1961),
    Chapters 11 and 13.
11. Salem, L., "The Molecular Orbital Theory of
    Conjugated Systems," Benjamin, New York (1966),
    Chapter 6.
12. It was shown by Hush, N.S. and Pople, J.A., Trans.
    Faraday Soc. (1955) 51, 600, that for alternant
    aromatic hydrocarbons the sum of electron
    affinity and ionization potential is constant.
13. Michl J. and Thulstrup, E.W., Tetrahedron (1976)
    32, 205.
14. Hammond, G.S., J. Am. Chem. Soc. (1955) 77, 334.

RECEIVED December 23, 1977.

Termination Reactions, Stable Radicals, and Radical Trapping

Self-Reactions of Alkylperoxy Radicals in Solution *(1)*

J. A. HOWARD

Division of Chemistry, National Research Council of Canada,
Ottawa, Ontario, Canada K1A 0R9

In 1967 Benson (2) noted that the most important area of disagreement, or perhaps uncertainty, which had been raised at the
International Oxidation Symposium in San Francisco (3-5) was the
nature of the termination reaction of alkylperoxy radicals. In
the ten years that have elapsed since that meeting numerous papers
have appeared on the kinetics and mechanisms of the self-reactions
·of these radicals and it is true to say that we are no closer to a
complete understanding of the mechanisms of these reactions than
we were in 1967.

Prior to 1957 the termination reaction for liquid-phase hydrocarbon autoxidation at oxygen pressures above *ca.* 100 torr was
generally written as

(1) $RO_2^{\bullet} + RO_2^{\bullet} \longrightarrow$ non-radical products.

Kinetic results were consistent with a bimolecular termination reaction whereas reaction products and mechanisms were some-
thing of a mystery. At that time it was known that the termination rate constant for autoxidation of cumene (6) is about three
orders of magnitude smaller than the termination rate constant for
autoxidation of tetralin (7). It was, however, generally accepted
that the termination rate constants for tertiary (8) and secondary
(9) alkylperoxy radicals are insensitive to the structure of the
hydrocarbon residue in the radical.

Russell (10) was the first to propose an acceptable mechanism
for the termination of primary and secondary alkylperoxy radicals
while Blanchard (11) made the important discovery that cumylperoxy
radicals are capable of undergoing non-terminating as well as
terminating interactions during autoxidation of ·cumene. These two
pieces of work stimulated a great deal of further research on the
self-reaction of alkylperoxy radicals. The results of this work,
which are reviewed here, have provided compelling evidence that
tertiary and secondary (and primary) alkylperoxy radicals can
terminate by different mechanisms in the liquid-phase. For this
reason these two types of radicals will be discussed separately.

© 0-8412-0421-7/78/47-069-413$05.00/0

Tertiary alkylperoxy radicals will be discussed first because the mechanism for these radicals is reasonably well understood. On the other hand the mechanism for self-reaction of secondary and primary alkylperoxy radicals is still in doubt and these radicals are considered in the second part of this review.

Tertiary Alkylperoxy Radicals

t-Butylperoxy. The simplest tertiary alkylperoxy radical is t-butylperoxy and there is considerable experimental evidence in support of the mechanism given in Scheme I for self-reaction of this radical.

Scheme I

(2) $2t\text{-BuO}_2{}^{\bullet}$ $\rightleftharpoons$ $t\text{-BuO}_4\text{Bu-}t$

(3) $t\text{-BuO}_4\text{Bu-}t$ $\xrightarrow{k_3}$ $\left[t\text{-BuO}^{\bullet}\text{O}_2{}^{\bullet}\text{OBu-}t \right]_{\text{cage}}$

(4) $\left[t\text{-BuO}^{\bullet}\text{O}_2{}^{\bullet}\text{OBu-}t \right]_{\text{cage}}$ $\xrightarrow{k_4}$ $t\text{-BuOOBu-}t + \text{O}_2$

(5) $\left[t\text{-BuO}^{\bullet}\text{O}_2{}^{\bullet}\text{OBu-}t \right]_{\text{cage}}$ $\xrightarrow{k_5}$ $2t\text{-BuO}^{\bullet} + \text{O}_2$

The alkoxy radicals formed in (5) may react with the solvent SH to give, in the presence of oxygen, solvent derived peroxy radicals or scavenge a $t\text{-BuO}_2{}^{\bullet}$,

(6) $t\text{-BuO}^{\bullet} + \text{SH}$ $\xrightarrow{\text{O}_2}$ $t\text{-BuOH} + \text{SO}_2{}^{\bullet}$

(7) $t\text{-BuO}^{\bullet} + t\text{-BuO}_2{}^{\bullet}$ $\longrightarrow$ $t\text{-BuO}_3\text{Bu-}t$

The initial reaction between $t\text{-BuO}_2{}^{\bullet}$ (2) must involve a head-to-head interaction because if peroxy radicals labelled with oxygen-18 ($t\text{-Bu}^{18}\text{O}^{18}\text{O}^{\bullet}$) are allowed to react with normal peroxy radicals ($t\text{-Bu}^{16}\text{O}^{16}\text{O}^{\bullet}$) the oxygen evolved has a total mass of 34 (12).

The existence of di-t-butyl tetroxide was deduced from studies of the influence of temperature on the concentration of t-butylperoxy radicals by electron spin resonance spectroscopy (13, 14). Thus it was shown that at temperatures below 193K the radical concentration can be increased by raising the temperature and decreased by lowering the temperature with no apparent loss in radical concentration. The influence of temperature on the concentration of $t\text{-BuO}_2^{\bullet}$ is shown in Figure 1. If the tetroxide is completely dissociated at the highest temperature equilibrium constants for reaction (2), K_2, can be calculated from

$$K_2 = \frac{[t\text{-BuO}_4\text{Bu-}t]}{[t\text{-BuO}_2^{\bullet}]^2}$$

because

$$[t\text{-BuO}_4\text{Bu-}t] = \tfrac{1}{2} \{[t\text{-BuO}_2^{\bullet}]_{max} - [t\text{-BuO}_2^{\bullet}]\}$$

where $[t\text{-BuO}_2^{\bullet}]$ is the measured radical concentration and $[t\text{-BuO}_2^{\bullet}]_{max}$ is the maximum radical concentration.

Equilibrium constants obtained at different temperatures are given in Table I. Plots of $\ln K_2$ against the reciprocal of the absolute temperature yielded values of ΔH_2° and ΔS_2° between -8.0 and -8.8 kcal mol^{-1} and -27 and -34 cal deg^{-1} mol^{-1}, respectively, depending on the method of radical preparation (14, 15). It would, therefore, appear that although ΔH_2° is known with a fair degree of accuracy there is some uncertainty about the magnitude of ΔS_2°, probably because of errors involved in measuring $[t\text{-BuO}_4\text{Bu-}t]$.

Table I   Equilibrium constants for $t\text{-BuO}_2^{\bullet}$ - $t\text{-BuO}_4\text{Bu-}t$ equilibrium

Temperature/K	183	173	163	153
$10^{-4}(K_2/M^{-1})$	0.1	0.4	1.8	10

Heats of formation of $t\text{-BuO}_4\text{Bu-}t$ and $t\text{-BuO}_2^{\bullet}$ have been estimated (16) to be -47 ± 8 and -21.5 ± 2.5 kcal mol^{-1}, respectively, which give a calculated $\Delta H_2^{\circ} \sim -5$ kcal mol^{-1} which is ~ 3 kcal mol^{-1} smaller than the measured value. This difference is probably because the tetroxide is more stable than was predicted (16).

Above 193K $t\text{-BuO}_2^{\bullet}$ decay irreversibly with second-order kinetics, i.e.,

$$(8) \quad \frac{-d[t\text{-BuO}_2^{\bullet}]}{dt} = 2k_a[t\text{-BuO}_2^{\bullet}]^2$$

The maximum value of $2k_a$ is $4k_3K_2k_5/(k_4 + k_5)$ and is obtained if all $t\text{-BuO}^{\bullet}$ are removed by (7). Values of this rate constant have been obtained for radicals prepared by photolysis of 2,2'-azoisobutane in oxygenated CF_2Cl_2 and by complete oxidation of $t\text{-BuOOH}$ with a large excess of Ce(IV) in CH_3OH using kinetic e.s.r. spectroscopy (17). Values of $2k_a$ at 303K from 2.5×10^2 to 2.5×10^4 M^{-1} s^{-1} have been reported with a "best" value of $\sim 10^4$ M^{-1} s^{-1}. The most reliable Arrhenius equation for this rate constant (14) appears to be

$$\log(2k_a/M^{-1}s^{-1}) = 9.7 - 8.7/\theta$$

where $\theta = 2.303$ RT kcal mol^{-1}.

In the presence of large concentrations of t-butyl hydroperoxide the t-butoxy radicals formed in (5) abstract the hydroperoxidic hydrogen to regenerate $t\text{-BuO}_2^{\bullet}$.

$$(9) \quad t\text{-BuO}^{\bullet} + t\text{-BuOOH} \longrightarrow t\text{-BuOH} + t\text{-BuO}_2^{\bullet}$$

Consequently irreversible radical decay is slower than it is in
the absence of t-BuOOH with a second-order rate constant k_b =
$2k_3K_2k_4/(k_4 + k_5)$.

There have been many determinations of this composite rate
constant by KESR and the hydroperoxide method ([17]). Absolute
values at 303K vary from 7×10^1 to 3×10^5 M^{-1} s^{-1}, values of
$\log(A_b/M^{-1} s^{-1})$ from 5.5 to 12, and activation energies from 4.5
to 10 kcal mol^{-1} ([17]). It has, however, been concluded that the
"best" Arrhenius equation is ([18]).

$$\log(2k_b/M^{-1} s^{-1}) = 9.2 - 8.5/\theta$$

Product and kinetic studies of the initiated decomposition of
t-BuOOH ([19],[20]) have been particularly useful in elucidating the
relative importance of (4) and (5). Induced hydroperoxide decomposition can be described by reactions (2) to (6) plus an initiation reaction such as the decomposition of di-t-butylperoxy-
oxalate (10).

(10) t-BuOOC(O)C(O)OOBu-t $\longrightarrow$ $2t$-BuO$\cdot$ + $2CO_2$

A kinetic analysis of this reaction gives

$$\frac{k_5}{k_4} = \frac{-d[t\text{-BuOOH}]/dt}{R_i} - 1 = \frac{2d[O_2]/dt}{R_i} - 1$$

where R_i is the rate of chain initiation. These equations have
been verified experimentally and Hiatt, Clipsham, and Visser ([19])
obtained values of $(-d[t\text{-BuOOH}]/dt)/R_i$ in the range 6-10 at 45°,
implying $k_5/k_4 \sim 7$. The ratio of t-BuOH to t-BuOOBu-t (corrected
for alcohol produced from the initiator) was consistent with this
ratio. Factor, Russell and Traylor ([20]) confirmed these results
and found that $d[O_2]/dt/R_i$ is ~10 in chlorobenzene.

The ratio k_5/k_4 increases with an increase in temperature
([18],[21]) with E_5-E_4 = 5.3 to 6.6 kcal mol^{-1} and $\log(A_5/A_4)$ = 4.6
to 5.2. If it is assumed that the reaction of t-BuO$\cdot$ in the cage
requires no activation energy, E_5, the energy required to diffuse
out of the cage, must be about 6 kcal mol^{-1}. The difference in A-
factors for (5) and (4) of about 4 orders of magnitude is consistent with a bimolecular cage reaction and first-order diffusion
out of the cage. These differences in activation parameters are,
however, not consistent with the "best" values for k_a and k_b given
above. Thus $2k_a/2k_b$ = $2k_5/k_4$ and $\log(k_a/k_b)$ = 0.5 and $E_5 - E_4$ =
0.2 kcal mol^{-1}. Clearly, there are discrepancies in the kinetic
data for self-reaction of t-BuO$_2\cdot$ which will require further
investigation before they are resolved.

Adamic, Howard and Ingold ([14]) have used the thermodynamic
parameters for the t-BuO$_2\cdot$-t-BuO$_4$Bu-t equilibrium and activation
parameters for irreversible radical decay to calculate activation
parameters for irreversible tetroxide decomposition. Decay

constants for t-BuO$_2^{\bullet}$ prepared by photolysis of 2,2'-azoisobutane in CF$_2$Cl$_2$ were used and it was assumed that $k_5/(k_4+k_5)$ is equal to 1 at the temperature of the experiments. The rate constant $2k_a$ is, therefore, equal to $4K_2k_3$ and

$$\ln A_a - \frac{E_a}{RT} = \ln 4 + \frac{\Delta S_2^o}{R} - \frac{\Delta H_2^o}{RT} + \ln A_3 - \frac{E_3}{RT}$$

Using $A_a = 10^{9.7}$ M^{-1} s^{-1}, $E_a = 8.7$ kcal mol^{-1}, $\Delta H_2^o = -8.8$ kcal mol^{-1}, and $\Delta S_2^o = -34$ cal deg^{-1} mol^{-1} values of $A_3 = 10^{16.6}$ s^{-1} and $E_3 = 17.5$ kcal mol^{-1} were calculated. It was concluded (14) from the magnitude of the A-factor that only one bond in the tetroxide is cleaved in the rate determining step.

(11) t-BuO$_4$Bu-t $\longrightarrow$ t-BuO$_3^{\bullet}$ + $^{\bullet}$OBu-t $\longrightarrow$ t-BuO$^{\bullet}$ + O$_2$ + $^{\bullet}$OBu-t

In support of the intermediacy of t-BuO$_3^{\bullet}$ there is some evidence that it can be prepared from t-butoxy and oxygen (22,23) while CF$_3$O$_3^{\bullet}$ has been unambiguously identified (24).

It should, however, be noted that Barlett and Guaraldi (13) and Mill and Stringham (25) obtained low A-factors ($10^{9.3}$ and 10^{12} s^{-1}, respectively) for irreversible decomposition of t-BuO$_4$ Bu-t and concluded that decomposition must be concerted.

There have been several reports of t-butylperoxy radicals undergoing self-reaction with first-order kinetics (21,26). Now it is well known that decay of certain radicals can be first-order if the radical is in equilibrium with a diamagnetic dimer and substantial concentrations of dimer are present at the decay temperature (27-28). t-Butylperoxy radicals do not fall into this category because the tetroxide is completely dissociated before irreversible radical decay occurs. Other less persistent t-RO$_2^{\bullet}$ do, however, decay irreversibly in the presence of tetroxide and in these cases first-order decay kinetics are observed (29) because radical decay monitors tetroxide decomposition.

Cumylperoxy. The initial reaction between cumylperoxy radicals (RO$_2^{\bullet}$) involves a head-to-head interaction to give dicumyl tetroxide (30)

(12) RO$_2^{\bullet}$ $\rightleftharpoons$ RO$_4$R

The equilibrium constants for this process, K_{12}, (17) fit the relation

$$\log(K_{12}/M^{-1}) = (-7 \text{ to } -10.5) + (9.2 \text{ to } 11.2)/\theta$$

Product analyses (11, 31, 32) have shown that the cumylperoxy radical undergoes non-terminating and terminating reactions during autoxidation of cumene.

(13) RO_4R $\longrightarrow$ $[RO^{\cdot} \ O_2 \ ^{\cdot}OR]_{cage}$

(14) $[RO^{\cdot} \ O_2 \ ^{\cdot}OR]_{cage}$ $\longrightarrow$ $ROOR + O_2$

(15) $[RO^{\cdot} \ O_2 \ ^{\cdot}OR]_{cage}$ $\longrightarrow$ $2RO^{\cdot} + O_2$

The cumyloxy radicals produced by non-terminating interactions either abstract a H-atom from cumene or undergo β-scission
at ambient temperatures to give acetophenone and methyl radicals,
the latter being converted to methylperoxy by reaction with
oxygen.

(16) $RO^{\cdot} + RH \longrightarrow ROH + R^{\cdot}$

(17) $RO^{\cdot} \xrightarrow{\ O_2\ } CH_3O_2^{\cdot} + C_6H_5C(O)CH_3$

Methylperoxy radicals either propagate autoxidation by reacting
with cumene or terminate the reaction by reacting with cumylperoxy radicals.

(18) $CH_3O_2^{\cdot} + RH \longrightarrow CH_3OOH + R^{\cdot}$

(19) $CH_3O_2^{\cdot} + RO_2^{\cdot} \longrightarrow ROH + CH_2O + O_2$

The absolute termination rate constant, $2k_t$, obtained from
rotating sector studies of autoxidation of neat cumene is, therefore, an *overall* rate constant and is given by

(20) $2k_t = 2fk_{13}K_{12} + 2(1-f)k_{13}K_{12}\left[\dfrac{k_{17}}{k_{17}+k_{16}[RH]}\right] \cdot$

$\dfrac{4k_{19}[RO_2^{\cdot}]}{4k_{19}[RO_2^{\cdot}] + k_{18}[RH]}$

where $f = k_{14}/(k_{14} + k_{15})$.

The magnitude of this rate constant depends on k_{13}, K_{12}, f, the
fraction of cumyloxy radicals which undergo β-scission, and the
fraction of methylperoxy radicals which are consumed in the
termination reaction (19).

The absolute value of $2k_t$ at $30° = 1.5 \times 10^4 \ M^{-1} \ s^{-1}$ with
$\log(2k_t/M^{-1} \ s^{-1}) = 10.1 - 9.2/\theta$ (17). The value of E_t is much
larger than the value of ~0 kcal mol^{-1} found by Melville and
Richards (6) and Thomas (21). This high activation energy has,
however, been confirmed by KESR (14).

Traylor and Russell (32) made the important discovery in 1965
that the rate of oxidation of cumene is increased by the addition
of cumene hydroperoxide. Thus the rate depends on the hydroperoxide concentration until a limiting rate is reached whereupon
addition of more hydroperoxide has no effect on the rate

(Figure 2). This increase in rate was attributed ($\underline{31}$) to reaction of cumyloxy and methylperoxy radicals with the hydroperoxide thus preventing $CH_3O_2^{\bullet}$ from undergoing chain termination reactions.

$$(21) \quad \left.\begin{array}{c} RO^{\bullet} \\ \\ CH_3O_2^{\bullet} \end{array}\right\} + ROOH \longrightarrow \left\{\begin{array}{c} ROH \\ \\ CH_3OOH \end{array}\right. + RO_2^{\bullet}$$

Measurement of the termination rate constant $2k_b$ under these conditions gave a value of 6×10^3 M^{-1} s^{-1} with $\log(2k_b/M^{-1}\ s^{-1})$ = $10.7 - 9.5/\theta$ ($\underline{17}$).

Values of A_b and E_b in conjunction with ΔS°_{12} and ΔH°_{12} have been used to calculate the activation parameters $\log(A_{13}/s^{-1})$=17.1 and $E_{13} = 16.5$ kcal mol^{-1} for irreversible decomposition of di-cumyl tetroxide ($\underline{14}$). Although there is little difference between the activation parameters for di-cumyl- and di-t-butyl-tetroxides it would appear that the former is somewhat less stable towards irreversible decomposition.

Fukuzumi and Ono have very recently concluded that the termination reaction for oxidation of cumene with manganese dioxide or cobalt oxide supported on silica ($\underline{33}$) and during autoxidation of cumene initiated by reaction of cumene hydroperoxide with lead oxide ($\underline{34}$) is strictly first-order with respect to the concentration of cumylperoxy radicals. These workers proposed an unprecedented 1,3-methyl shift followed by O–O bond cleavage to account for these unusual kinetics,

$$RO_2^{\bullet} \longrightarrow C_6H_5(CH_3)\overset{\bullet}{C}OOCH_3 \longrightarrow C_6H_5C(O)CH_3 + CH_3O^{\bullet}$$

whereas a pseudo-first order reaction is more plausible.

Other t-alkylperoxys. The self-reactions of a wide variety of other t-alkylperoxy radicals have been examined by hydrocarbon autoxidation and KESR ($\underline{14}$, $\underline{17}$, $\underline{35}$, $\underline{36}$) and they all exist in equilibrium with t-RO$_4$R-t. Unfortunately, most of these radicals are less persistent than t-BuO$_2^{\bullet}$ and values of ΔS° could not be determined with any degree of accuracy because [t-RO$_4$R-t] could not be measured. Estimates of ΔS° and ΔH° have, however, been made ($\underline{17}$, $\underline{36}$) and it has been concluded that the nature of R has very little influence on the magnitude of these parameters. In addition, the nature of R appears to have little or no influence on the ratio of rate constants for non-terminating and terminating interactions ($\underline{37}$).

Absolute values of overall termination rate constants ($2k_t$) and rate constants for self-termination ($2k_b$) have been measured and typical values are presented in Table II.

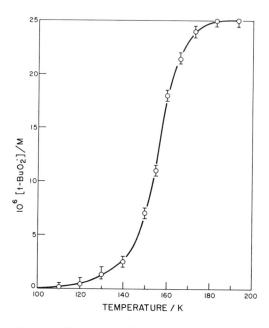

Figure 1. Variation of the concentration of $(CH_3)_3$-$CO_2 \cdot$ with temperature in the range where irreversible decay does not occur

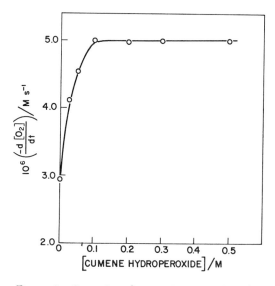

Figure 2. Rate of oxidation of cumene (3.6M) in chlorobenzene as a function of cumene hydroperoxide concentration at 330.2 K (31)

Table II   Overall and self-termination rate constants for some
tertiary alkylperoxy radicals (17).

Peroxy radical	$10^{-4}(2k_t)^a$ $(M^{-1} s^{-1})$	$10^{-4}(2k_b)^b$ $(M^{-1} s^{-1})$
Cumylperoxy	1.5	0.6 - 0.8
2-Phenyl-2-butylperoxy	18	3.2
1,1-Diphenylethylperoxy	13	6.4
1-Methylcyclopentylperoxy	-	0.6 - 23
2-Cyano-2-propylperoxy	-	100

[a] From rotating sector studies of hydrocarbon autoxidation.

[b] From rotating sector studies of hydrocarbon autoxidation in the
presence of hydroperoxide or by KESR in the presence of hydroperoxide.

Values of $2k_t$ for these radicals will be given by an equation
similar to (20) and as might be expected $2k_t$ depends on the nature
of R because the t-RO$^{\bullet}$ produced by non-terminating interactions
exhibit different susceptibilities to β-scission. Somewhat more
surprisingly values of $2k_b$ also depend on the nature of R. For
instance the self-termination rate constant for 1,1-diphenylethyl-
peroxy is over an order of magnitude larger than $2k_b$ for t-BuO$_2^{\bullet}$.
Since K and f do not depend on R it has been concluded that
differences in k_b are due to differences in the rate constants for
unimolecular decomposition of t-RO$_4$R-t.

Acylperoxy radicals

Traylor and co-workers have recently provided evidence from
autoxidation of acetaldehyde (38-40) and induced decomposition of
peracetic acid (41) for a mechanism for the self-reaction of the
acetylperoxy radical which involves a non-terminating interaction
via a tetroxide.

(22) $2CH_3C(O)O_2^{\bullet} \longrightarrow CH_3C(O)O_4C(O)CH_3 \longrightarrow 2CH_3^{\bullet} + 2CO_2 + O_2$

Acetyl peroxide, methyl acetate, dimethyl peroxide and ethane are
not produced in these reactions indicating that there is no cage
collapse of the radicals. Instead the methyl radicals diffuse out
of the cage and react with oxygen to give methylperoxy. This means
that autoxidation of acetaldehyde is terminated either by reaction
of methylperoxy with acetylperoxy or by self-reaction of methylperoxy (23) and (24).

(23) $CH_3O_2^{\bullet} + CH_3C(O)O_2^{\bullet} \longrightarrow CH_3C(O)OH + CH_2O + O_2$

(24) $2CH_3O_2^{\bullet} \longrightarrow CH_2O + CH_3OH + O_2$

Termination rate constants for autoxidation of some aldehydes (42) are given in Table III. The value for acetaldehyde must be a composite one containing contributions from $2k_{23}$ and $2k_{24}$. If reactions (22), (23), and (24) apply to the other aldehydes all the rate constants in this table will be composite ones.

Benzoyloxy radicals do not undergo decarboxylation as readily as other acylperoxy radicals. The termination rate constant is, however, close to the diffusion controlled limit. This apparent anomaly has been attributed (42) to irreversible formation of the tetroxide followed by complete dimerization of benzoyloxy radicals in the cage.

(25) $2C_6H_5C(O)O_2^{\bullet} \longrightarrow C_6H_5C(O)O_4C(O)C_6H_5 \longrightarrow$

$C_6H_5C(O)O^{\bullet} \; O_2 \; {}^{\bullet}OC(O)C_6H_5 \longrightarrow C_6H_5C(O)O_2C(O)C_6H_5$

Table III. Termination rate constants for aldehyde autoxidation[a]

Aldehyde	$10^{-7}(2k_t)$ /$M^{-1} s^{-1}$
Acetaldehyde	10.4
Heptaldehyde	5.4
Octaldehyde	7.0
Cyclohexanecarboxaldehyde	0.7
Pivaldehyde	0.7
Benzaldehyde	176

[a] At $273^{\circ}K$.

Primary and secondary alkylperoxy radicals.

Two mechanisms have been considered for the self-reaction of primary and secondary alkylperoxy radicals, the Russell mechanism (10) reactions (26), (27), and (28) and a mechanism involving the intermediacy of alkoxy radicals (43, 44) reactions (29), (30), (31), and (32). The reactions involved in these two mechanisms are presented in Scheme II.

Scheme II

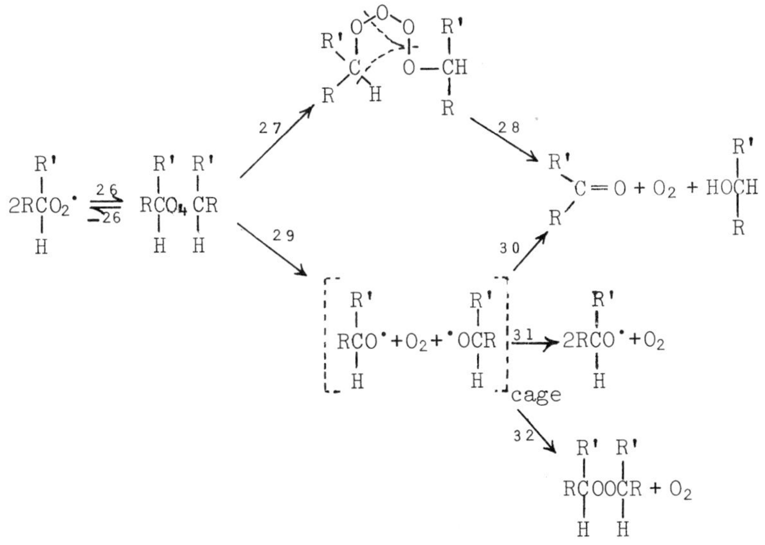

$$2R\dot{C}O_2 \underset{-26}{\overset{26}{\rightleftharpoons}} \underset{\underset{H\ H}{|\ \ |}}{R\overset{R'\ R'}{CO_4}CR}$$

(27), (28), (29)

$$\underset{\overset{|}{R'}}{R'}C = O + O_2 + HO\underset{\overset{|}{R}}{CH}$$

cage:
$$R\dot{C}O^{\bullet}+O_2+{}^{\bullet}O CR \xrightarrow{31} 2R\dot{C}O^{\bullet}+O_2$$

(30), (32)

$$\underset{\overset{|}{H}\ \overset{|}{H}}{R\overset{R'\ R'}{C}OOCR} + O_2$$

Russell (10) suggested that the bimolecular self-reaction of s-$RO_2^{\bullet}$ involves the concerted decomposition of a cyclic tetroxide formed by combination of the radicals. This mechanism was deduced from a consideration of the results of a kinetic and product study of the autoxidation of ethylbenzene. Thus Russell found that almost one molecule of acetophenone is produced per two kinetic chains and that $C_6H_5CH(CH_3)O_2^{\bullet}$ interact to form non-radical products nearly twice as fast as $C_6H_5CD(CH_3)O_2^{\bullet}$. The former result is only compatible with (29) if all the alkoxy radicals disproportionate in the solvent cage (30) while the deuterium isotope effect requires a H-atom transfer reaction to be rate controlling, which is unlikely for the radical pathway.

Support for the Russell mechanism came from Howard and Ingold (45) who found that some of the oxygen evolved from self-reaction of s-butylperoxy radicals is in the electronically excited singlet delta state ($^1\Delta g$) as required by the Wigner spin conservation rule for concerted decomposition to give singlet ketone and singlet alcohol. Kellogg (46) has, however, suggested that tetroxide decomposition gives triplet ketone and triplet oxygen and that singlet oxygen is formed by cage encounters of these two species. On the other hand Beutel (47) has argued that ketone can be produced either in a vibrationally excited singlet state from which it may undergo an inefficient adiabatic transition to the triplet manifold or in an electronically excited singlet state followed either by fluorescence or intersystem crossing. Although the exact mechanism for the production of chemiluminescence during autoxidation is debatable there is no doubt that emission spectra from triplet ketones have been observed and the intensity of chemiluminescence during non-stationary state autoxidation has been used to provide kinetic data for termination reactions (17).

Production of singlet oxygen (both $^1\Sigma g^+$ and $^1\Delta g$) by self-reaction of s-butylperoxy radicals and the peroxy radicals derived from linoleic acid has very recently been confirmed by analysis of the emission spectra observed during ceric ammonium nitrate oxidation of the appropriate hydroperoxide (48).

Howard and Ingold (49) confirmed Russell's conclusion that the rate constant for termination of s-alkylperoxy radicals depends on the strength of the α-C-H bond. Thus the average isotope effect for this reaction at 303K, $(2k_t)_H/(2k_t)_D$, is 1.37 ± 0.14 which provides compelling evidence for abstraction of this hydrogen in the rate controlling step (Table IV). Furthermore $2k_t$ for cyclohexenylperoxy is 2.8 times larger than $2k_t$ for cyclohexylperoxy which is consistent with the former radical having the weaker α-C-H bond.

Table IV. Deuterium isotope effects on rate constants for self-reaction of s-RO$_2$$^\bullet$

Peroxy radical	Temperature/K	$(2k_t)_H/(2k_t)_D$
$(C_6H_5)_2CH(D)O_2^\bullet$	303	1.36
$C_2H_5(CH_3)CH(D)O_2^\bullet$	303	1.37
cyclo-$C_5H_9(D_9)O_2^\bullet$	180	4.0

Hiatt and Zigmund (50) provided further evidence for the Russell mechanism when they discovered that the interaction of two s-butylperoxy radicals does not give di-s-butyl peroxide at 318K, a product that might be expected if the reaction involves the intermediacy of s-butoxy radicals. Diaper (51), however, reported that 1-methoxy- and 1-t-butoxy-nonane-1-hydroperoxides are oxidized to di-(1-alkoxyalkyl) peroxides by 1 equiv of Ce (IV) in methanol at 273K. Since Ce(IV) oxidizes alkyl hydroperoxides to the corresponding alkylperoxy radical in high yield these results are not compatible with Hiatt and Zigmund's findings or the Russell mechanism.

Other product studies (25, 52-54), mainly at ambient temperatures, have shown that almost equal yields of alcohol and ketone are formed from the self-reaction of s-RO$_2$$^\bullet$ as predicted by (27) and (28) or by (29) and (30).

In addition Hiatt and co-workers (55) found chain lengths of 0.7 to 1.0 for di-t-butyl peroxyoxalate induced decomposition of n-butyl-, s-butyl-, and α-tetralyl-hydroperoxides at 318K indicating that interactions of the s-RO$_2$$^\bullet$ from these hydroperoxides are almost always terminating at this temperature.

Lindsay *et al* (56) made a thorough study of the products of the self-reaction of 1-ethoxyethylperoxy and 1,2-diphenylethylperoxy radicals at ambient temperatures, radicals which would be expected to give the products outlined in Schemes III and IV.

Scheme III

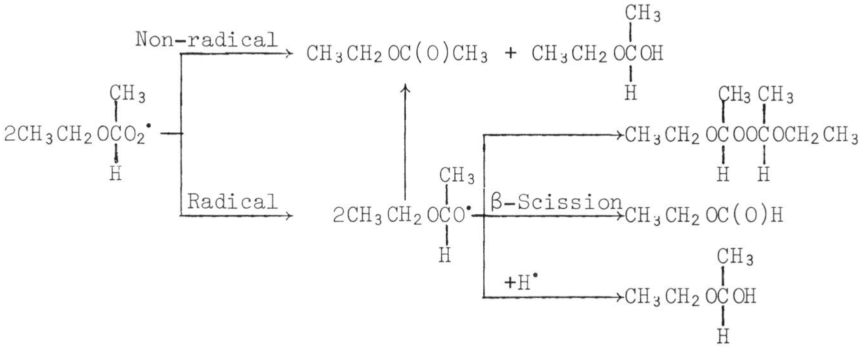

Scheme IV

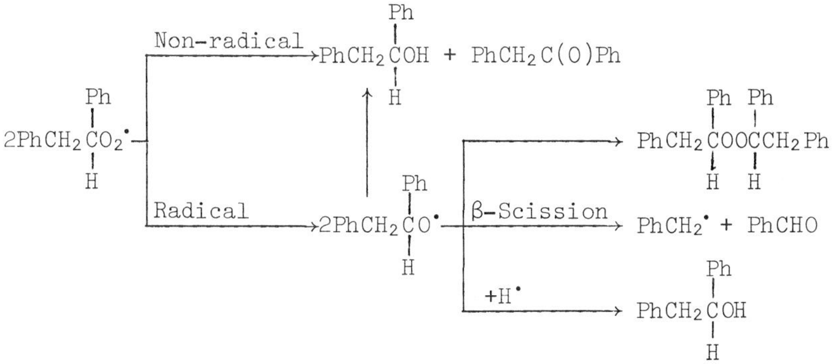

1-Ethoxyethylperoxyls were prepared by a variety of methods and react to give ethyl acetate, ethanol, acetaldehyde and ethyl formate. (Table V). There was no evidence for the formation of di-(1-ethoxyethyl) peroxide, a result that is not entirely incompatible with Diaper's work in view of the large difference in the size the peroxy radicals studied by the two groups of workers. The absolute yields of the products were difficult to ascertain with any degree of certainty because of secondary reactions. The high yield of ethyl formate is, however, diagnostic for the intermediacy of alkoxy radicals and indicates that at least 20% of the self-reactions of 1-ethoxyethylperoxyls occur via the radical mechanism.

Table V. Products[a] from oxidation of 1-ethoxyethyl hydroperoxide

Oxidant	Solvent	Ethyl formate	Ethyl acetate	Ethyl alcohol	Acet- aldehyde	O_2
Ce(IV)	CH_3OH	0.20	0.3	0.46	0.23[b]	0.4
t-BuO·[c]	CCl_4-C_6H_6	0.04-0.32	0.43	0.18	0.2	-
Ag_2O	C_6H_6	0.12	0.12	0.52	0.36	-

[a] mol per mol of hydroperoxide, [b] as 1,1-dimethoxyethane, [c] from di-t-butyl hyponitrite.

1,2-Diphenylethylperoxy radicals undergo self-reaction to give almost equal yields of benzaldehyde, benzyl alcohol, 1,2-diphenylethanol, and benzoin. Benzaldehyde is produced by β-scission of 1,2-diphenylethoxyls while 1,2-diphenylethanol and benzoin are formed by a non-radical process.

The work of Lindsay et al provides good evidence that not all s-RO_2· undergo bimolecular self-reaction at ambient temperatures entirely by the Russell mechanism. It would, however, appear from their work that the yield of alkoxy radicals is very dependent on the structure of the peroxy radical. This lead these workers (56) to propose that the tetroxide may decompose by multiple bond scission by a concerted but non-cyclic mechanism.

$$\underset{\substack{| \\ R'}}{\overset{\substack{R'' \\ |}}{R}}C O_4 \underset{\substack{| \\ R'}}{\overset{\substack{R'' \\ |}}{C}}R \xrightarrow{\text{slow}} R· + \underset{\substack{| \\ R'}}{\overset{\substack{R'' \\ |}}{C}}=O + \left(\underset{\substack{| \\ R'}}{\overset{\substack{R'' \\ |}}{R}}CO_3· \right) \longrightarrow \text{Reaction products}$$

Structural factors may, therefore, have a profound influence not only on the rate constants for termination of s-RO_2· but also on the reaction mechanism.

Product studies have indicated (57) that the fraction of s-RO_2· interactions that terminate during autoxidation of neat n-butane is 0.3-0.6 at 373K and 0.18-0.35 at 398K. Furthermore, the non-radical concerted reaction provides most of the termination for s-BuO_2· at 373K whereas at the higher temperature a major fraction of the interactions give free s-BuO·. These results are quite compatible with all s-BuO_2· self-reactions being terminated via a concerted non-radical process at 303K.

Very recently Bennett and Summers (58) reported a product study of the low temperature self-reaction of s-butylperoxy, s-hexylperoxy cycloheptylperoxy, and cyclopentylperoxy radicals. Although alcohol to ketone ratios close to 1.0 are produced at ambient temperatures this ratio $decreases$ as the temperature is reduced. For instance the ratio of cyclopentanol to cyclopentanone is 0.06 at 173K. Furthermore significant yields of hydrogen peroxide are obtained. These results prompted Bennett and Summers to propose an alternative transition state to the Russell mechanism which increases in importance as the temperature is lowered.

$$2R_2CHO_2^{\cdot} \; \rightleftharpoons \; R_2-C\underset{O-O\cdots H}{\overset{H\cdots O-O}{\diagup\vert\diagdown}}C-R_2 \; \longrightarrow \; R_2C{=}O \; + \; H_2O_2$$

Kinetic data for self-reaction of p-RO$_2^{\cdot}$ and s-RO$_2^{\cdot}$ have been obtained by EPR spectroscopy and rotation sector studies on hydrocarbon autoxidation (17). These reactions usually obey the kinetic expression

$$\frac{-d[RO_2^{\cdot}]}{dt} = 2k_t[RO_2^{\cdot}]^2$$

where $2k_t$ is the bimolecular termination rate constant.

It has generally been found that p-RO$_2^{\cdot}$ have termination rate constants in excess of 10^8 M^{-1}s^{-1} while s-RO$_2^{\cdot}$ exhibit a wide variation in $2k_t$ which, with a few exceptions, can be classified according to whether the radical is derived from a benzylic, allylic, cyclic, or alkyl system. In the benzylic group are compounds such as ethylbenzene, n-butylbenzene, and styrene, with $2k_t$'s of $2{-}6 \times 10^7$ M^{-1}s^{-1}. Secondary alkylperoxy radicals from alkenes undergo termination more slowly than benzylic peroxy radicals. This is partly due to steric effects associated with the alkyl moiety of the peroxy radical since $2k_t$'s increase as the size of the olefin is decreased and partly due to an increase in the strength of the α-C-H bond. Cyclic hydrocarbons such as tetralin and cyclohexene have termination rate constants in the range $2{-}8 \times 10^6$ M^{-1}s^{-1}. Addition of secondary alkyl hydroperoxides, e.g., α-tetralin hydroperoxide to the parent hydrocarbon has no effect on the magnitude of $2k_t$ indicating that in these cases non-terminating self-reactions are not significant. Values of $2k_t$ for some p-RO$_2^{\cdot}$ and s-RO$_2^{\cdot}$ are given in Table VI.

Table VI. Termination rate constants at 303K for some typical primary and secondary alkylperoxy radicals.

Alkylperoxy radical.	$10^{-6}(2k_t)/M^{-1}s^{-1}$	Reference
Methylperoxy	500^a	59
Ethylperoxy	80^a	59
n-Butylperoxy	40^c, 300^b	17,59
2-Propylperoxy	3^b	59
2-Butylperoxy	1.5^c, 9^b	17,59
Cyclohexylperoxy	2.0^c, 10^b	17,59
Cyclohexenylperoxy	5.6^c	17
α-Tetralylperoxy	7.6, 7.2^d	17
Benzylperoxy	300^c	17
1-Phenylethylperoxy	40^c	17
Poly(peroxystyrylperoxy)	42^c	17
Tetrahydrofuranylperoxy	31^c	17

[a]Determined in the gas phase using molecular modulation spectroscopy (59). [b]Determined in the liquid-phase by KESR (59). [c]Rotating sector study of autoxidation (17). [d]In the presence of 0.2 M α-tetralin hydroperoxide (17).

Rate constants for methylperoxy and ethylperoxy were obtained in the gas-phase by molecular modulation spectroscopy (59) and it was concluded in this work that changing from the liquid to the gas phase has only a minor influence on the magnitude of $2k_t$. The order of reactivity, $CH_3O_2^{\cdot} > C_2H_5O_2^{\cdot} > (CH_3)_2CHO_2^{\cdot}$ is contrary to the previous conclusion that $2k_t$ increases as the α-hydrogen becomes weaker. It would, therefore, appear for these radicals that the relative stability of the incipient carbonyl compound overshadows a bond strength effect.

It should be noted that there is a significant discrepancy between the values of $2k_t$ determined by the rotating sector method and KESR for some of the radicals in Table VI. At the present time we are inclined to favour the values obtained from hydrocarbon autoxidation because values of the propagation rate constant are also obtained by this method, the magnitude of which gives an internal check on the value of $2k_t$.

Accurate Arrhenius parameters have been determined for a few primary and secondary $RO_2^{\cdot}$'s and examples are given in Table VII.

Table VII. Arrhenius parameters for self-reaction of some primary and secondary alkylperoxy radicals.

Alkylperoxy	$\log(A_t/M^{-1}s^{-1})$	E_t/kcal mol^{-1}	Reference
Methylperoxy	8.7	0	<u>59</u>
Ethylperoxy	7.9	~0.5	<u>59</u>
s-Butylperoxy	9.0	2.7	<u>60</u>
Cyclopentylperoxy	10.0	3.1	<u>60</u>

The pre-exponential factors for these radicals are close to the normal values for liquid-phase bimolecular reactions while the activation energies are small and *positive*.

There is evidence from e.p.r. studies (<u>29</u>,<u>36</u>) that s-RO$_2$· exist in equilibrium with a tetroxide below 173K and in this respect behave analogously to t-RO$_2$·. Unfortunately s-RO$_2$· decay irreversibly before the tetroxide is completely dissociated and values of the equilibrium constant cannot be estimated. It would, however, appear that the thermodynamic parameters for s-RO$_2$· are similar to those for t-RO$_2$·, e.g., $\Delta H° \sim -7$ kcal mol^{-1} and $\Delta S° < -25$ cal deg^{-1} mol^{-1} for isopropylperoxy.

The difference in termination rate constants between secondary (and primary) alkylperoxys and tertiary alkylperoxys is, therefore, almost entirely due to differences in the rate constants for irreversible tetroxide decay. Thus at 303K 2k_b for t-BuO$_2$·= 1.2 × 10^3 M^{-1}s^{-1} and 2k_t for s-BuO$_2$· = 10^7 M^{-1}s^{-1}. That is $k_t/k_b = 8.3 × 10^3$. Now Ingold (<u>61</u>) has suggested that $\Delta S°$ for (27) ≈ -14.4 cal deg^{-1}mol^{-1} because of the entropy loss from four hindered internal rotors (4 × 3.6 cal deg^{-1}mol^{-1}). Thus A_t/A_b and E_b-E_t should be 10^{-3} and 9.6 kcal mol^{-1}, respectively, for t-BuO$_2$· and s-BuO$_2$·, i.e., E_t should be *ca* 1 kcal mol^{-1} *negative*. Neither of these predictions are observed experimentally and we are forced to conclude that kinetic data for self-reaction of s-RO$_2^•$ provides further evidence against the complete acceptance of the Russell mechanism.

Conclusions

There is no doubt that all alkylperoxy radicals interact to give a tetroxide which decomposes to give either radical or non-radical products. Furthermore, it would appear that the structure of the tetroxide determines the overall rate and mechanism of the reaction. Di-t-alkyl tetroxides decompose either by a concerted or two step process to give t-alkoxy radicals, a fraction of which combine in the cage. This reaction pathway is also available to primary and secondary alkylperoxy radicals but seems to be preferred at higher temperatures. At temperatures below 373K these radicals appear to react principally by a non-radical

concerted process (Russell mechanism) although reaction products
and kinetics are not entirely consistent with this mechanism.

Literature cited

1. Issued as NRCC No. 16304.
2. Benson, S.W., Advan. Chem. Ser. (1968) 76, 143.
3. Oxidation of Organic Compounds -I, Advan. Chem. Ser. (1968),
   75.
4. Oxidation of Organic Compounds -II, Advan. Chem. Ser. (1968),
   77.
5. Oxidation of Organic Compounds -III, Advan. Chem. Ser.
   (1968), 77.
6. Melville, H.W. and Richards, S., J. Chem. Soc. (1954) 944.
7. Bamford, C.H. and Dewar, M.J.S., Proc. Roy. Soc. (London)
   (1949) 198A, 252.
8. Russell, G.A., J. Am. Chem. Soc. (1956) 78, 1047.
9. Bolland, J.L., Quart. Revs. (London) (1949) 3, 1.
10. Russell, G.A., J. Am. Chem. Soc. (1957) 79, 3871.
11. Blanchard, H.S., J. Am. Chem. Soc. (1959) 81, 4548.
12. Bennett, J.E. and Howard, J.A., J. Am. Chem. Soc. (1973) 95,
    4008.
13. Bartlett, P.D. and Guaraldi, G., J. Am. Chem. Soc., (1967)
    89, 4799.
14. Adamic, K., Howard, J.A., and Ingold, K.U., Can. J. Chem.
    (1969) 47, 3803.
15. Benson, S.W., Organic Peroxides, Vol. I, D. Swern, Ed.,
    Interscience, New York (1971) p.129.
16. Benson, S.W., J. Am. Chem. Soc. (1964), 86, 3922.
17. Howard, J.A., Advan. Free Radical Chem. (1972) 4, 49.
18. Korcek, S., Chenier, J.H.B., Howard, J.A., and Ingold, K.U.,
    Can. J. Chem. (1972) 50, 2285.
19. Hiatt, R., Clipsham, J., and Visser, T., Can. J. Chem. (1964)
    42, 2754.
20. Factor, A., Russell, C.A., and Traylor, T.G., J. Am. Chem.
    Soc. (1965), 87, 3692.
21. Thomas, J.R., J. Am. Chem. Soc. (1965) 87, 3935.
22. Weiner, S.A., and Hammond, G.S., J. Am. Chem. Soc. (1969) 91,
    2182.
23. Symons, M.C.R., J. Am. Chem. Soc. (1969) 91, 5924.
24. Fessenden, R.W., J. Chem. Phys. (1968) 48, 3725.
25. Mill, T. and Stringham, R.S., J. Am. Chem. Soc. (1968) 90,
    1062.
26. Ingold, K.U. and Morton, J.R., J. Am. Chem. Soc. (1964) 86,
    3400.
27. Weiner, S.A. and Mahoney, L.R., J. Am. Chem. Soc. (1972), 94,
    5029.
28. Brokenshire, J.L., Roberts, J.R. and Ingold, K.U., J. Am.
    Chem. Soc. (1972) 94, 7040.
29. Furimsky, E. and Howard, J.A., Unpublished results.

30. Bartlett, P.D. and Traylor, T.G., J. Am. Chem. Soc. (1963) 85, 2407.
31. Thomas, J.R., J. Am. Chem. Soc. (1967) 89, 4872.
32. Traylor, T.G. and Russell, C.A., J. Am. Chem. Soc. (1965) 87, 3698.
33. Fukuzumi, S., and Ono, Y., J. Chem. Soc. Perkin 2 (1977) 622.
34. Fukuzumi, S., and Ono, Y., J. Chem. Soc. Perkin 2 (1977) 784.
35. Bennett, J.E., Brown, D.M., and Mile, B., Trans. Faraday Soc. (1970) 66, 386.
36. Bennett, J.E., Brown, D.M., and Mile, B., Trans. Faraday Soc. (1970) 66, 397.
37. Howard, J.A. and Ingold, K.U., Can. J. Chem. (1969) 47, 3797.
38. Clinton, N.A., Kenley, R.A., and Traylor, T.G., J. Am. Chem. Soc. (1975) 97, 3746.
39. Clinton, N.A., Kenley, R.A., and Traylor, T.G., J. Am. Chem. Soc. (1975) 97, 3752.
40. Clinton, N.A., Kenley, R.A., and Traylor, T.G., J. Am. Chem. Soc. (1975) 97, 3757.
41. Kenley, R.A. and Traylor, T.G., J. Am. Chem. Soc. (1975) 97, 4700.
42. Zaikov, G.E., Howard, J.A., and Ingold, K.U., Can. J. Chem. (1969) 47, 3017.
43. Bell, E.R., Raley, J.H., Rust, F.F., Seubold, Jr., F.H. and Vaughan, W.E., Disc. Faraday Soc., (1951), 10, 242.
44. Seubold, Jr., F.H., Rust, F.F. and Vaughan, W.E., J. Am. Chem. Soc., (1951) 73, 18.
45. Howard, J.A. and Ingold, K.U., J. Am. Chem. Soc. (1968) 90, 1056.
46. Kellogg, R.E., J. Am. Chem. Soc. (1969) 91, 5433.
47. Beutel, J., J. Am. Chem. Soc. (1971) 93, 2615.
48. Nakano, M., Takayama, K., Shimizu, Y., Tsuji, Y., Inaba, H., and Migita, T., J. Am. Chem. Soc. (1976) 98, 1974.
49. Howard, J.A. and Ingold, K.U., J. Am. Chem. Soc. (1968) 90, 1058.
50. Hiatt, R. and Zigmund, L., Can. J. Chem. (1970) 48, 3967.
51. Diaper, D.G.M., Can. J. Chem. (1968) 46, 3095.
52. McCarthey, R.L. and MacLachlan, A., J. Chem. Phys. (1961) 35, 1625.
53. Cramer, W.A., J. Phys. Chem. (1967) 71, 1171.
54. MacLachlan, A., J. Am. Chem. Soc. (1965) 87, 960.
55. Hiatt, R., Mill, T., Irwin, K.C., and Castleman, J.K., J. Org. Chem. (1968), 33, 1428.
56. Lindsay, D., Howard, J.A., Horswill, E.C., Iton, L., and Ingold, K.U., Can. J. Chem. (1973) 51, 870.
57. Mill, T., Mayo, F., Richardson, H., Irwin, K., and Allara, D.L., J. Am. Chem. Soc. (1972) 94, 6802.
58. Bennett, J.E. and Summers, R., Can. J. Chem. (1974) 52, 1377.
59. Bennett, J.E. and Parkes, D.A., Verbal communication to the EUCHEM conference on "Organic Free Radicals", Schloss Elmau, Germany, Oct. 19-24, (1975).

60.   Howard, J.A. and Bennett, J.E., Can. J. Chem. (1972) 50, 2374.
61.   Bowman, D.F., Gillan, T., and Ingold, K.U., J. Am. Chem. Soc.,
      (1971) 93, 6555.

RECEIVED December 23, 1977.

The Spin Trapping Reaction

EDWARD G. JANZEN, C. ANDERSON EVANS (1), and EDWARD R. DAVIS

Department of Chemistry, Guelph Waterloo Center for Graduate Work in
Chemistry, University of Guelph, Guelph, Ontario

A number of different kinds of radicals add to the nitrone
function to produce esr detectable (2) and sometimes isolable
nitroxides (3):

$$R\cdot + \underset{+}{\overset{O-}{\underset{\;}{\;}}C{=}N} \longrightarrow R{-}\underset{|}{\overset{O-}{\underset{+}{C}}}{-}N\cdot \longleftrightarrow R{-}\overset{O\cdot}{\underset{\cdot\cdot}{C}}{-}N$$

The α-phenyl N-t-butyl nitrones (PBN's) have been studied in
greatest detail (4):

$$R\cdot + \underset{+}{PhCH{=}N}{-}CMe_3 \longrightarrow R{-}\overset{H}{\underset{Ph}{C}}{-}\overset{O\cdot}{N}{-}CMe_3$$

The magnitude of the β-hydrogen hyperfine splitting depends mainly
on the bulk of R and indirectly on the group electronegativity of
R. Since the spin density on the nitrogen atom and thus the g-
value of the nitroxide is sensitive to the polar character of R
the β-hydrogen splitting will vary depending on the group
electronegativity of R. If R contains atoms with nuclear spin
additional hyperfine splitting can be expected. Thus in principal
each trapped radical should produce a unique spin adduct spectrum.
This technique has been named spin trapping.

Quantitative Studies

It is of some interest to know what fraction of radicals can
be trapped by nitrones at suitable concentrations. The first
quantitative measurements were made with benzoyl peroxide in
benzene at room temperature (5).

$$\overset{O}{\underset{\;}{Ph}}\overset{O}{\underset{\;}{COOC}}Ph \longrightarrow 2\;Ph\overset{O}{\underset{\;}{C}}{-}O\cdot$$

$$
\underset{\substack{\| \\ O}}{PhC}-O\cdot \; + \; \underset{\substack{+ \\ }}{PhCH=}\overset{O^-}{\underset{|}{N}}-CMe_3 \;\longrightarrow\; PhCH-\overset{O\cdot}{\underset{|}{N}}-CMe_3 \\
\hspace{6.5cm} \underset{\substack{\| \\ O}}{Ph-C-O}
$$

Initial slopes of the build-up of the benzoyloxy adduct of PBN are
linear as a function of time for 5-15 minutes at 32-48° when
PBN=0.001-1M for 0.1M benzoyl peroxide or when PBN=0.1M and
benzoyl peroxide is varied between 0.05-0.5M. Rates obtained from
these initial slopes show that adduct formation is 1st order in
benzoyl peroxide and almost zeroth order in PBN as long as PBN >
0.075M. At lower concentrations of PBN decarboxylation of benzoyloxy radicals competes with trapping and phenyl radicals are
detected:

$$PhCO_2\cdot \;\longrightarrow\; Ph\cdot \; + \; CO_2$$

$$
Ph\cdot \; + \; \underset{+}{PhCH=}\overset{O^-}{\underset{|}{N}}-CMe_3 \;\longrightarrow\; Ph_2CH-\overset{O\cdot}{\underset{|}{N}}-CMe_3
$$

Benzene as solvent also competes for phenyl radicals at low concentrations of PBN to produce non-detected phenylcyclohexadienyl
radicals:

Ph· + ⟨benzene⟩ ⟶ ⟨phenylcyclohexadienyl with H and Ph⟩

It was also shown that the rate constants for thermal decomposition
of benzoyl peroxide as a function of temperature obtained by esr
are in good agreement with those previously reported in the
literature. Thus we conclude that benzoyloxy radicals are trapped
quantitatively by PBN at suitable concentrations in benzene at
room temperature and that induced decomposition of benzoyl peroxide
by PBN or the spin adducts of PBN is not an important part of the
spin trapping process.

Similar results were obtained with di-t-butyl peroxalate
(DBPO) (6):

$$
\underset{\substack{\|\; \| \\ O\; O}}{Me_3CO-OC-CO}-OCMe_3 \;\longrightarrow\; 2\; Me_3CO\cdot \; + \; 2\; CO_2
$$

$$
Me_3CO\cdot \; + \; \underset{+}{PhCH=}\overset{O^-}{\underset{|}{N}}-CMe_3 \;\longrightarrow\; PhCH-\overset{O\cdot}{\underset{|}{N}}-CMe_3 \\
\hspace{6.8cm} Me_3CO
$$

The reaction is 1st order in DBPO and zeroth order in PBN over the
range of concentration studied: DBPO=0.001-0.1M and PBN=0.02-0.2M.

Initial slopes of the build-up of the t-butoxy spin adduct were
used for this study. The rate constants for thermal decomposition
of the peroxalate obtained by esr as a function of temperature are
in good agreement with those reported previously. At low
concentrations of PBN some cleavage of t-butoxy radicals takes
place and methyl radicals are detected:

$$Me_3CO\cdot \longrightarrow Me\cdot + MeCOMe$$

$$
\underset{\underset{+}{Me\cdot} + PhCH=N-CMe_3}{\overset{O^-}{|}} \longrightarrow \underset{\underset{Ph}{|}}{\overset{O\cdot}{|}} MeCH-N-CMe_3
$$

Thus in the thermal decomposition of DBPO all t-butoxy radicals
are trapped at appropriate concentrations of PBN and induced
decomposition of the peroxalate by PBN or the spin adducts can not
be detected.

The thermal decomposition of phenylazotriphenylmethane (7)
(PAT) in benzene containing PBN produces the phenyl spin adduct
and triphenylmethyl radicals (detectable after longer periods of
time).

$$Ph-N=N-CPh_3 \longrightarrow Ph\cdot + N_2 + Ph_3C\cdot$$

The kinetics are complicated by the competition of benzene for
phenyl radicals. An analysis of the data as a function of [PBN]
in benzene gives a rate constant for the thermal decomposition
of PAT which is in good agreement with literature values. Thus
it appears that in an inert solvent phenyl radicals would be
trapped quantitatively by PBN.

Spin Trapping Rate Constants of PBN

If at appropriate concentrations PBN can trap radicals
quantitatively the rate constant of trapping must be quite large.
By means of competitive kinetics estimates of the spin trapping
rate constants have been obtained:

$$
\underset{\underset{+}{R\cdot} + PhCH=N-CMe_3}{\overset{O^-}{|}} \overset{k_R}{\longrightarrow} \underset{\underset{R}{|}}{\overset{O\cdot}{|}} PhCH-N-CMe_3
$$

$$R\cdot + SH \overset{k_S}{\longrightarrow} RH + S\cdot$$

$$
\underset{\underset{+}{S\cdot} + PhCH=N-CMe_3}{\overset{O^-}{|}} \longrightarrow \underset{\underset{Ph}{|}}{\overset{O\cdot}{|}} S-CH-N-CMe_3
$$

Thus when R· is t-butoxy radical and SH is cyclohexane the ratio

of the initial slopes of the build-up of the two signals due to
the spin adducts is equal to the ratio of the appropriate rate
constants times the ratio of the concentrations of PBN and
cyclohexane:

$$
\left[\frac{\dfrac{d(SA\text{-}PBN)_R}{dt}}{\dfrac{d(SA\text{-}PBN)_S}{dt}} \right]_{t=o} = \frac{k_R[PBN]_o}{k_S[SH]_o}
$$

Since the rate constant for hydrogen atom abstraction from cyclohexane by t-butoxy radicals has been estimated (8) (namely 1×10^5
M^{-1} sec^{-1} at ambient temperatures) a rate constant of 5.5×10^6
M^{-1} sec^{-1} for the trapping of t-butoxy radicals can be determined.
(Actually a cyclic nitrone trap, 5,5-dimethylpyrroline-1-oxide
(DMPO) was used for this study since a serious peak overlap
problem was encountered with the two spin adducts of PBN (6).
The rate constant for PBN was related to that of DMPO by further
competitive experiments).

A similar approach was used to obtain the rate constant of
trapping phenyl radicals (7). In this case the reported rate
constant of hydrogen atom abstraction from methanol by p-methyl-
phenyl radicals was used to obtain the rate constant of trapping
phenyl radicals by PBN. A value of 1.2×10^7 M^{-1} sec^{-1} was obtained.

By combining some decarboxylation rate data of benzoyloxy
radicals with those of phenylation of benzene an estimate of
4×10^7 M^{-1} sec^{-1} for the rate constant of trapping benzoyloxy
radicals by PBN is obtained (7).

Ingold and Schmid (9) have obtained a rate constant for spin
trapping 5-hexenyl radical by PBN equal to 1.7×10^5 M^{-1} sec^{-1} at
40° in benzene.

$$
CH_2=CH\text{-}(CH_2)_3\text{-}CH_2 \cdot \; + \; PhCH=\overset{O-}{\underset{+}{N}}\text{-}CMe_3 \; \longrightarrow \; CH_2=CH\text{-}(CH_2)_3\text{-}CH_2\text{-}\underset{Ph}{\overset{O\cdot}{CH}}\text{-}N\text{-}CMe_3
$$

We thus have the following summary of PBN spin trapping rate
constants in benzene at room temperature.

$CH_2=CH\text{-}(CH_2)_3\text{-}CH_2 \cdot$ 1.8×10^5 M^{-1} sec^{-1}

$Me_3CO \cdot$ 5.5×10^6

$Ph \cdot$ 1.2×10^7

$PhCO_2 \cdot$ 4×10^7

It should be pointed out that the latter three values are based
on estimates with large possible errors. The numbers may be
correct only within an order of magnitude.

Spin Trapping Rate Constants of Other Nitrones

By competitive experiments spin trapping rate constants of
other nitrones were obtained for t-butoxy and benzoyloxy radicals
(5, 6). The values found are given in Table 1. Two features of
the rate data are noteworthy: (1) Phenyl or methyl substitution
on the carbon atom of the nitronyl function slows the reaction;
(2) For benzoyloxy radical electron-withdrawing substituents on
the phenyl group decrease the trapping rate constant whereas
electron-donating substituents increase the rate constant although
the effect is small. For t-butoxy radical the substituent effect
is not clear.

The slowing effect of the phenyl group in the α-position when
methyl radical adds to substituted ethylenes is well known from
work of Szwarc (10):

Me· addition to	abstraction/addition	
$CH_2=CH_2$	34	(1)
$CH_2=CH-Ph$	796	(23)
$CH_2=CPh_2$	1590	(47)
PhCH=CHPh (trans)	104	(3)
(cis)	29	(0.9)
PhCH=CPh_2	48	(1.4)
$Ph_2C=CPh_2$	8	(0.2)

The rate enhancement in styrene or 1,1-diphenylethylene as
compared to ethylene is easily accounted for by increased
resonance stabilization of the radical in the addition products.
However in <u>trans</u> stilbene the second phenyl group goes to stabilize the ground state and doesn't help to declocalize the
unpaired electron in the addition product. Simple HMO calculations provide the following picture:

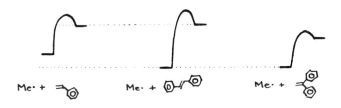

Table I

with $Me_3CO\cdot$ (M^{-1} sec^{-1})		with $PhCO_2\cdot$ (M^{-1} sec^{-1})	

$$CH_2=\overset{\overset{\displaystyle O^-}{|}}{\underset{+}{N}}-CMe_3$$ $3-5\times10^8$ (100)

$$PhCH=\overset{\overset{\displaystyle O-}{|}}{\underset{+}{N}}-CMe_3$$ 5.5×10^6 (1) 4×10^7 (7.3)

$$p-NO_2PhCH=\overset{\overset{\displaystyle O-}{|}}{\underset{+}{N}}-CMe_3$$ 9×10^6 (1.6) 1.8×10^7 (3.3)

$$p-ClPhCH=\overset{\overset{\displaystyle O-}{|}}{\underset{+}{N}}-CMe_3$$ 6.5×10^6 (1.2) 3×10^7 (5.5)

$$p-MePhCH=\overset{\overset{\displaystyle O-}{|}}{\underset{+}{N}}-CMe_3$$ 3.4×10^6 (1.6) 6.3×10^7 (11)

$$p-MeO\ PhCH=\overset{\overset{\displaystyle O}{|}}{\underset{+}{N}}-CMe_3$$ 5.5×10^6 (1.0) 9.3×10^7 (17)

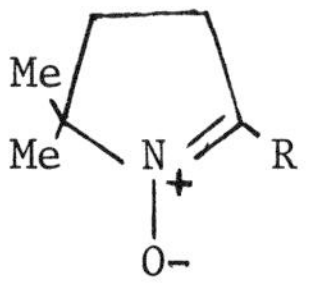

R=H 5×10^8 (100)

R=Ph 1.3×10^7 (2.4)

R=Me 9×10^6 (1.6)

DMPO

Thus resonance of the phenyl group can slow down the radical
addition reaction in certain cases.

However if this were the only effect one would expect the cis
isomer to give about the same rate of addition as the trans isomer.
There must also be a steric effect to the addition of the radical.
It is generally accepted that the transition state for radical
addition is rather loose and resembles a σ-complex more than it
does a π-complex (11, 12). The phenyl rings in cis-stilbene will
have an ortho hydrogen blocking easy approach to the appropriate
p-orbital of the ethylenic function. Perhaps some steric hindrance to approach also exists in the addition to trans-stilbene
because the phenyl rings may not be entirely planar in this isomer.

In applying the above conclusions to nitrone spin trapping we
have the same α-phenyl substitution pattern when the unsubstituted
nitrone is compared with PBN and DMPO with 2-phenyl-DMPO:

relative   1   0.01   1   0.03
scale

Probably a major factor is resonance although some steric hindrance
to approach may also exist depending on the strucutre of these
nitrones. X-ray diffraction studies of PBN crystals are planned.
It would of course be of interest to compare the spin trapping
rate constant of N-t-butyl nitrone with N-phenyl nitrone but this
compound has never been synthesized:

We may have to be satisfied with comparing PBN with α-phenyl
N-phenyl nitrone.

There may be a third effect influencing the rate of radical
addition, namely steric hindrance to product formation associated
with resistance of groups to move into the direction of the initial
conformation of the addition product. Examples of this effect
appear in the addition of methyl radical to methyl substituted
ethylenes as seen in Table II (13). The relative rates are
ethylene (1.0), propylene (0.7), cis- and trans- 2-butene
(0.23 and 0.33), 1,1-dimethylethylene (1.1), trimethylethylene

Table II

Relative Rate Constants for Reaction of Radicals With Olefins($\underline{13}$)

OLEFIN	$CH_3\cdot$ ($453^{\circ}K$)	$H\cdot$ ($298^{\circ}K$)	$CF_3\cdot$ ($338^{\circ}K$)	$\cdot NF_2$ ($373^{\circ}K$)	$\cdot OH$[14] ($298^{\circ}K$)	$\cdot OH$[15] ($305^{\circ}K$)
$F_2C=CF_2$	9.51		0.12			
$Cl_2C=CCl_2$						0.3
$Cl_2C=CHCl$						0.6
$H_2C=CH_2$	1	1	1	1	1	1
$CH_3CH=CH_2$	0.7	1.8	1.2	4.4	10	3.3
$CH_3CH=CHCH_3$ (CIS)	0.23	0.8	0.9	10.3	36	7.5
$CH_3CH=CHCH_3$ (TRANS)	0.33	1.1	1.0	10.4	42	
$(CH_3)_2C=CH_2$	1.1	4.4	3.7	20	38	
$(CH_3)_2C=CH(CH_3)$	0.32	–	–	34	70	
$(CH_3)_2C=C(CH_3)_2$	0.20	1.5	1.3	99	90	
$CH_2=CH-CH=CH_2$	12.0	8.7	8.7	–	–	9

(0.32) and tetramethylethylene (0.20). Consider the reaction course from the first formed σ-complex through the initial conformation to the final conformation for the various ethylenes. The relative rate for addition to 1,1-dimethylethylene is the same as for ethylene probably because no serious new steric hindrance to the approach of the radical exists and the initial conformation is also the most stable final conformation for the $CH_3CH_2\overset{\cdot}{C}(CH_3)_2$ radical. The gain in producing a more stable tertiary radical as a product must be approximately offset by the fact that alkyl-substituted ethylenes are more nucleophilic olefins than ethylene. The addition to propylene is slower than to ethylene or 1,1-dimethylethylene probably because the initial conformation is not a particularly stable one – the barrier between the σ-complex and the final conformation of the addition product is greater in this case and the reaction is slower. The addition rate to trans-2-butene is much slower than the rate of addition to propylene although the initial conformation produced in trans-2-butene suffers from methyl interaction essentially in the same way as does the initial conformation produced from propylene. The slower rate in this case can be attributed to steric hindrance of approach to the appropriate p-orbital for σ-complex formation. The rates of addition in the case of cis-2-butene, trimethylethylene and tetramethylethylene are all slow probably because of a combination of the above effects.

In the case of nitrones when t-butoxy radicals adds the initial conformation is the most stable one for N-t-butyl nitrone (or very close to the most stable) but not for PBN. The same problem exists in the addition of t-butoxy to DMPO and 2-phenyl DMPO. In the latter case the steric resistance of the 5,5-dimethyl group contributes to the slower rate of addition.

Polar effects also appear to exist in the addition reactions of certain radicals to substituted ethylenes. It is clear from Table II that for electronegative radicals like difluoroamino or hydroxy and to a lesser extent trifluoromethyl radical and hydrogen atom, substitution with donor groups like methyl increases the rate of addition. Substitution with electronegative atoms like fluorine or chlorine has the effect of slowing the rate of addition by hydroxy or trifluoromethyl radicals but the rate of addition of methyl radicals is increased.

Similar rate constants for trifluoromethyl radical are obtained in solution phase as in the gas phase when the solvent is a hydrocarbon (16). However in water the addition seems to be faster:

$$CF_3\cdot + CH_2 = CH_2 \rightarrow CF_3CH_2CH_2\cdot$$

k=3.5x10^6 (gas phase) (17), 3x10^6 (cyclohexane or heptane) (18), 4.0x10^7 (water) (19). The substituted effect and the solvent effect both point to a transition state where some charge transfer is occurring: e.g.

$$HO\cdot + \quad \begin{matrix} Me \\ Me \end{matrix} C=C \begin{matrix} Me \\ Me \end{matrix} \quad \leftrightarrow \quad HO; \begin{matrix} Me \\ (-) \\ Me \end{matrix} C \overset{(+)}{\text{---}} C \begin{matrix} Me \\ Me \end{matrix}$$

$$CH_3\cdot + \quad \begin{matrix} F \\ F \end{matrix} C=C \begin{matrix} F \\ F \end{matrix} \quad \leftrightarrow \quad CH_3 \begin{matrix} F \\ (+) \\ F \end{matrix} C \overset{(-)}{\text{---}} C \begin{matrix} F \\ F \end{matrix}$$

Unfortunately the same radicals have not as yet been studied quantitatively by spin trapping. However if one assumes that t-butoxy, benzoyloxy and phenyl radicals are similar to hydroxy and trifluoromethyl radicals comparisons can be made. In PBN the Hammett slope is slightly positive with benzoyloxy radical, i.e. some charge transfer is indicated in the transition state:

$$RO\cdot \quad \begin{matrix} H \\ \phi \end{matrix} C=N \begin{matrix} CMe_3 \\ O \end{matrix} \quad \leftrightarrow \quad RO^- \quad \begin{matrix} H \\ + \\ \phi \end{matrix} C\overset{+}{\text{---}}N \begin{matrix} CMe_3 \\ O \end{matrix}$$

Although the effect is relatively small it is in the same direction as found for electronegative radicals adding to substituted ethylene.

t-Butoxy radical seems to behave both as an electronegative and as a nucleophilic radical with p-substituted PBN's. However the effect is very small. Examples of similar behaviour are available among certain haloalkyl radical additions to fluoroethylenes (12): e.g.

$$\cdot CHF_2 + CH_2 = CH_2 \quad 1.0 \quad \text{(relative rate constant)}$$

$$\cdot CHF_2 + CH_2 = CHF \quad 0.38$$

$$\cdot CHF_2 + CH_2 = CF_2 \quad 0.12$$

$$\cdot CHF_2 + FCH = CF_2 \quad 0.29$$

$$\cdot CHF_2 + F_2C = CF_2 \quad 1.10$$

Similar trends but with less marked changes in direction are found in the addition of $\cdot CH_2Cl$, $\cdot CFBr_2$ and $\cdot CF_3$ to these fluoro-ethylenes (12).

Thus in the addition to PBN of benzoyloxy or t-butoxy radicals, both electronegative radicals, one gets the picture that the nitronyl function is rather electrophilic itself or the site of radical attack is quite elctrophilic, for if it were otherwise one might expect to see a larger polar effect. It would be interesting to investigate the same substitutent effect using an alkyl radical. Perhaps the Hammett slope would be positive in this case indicating charge acceptance by the nitronyl function. If charge transfer in the transition state is important the rates could be faster in aqueous media.

We explored the question of charge transfer in the transition state by a simple HMO approach. If one imagines tearing away carbon-2 from vinyl nitroxide the reverse process would resemble at the initial stages the addition of a kind of methyl radical (i.e. a radical with an unpaired electron in a p-orbital centered on carbon) to the nitrone function:

$$
\begin{array}{ccc}
C & \longrightarrow & C\text{-}N\text{-}O & \longrightarrow & C_2\text{-}C_1\text{-}N\text{-}O
\end{array}
$$

Nitrone   Vinyl Nitroxide

The spin and π-electron densities were calculated for vinyl nitroxide using simple HMO theory and coulomb integral values for nitrogen and oxygen (α_N and α_O) which provided best fits for nitroxides in previous studies (20). To simulate the approach of the methyl radical β for $C_1\text{-}C_2$ was increased from 0.0 to 1.0.

Of interest are the simulations of the addition of an oxy radical to the nitronyl function. Here instead of using the parameters for a carbon atom for C_2 an atom with $\alpha_O = \alpha_C + 1.0\beta$ was used. At $\beta_{1-2} = 0.0$ the approaching radical has no spin but instead a full negative charge. Thus HMO predicts charge transfer when an oxy radical approaches a nitrone function:

$$
\begin{array}{ccc}
O + C\text{-}N\text{-}O & \longrightarrow & \overset{(-)}{O} \quad \overset{(+)}{C\text{-}N\text{-}O}
\end{array}
$$

Similarly if a donor radical with the unpaired electron centered on M is used where $\alpha_M = \alpha_C - 1.2\beta$ again electron transfer is indicated but now to the nitronyl function (accidental degeneracy exists at $\alpha_M = \alpha_C - 1.0\beta$ and at $\alpha_M = \alpha_C - 0.8\beta$ no charge transfer occurrs):

$$
\begin{array}{ccc}
M + C\text{-}N\text{-}O & \longrightarrow & \overset{(+)}{M} \quad \overset{(-)}{C\text{-}N\text{-}O}
\end{array}
$$

A comparison of the magnitudes of the absolute rate constants of addition is also of interest. It would seem at first sight that the rapid rate of radical addition to the nitronyl function is unusual since unsaturated functions containing heteroatoms are not known to accept radicals readily. Radicals preferentially

abstract the "vinyl" hydrogen in aldehydes and imines (21) but
no evidence for vinyl hydrogen abstraction could be found in
nitrones (22):

$$\text{RO}\cdot + \text{R}-\overset{\overset{\text{H}}{|}}{\text{C}}=\text{O} \longrightarrow \text{R}-\overset{\bullet}{\text{C}}=\text{O} + \text{ROH}$$

$$\text{RO}\cdot + \text{R}-\overset{\overset{\text{H}}{|}}{\text{C}}=\text{N}-\text{R}^1 \longrightarrow \text{R}-\overset{\bullet}{\text{C}}=\text{N}-\text{R}^1 + \text{ROH}$$

$$\text{RO}\cdot + \text{R}-\overset{\overset{\text{H}\;\;\text{O}}{|\;\;|}}{\text{C}}=\text{N}-\text{R}^1 \overset{-\text{X}}{\longrightarrow} \text{R}-\overset{\overset{\text{O}}{|}}{\underset{\bullet}{\text{C}}}=\text{N}-\text{R}^1 + \text{ROH}$$

Although a few examples of radical additions to carbonyl compounds
are known (23, 24), none are known to the authors for imines (25):

$$\text{R}\cdot + \overset{\diagdown}{\underset{\diagup}{\text{C}}} = \text{O} \longrightarrow \text{R}-\overset{|}{\underset{|}{\text{C}}}-\text{O}\cdot \quad \text{OR} \quad \cdot\overset{|}{\underset{|}{\text{C}}}-\text{OR}$$

$$\text{R}\cdot + \overset{\diagdown}{\underset{\diagup}{\text{C}}}=\text{N}- \longrightarrow \text{R}-\overset{|}{\underset{|}{\text{C}}}-\overset{\bullet}{\text{N}}- \quad \text{OR} \quad \cdot\overset{|}{\underset{|}{\text{C}}}-\overset{|}{\underset{|}{\text{N}}}-\text{R}$$

The radical addition to a nitrone is actually more analogous
to the addition of a radical to butadiene because in both cases
the unpaired electron in the addition product is stabilized by
delocalization in a π-orbital over two or three atoms. The 9-12
fold increase in rate constant of addition produced by attaching
the vinyl substituent to ethylene (butadiene) is much greater when
an oxygen atom is attached to the nitrogen of an imine to make a
nitrone:

$$\text{R}\cdot + \text{CH}_2=\text{CH}-\text{CH}=\text{CH}_2 \quad \text{addition } 9\text{-}12\text{X faster than in } \text{R}\cdot + \text{CH}_2=\text{CH}_2$$

$$\text{R}\cdot + \text{CH}_2=\overset{\overset{\text{O}-}{|}}{\underset{+}{\text{N}}}-\text{R} \quad \text{addition perhaps } 10^5\text{X faster than in } \text{R}\cdot + \text{CH}_2=\text{N}-\text{R}$$

(when R· is t-butoxy radical in benzene at room temperature (22).
The stabilization of the unpaired electron by delocalization over
the two heteroatoms of the nitroxyl function in the addition
product produces a rapid rate of addition which otherwise is very
slow.
    The absolute rate constants of radical addition to butadiene
are 1.25×10^6 M^{-1} sec^{-1} for methyl (26) and 4.0×10^7 M^{-1} sec^{-1} for
trifluoromethyl (19) both in aqueous solution. These values are
of the same order of magnitude as estimated for additions to
nitrones in benzene.

Spin Trapping Rate Constants with Nitroso Compounds

Carbon, oxygen and nitrogen centered radicals are known to add to the nitroso function to produce nitroxides:

$$R\cdot \ + \ \overset{|}{N}{=}O \ \longrightarrow \ R{-}\overset{|}{N}{-}O\cdot$$

Alkoxy nitroxides are not particularly stable but can be detected by ESR. The rate constants of addition for certain nitroso compounds are given here for completeness.

Reference

$(Me)_3CN{=}O$	MeO·	$1.3 \times 10^8 \ M^{-1} \ sec^{-1}$	27
$(Me)_3CN{=}O$	$Me_3CO\cdot$	$1.5 \times 10^6 \ M^{-1} \ sec^{-1}$	6
$(Me)_3CN{=}O$	$CH_2{=}CHCH_2CH_2CH_2\cdot$	$8.8 \times 10^6 \ M^{-1} \ sec^{-1}$	9

$-N{=}O$ $CH_2{=}CHCH_2CH_2CH_2\cdot$ $5.9 \times 10^5 \ M^{-1} \ sec^{-1}$ 9

Literature Cited

1. Present address, Varian Associates, Florham Park, New Jersey.
2. Janzen, E.G., Accounts Chem. Res., (1971), 4, 31.
3. Iwamura, M. and Inamoto, N., Bull. Chem. Soc. Jap., (1967), 40, 702, 703; (1970), 43, 856, 860.
4. Janzen, E. G., and Blackburn, B. J., J. Amer. Chem. Soc., (1968) 90, 5909; (1969) 91, 4481.
5. Janzen, E. G., Evans, C. A. and Nishi, Y., J. Amer. Chem. Soc., (1972), 94, 8236.
6. Janzen, E. G. and Evans, C. A., ibid., (1973), 95, 8205.
7. Janzen, E. G. and Evans, C. A., ibid., (1975), 97, 205.
8. Howard, J. A., Advan. Free Rad. Chem., (1971) 4, 49; the value $1 \times 10^5 \ M^{-1} \ sec^{-1}$ is an average of two values obtained by two different groups, namely Ingold's and Walling's and as such has a substantial uncertainty itself.
9. Schmid, P. and Ingold, K. U., J. Amer. Chem. Soc., submitted.
10. From Table 17 in Ingold, K. U., "Rate Constants for Free Radical Reactions in Solution" ch. 2 in Kochi, J. K.. "Free Radicals", Volume 1, Wiley-Interscience, New York, N.Y. 1973.
11. Owen, G. E., Pearson, J. M., and Szwarc, M. ibid., (1965) 61, 1722.
12. Tedder, J. M. and Walton, J. C., Accounts of Chem. Res., (1976) 9, 183.
13. Kerr, J. A., "Rate Processes in the Gas Phase", Chaper 1, page

1 of Kochi, J. K. "Free Radicals", Volume 1, J. Wiley and Sons, New York, N.Y. 1973.

14. Morris, Jr., E. D., and Niki, H., J. Phys. Chem., (1971) 75, 3640; J. Amer. Chem. Soc., (1971), 93 3570.

15. Winer, A. M., Lloyd, A. C., Darnall, K. R., and Pitts, Jr., J. N., J. Phys. Chem., (1976), 80, 1635; Lloyd, A. C., Darnall, K. R., Winer, A. M., and Pitts, Jr., J. N., ibid., (1976) 80, 789 and references therein; see also Meagher, J. F. and Heicklen, J., ibid., (1976) 80, 1645.

16. Dixon, P. S., and Szwarc, M., Trans. Faraday Soc., (1963) 59, 112.

17. Sangster, J. M. and Thynne, J. C. J., J. Phys. Chem., (1969), 73, 2746.

18. Weir, R. A., Infelta, P.P., and Schnuler, R. H., J. Phys. Chem. (1970) 74, 2596.

19. Bullock, G. and Cooper, R., Trans. Faraday Soc., (1970) 66, 2055; (a value of 7×10^8 M^{-1} sec^{-1} was estimated for this reaction in water) Balkas, T. I., Fendler, J. H., and Schuler, R. H., J. Phys. Chem., (1971) 75, 455; see however Lilie, J., Dehar, D., Snjdak, R. J. and Schuler, R. H., ibid., (1972), 76, 2517.

20. Janzen, E. G., and Happ, J. W., J. Phys. Chem., (1969) 73, 2335.

21. Danen, W. E., and West, C. T., J. Amer. Chem. Soc., (1973) 95, 6872.

22. Nutter, Jr., D. E., "Two Applications of ESR Spin Trapping: Investigation of Reaction of t-Butoxy Radicals with Benzald-imines and Photolysis of Alcohol/Nitrone Solutions", Ph.D. dissertation, University of Georgia, Athens, Georgia, 1974.

23. Huyser, E. S., "Free-Radical Chain Reactions", p. 89, Wiley-Interscience, New York, N.Y. 1970; three examples are given: addition of (1) benzoyl radicals to benzaldehyde (Rust, F. F., Seubold, F. H., and Vaugham, W. E., J. Amer. Chem. Soc., (1948) 70, 3258) (2) cycloalkyl radicals to formaldehyde (Fuller, G. and Rust, F. F., ibid., (1958) 80, 6148) (3) n-butyroyl radicals to biacetyl (Urry, W. H., Pai, M. H., and Chen, C. Y., ibid., (1964) 86, 5342).

24. Ingold, K. U. and Roberts, B. P. "Free-Radical Substitution Reactions", p. 86-89, Wiley-Interscience, New York, N.Y., 1971; two more examples of radical additions to biacetyl in solution are given (Bentrude, W. G., and Darnell, K. R., Chem. Comm., 810 (1968); J. Amer. Chem. Soc., 90, 3588 (1968)) and other examples are cited from gas phase studies.

25. Khaba, R. A., Griller, D., and Ingold, K. U., J. Amer. Chem. Soc. (1974), 96, 6202; Brunton, G., Taylor, J. F., and Ingold, K. U., ibid., (1976), 98, 4879.

26. Thomas, J. K., J. Phys. Chem., (1967), 71, 1919.

27. Sargent, F. P., J. Phys. Chem., 81, 89 (1977).

RECEIVED December 23, 1977.

Pyridinyl Paradigm Proves Powerful

EDWARD M. KOSOWER

Department of Chemistry, Tel–Aviv University, Tel–Aviv, Israel and
Department of Chemistry, State University of New York, Stony Brook, NY 11794

Radicals stable enough to be transferred from one environment
to another have been useful for (a) analyzing the effect of the
molecular milieu on the reactions of radicals and (b) probing the
viscosity and polarity of the domain in which the radical is located. The simplicity and chemical reactivity of pyridinyl radicals,
generated through chemical reduction of pyridinium halides (Eq.1),
has provided the opportunity for studies on the mechanisms of radical reactions and on the properties of π-complexes of radicals.

(1)
$$\text{(pyridinium-COOCH}_3, \text{N}^+\text{-R)} \xrightarrow[\text{CH}_3\text{CN}]{\text{Na(Hg)}} \text{(pyridinyl-COOCH}_3, \text{N-R)} \quad R = Me, Et, i\text{-}Pr, t\text{-}Bu \quad (Py\cdot)$$

Halocarbon Reactions

Pyridinyl radicals (in most cases, this term will refer to a
4-carbomethoxy-1-alkylpyridinyl radical as shown in Eq.1) react
with halocarbons in revealing ways. The kinetic constants shown in
Table 1 for the reactions with dichloromethane, bromochloromethane
and iodochloromethane illustrate (a) discrimination between C-Cl,
C-Br and C-I bonds (b) rates which are convenient to measure by
conventional techniques. The overall reaction is given in Eq. 2.

(2) $2Py\cdot + RX \longrightarrow Py^+X^- + PyR$

In order to establish the mechanism of the reaction, the solvent
effect on the reaction of pyridinyl radical with dibromomethane
was investigated. As the results listed in Table 1 show, there is
no solvent effect on the rate of the reaction. How could one reconcile the formation of a salt with the lack of solvent polarity
effect on the rate? Since the initial state ($Py\cdot + RX$) is not
very polar (pyridinyl radical with a 1-ethyl group is soluble in
n-hexane, $BrCH_2Br$ has a dipole moment of ca. 1 Debye), the lack of

© 0-8412-0421-7/78/47-069-447$05.00/0

solvent effect requires that the transition state for the rate
limiting step be also weakly polar. The only possible reaction
which could account for this result is that of atom-transfer, in
which a bromodihydropyridine is formed as a first intermediate.
(Eq. 3).

To explore the mechanism of the reaction further, the rates
of reaction of pyridinyl radical with a series of substituted
benzyl chlorides were measured. The idea was that the Hammett
ρ-value would help us to understand the possible charge separation
in the transition state. The rate constants (Table 1) indicated
that the effect of substituent was very small except in the case
of the 4-nitrobenzyl chloride-pyridinyl radical reaction. The rate
constant for the latter reaction was so high and so different from
those for the other benzyl chlorides that a change in mechanism
was likely, with an electron-transfer reaction being a reasonable
possibility in view of the electron-withdrawing character of the
nitro-group. This could be confirmed by a study of the effect of
solvent polarity on the rate constants for the reaction of pyridinyl radical with 4-nitrobenzyl chloride, for which a very large
effect was found. (Table 1).The criterion of solvent polarity
used was the Z-value, an empirical solvent polarity parameter
which I introduced in 1958, and which is based on the charge-
transfer light absorption of 1-ethyl-4-carbomethoxy-pyridinium
iodide. (Eq.4). The theoretical model for the solvent effect
underlying the Z-values suggests that a reasonable estimate for
the transfer of ion-pairs from one solvent to another may be made
by dividing the difference in Z-values by two. We could then
compare the ½ ΔZ thus obtained with the ΔΔG values derived from
the transition state energies for the rate constants for the
4-nitrobenzyl chloride-pyridinyl radical reaction. The similarity
of the two numbers for the solvents, 2-methyltetrahydrofuran and
acetonitrile, shows that the interpretation of the reaction as an
electron-transfer reaction is not only qualitatively valid but is
quantitatively consistent with such a formulation. (Eq. 5).

Further details on the halocarbon reaction with pyridinyl
radicals may be found in a review (1) and in the original papers
with Schwager (2) and Mohammad (3).

Bimolecular Reactions of Pyridinyl Radicals in Water

1-Alkyl-3-carbamidopyridinyl radicals (e.g. 3˙) are of considerable significance biologically, since such radicals represent
potential intermediates in biochemical reactions involving the
coenzyme, NAD˙ (NADH). The 3˙ radicals are, in general, not sufficiently stable to be isolated and transferred from one medium to
another, but can readily be produced by pulse radiolysis (4) of
1-alkyl-3-carbamidopyridinium ions in water. The 3˙ radicals
disappear in bimolecular reactions at rates which are close to
diffusion-controlled and independent of pH. (5,6). The product of
3˙ reaction with 3˙ is the dimer. (Eq. 6).

TABLE I. Rate Constants for the Reaction of 1-Ethyl-4-carbomethoxy-pyridinyl Radical with Halides and Polyhalides at 25°C

Halide	Solvent	Rate constant $(M^{-1}sec^{-1})$	Ref.
Group a: Element effect—atom-transfer reactions			
$ClCH_2Cl$	CH_3CN	2.6×10^{-8a}	2
$BrCH_2Cl$	CH_3CN	5.0×10^{-5}	2
ICH_2Cl	CH_3CN	1.3×10^{-1}	2
ICH_3	CH_3CN	5.0×10^{-6}	2
Group b: Substituent effect—aromatic ring			
$ClCH_2C_6H_4OCH_3$	CH_3CN	11.3×10^{-4}	17
$ClCH_2C_6H_4CH_3$	CH_3CN	3.68×10^{-4}	17
$ClCH_2C_6H_5$	CH_3CN	3.31×10^{-4}	17
$ClCH_2C_6H_4Cl$	CH_3CN	6.5×10^{-4}	17
$ClCH_2C_6H_4NO_2$	CH_3CN	$240,000,000 \times 10^{-4}$	17
Group c-1: Solvent effect—atom transfer			
$BrCH_2Br$	CH_3CN	1×10^{-4}	2
	CH_2Cl_2	0.48×10^{-4}	2
	i-PrOH	0.94×10^{-4}	2
	EtOH	1.7×10^{-4}	2
Group c-2: Solvent effect—electron transfer			
$ClCH_2C_6H_4NO_2$	$MTHF^b$	1.62	3
	DME^c	8.3	3
	CH_2Cl_2	75	3
	CH_3COCH_3	450	3
	DMF^d	1.2×10^4	3
	CH_3CN	2.4×10^4	3

[a] Estimated from data at higher temperatures.
[b] 2-Methyltetrahydrofuran.
[c] 1,2-Dimethoxyethane.
[d] N,N-Dimethylformamide.

COOCH$_3$ + BrCH$_2$Br $\longrightarrow$ COOCH$_3$ (Br, H, N, CH$_2$CH$_3$) + $\cdot$CH$_2$Br

(3)   COOCH$_3$ (N, CH$_2$CH$_3$) + $\cdot$CH$_2$Br $\longrightarrow$ BrCH$_2$ COOCH$_3$ (N, CH$_2$CH$_3$) + COOCH$_3$ (CH$_2$Br, H, N, CH$_2$CH$_3$)

COOCH$_3$ (Br, H, N, CH$_2$CH$_3$) $\longrightarrow$ COOCH$_3$ (+N, CH$_2$CH$_3$) + Br$^-$

(4)   COOCH$_3$ (N, i$^-$, +CH$_2$CH$_3$) $\xrightarrow{h\nu}$ COOCH$_3$ (N, i$\cdot$, CH$_2$CH$_3$)

(5)   COOCH$_3$ (N, CH$_2$CH$_3$) + NO$_2$ (CH$_2$Cl) $\longrightarrow$ CH$_3$OOC (+N, CH$_3$CH$_2$) , NO$_2$ (i$^-$, CH$_2$Cl)
$\longrightarrow$ COOCH$_3$ (+N, Cl$^-$, CH$_2$CH$_3$) + NO$_2$ (CH$_2\cdot$)

COOCH$_3$ (N, CH$_2$CH$_3$) + NO$_2$ (CH$_2\cdot$) $\longrightarrow$ COOCH$_3$ (CH$_2$, H, N, CH$_2$CH$_3$)—CH$_2$—NO$_2$

(6)

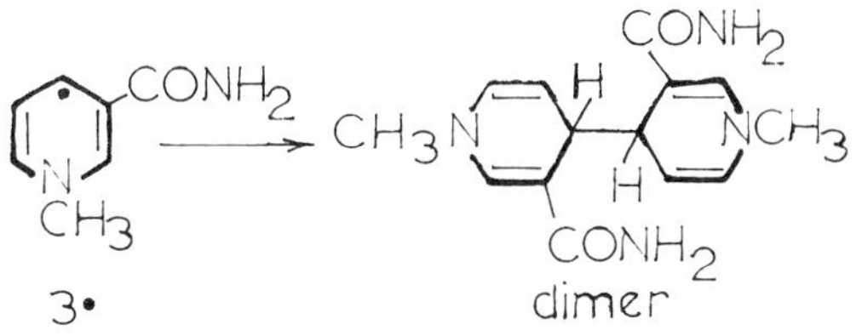

$$\text{(structures: } 3\cdot \longrightarrow CH_3N \ldots \text{dimer)}$$

3·   dimer

1-Alkyl-4-carbamidopyridinyl radicals (4·) react with one
another in a pH-dependent reaction (7). A protonated radical can
be demonstrated at low pH and its pK_a has been estimated as 2.3
for the 1-hexyl derivative (8). The disappearance of 4· can be
described as a reaction between an electron-transfer reaction
between 4· and 4H·$^+$. (Eq. 7).

(7)

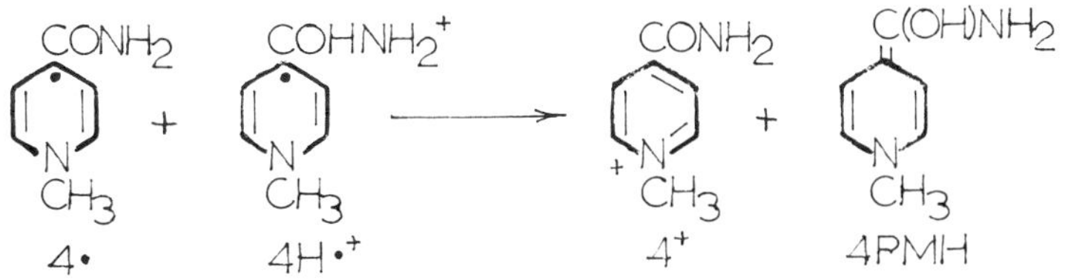

$$4\cdot \quad + \quad 4H\cdot^+ \quad \longrightarrow \quad -4^+ \quad + \quad 4PMH$$

The 4PMH is not stable in aqueous solution. After hydration,
another proton is consumed, yielding a product which slowly hydro-
lyzes to an alkylamine. Methylamine has been detected after
radiolysis of 1-alkyl-4-carbamidopyridinium ions. (Eq. 8).

(8)

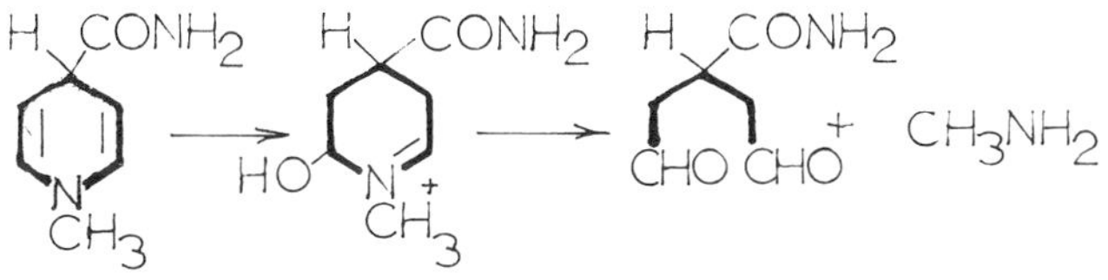

$$\ldots \longrightarrow \ldots \longrightarrow \ldots + CH_3NH_2$$

We have demonstrated a particularly clear and simple technique
for evaluating the reactions of pyridinyl radicals in water
through measurement of the pH changes resulting from radiolysis of
pyridinium ions in unbuffered aqueous solutions containing isopropyl alcohol (6). Radiolysis generates one proton for each
pyridinyl radical, as set forth in Eq. 9. (4).

$$3 \ H_2O \ \xrightarrow{\gamma\text{-radiation}} \ 3 \cdot OH \ + \ 3H^+ \ + \ 3e^-$$

(9) $$3e^- \ + \ 3Py^+ \ \longrightarrow \ 3Py\cdot$$

$$3 \cdot OH \ + \ 3 \ i\text{-PrOH} \ \longrightarrow \ 3 \ H_2O \ + \ 3 \ (CH_3)_2C(OH)\cdot$$

$$3 \ (CH_3)_2C(OH)\cdot \ + \ 3Py^+ \ \longrightarrow \ 3Py\cdot \ + \ 3H^+ \ + \ 3 \ (CH_3)_2CO$$

$$\overline{3 \ H_2O \ + \ 6 \ Py^+ \ + \ 3 \ i\text{-PrOH} \ \longrightarrow 6Py\cdot \ + \ 6H^+ \ + \ 3 \ (CH_3)_2CO}$$

Each 1-methyl-x-carbamidopyridinium ion behaves differently upon radiolysis, with the results being summarized in Eq. 10-12.

(10) $$3\cdot \ + \ 3\cdot \ \longrightarrow \ 3\text{-}3 \ \text{dimer (No } H^+ \text{ consumption)}$$

(11) $$2\cdot \ + \ 2\cdot \ H^+ \ \longrightarrow \ 2^+ \ + \ 2H \ (1 \ H^+ \text{ consumed per2Py}\cdot\text{formed)}$$

(12) $$4\cdot \ + \ 4\cdot + \ 2H^+ \ \longrightarrow 4^+ \ + \ 4HH^+ \ (1 \ H^+ \text{ consumed per Py}\cdot \text{ formed)}$$

Pyridinyl Diradicals and Their Magnesium Complexes

The stability evident for 1-alkyl-4-carbomethoxypyridinyl radicals stimulated the preparation of a series of pyridinyl diradicals, with the general formula shown below (9,10). In many cases, as for the case of magnesium, calcium or other strongly reducing metals, a diradical-metal ion complex is produced, from which the metal ion is removed with great difficulty.

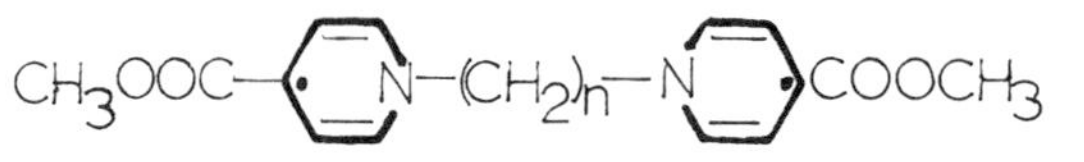

$$CH_3OOC\text{---}\langle \ \cdot \ N\text{---}(CH_2)_n\text{---}N \ \rangle COOCH_3 \qquad n = 3\text{--}10$$

The most striking property of the diradicals, most evident in their magnesium complexes, is that of π-merization. π-mers are defined as charge-transfer complexes between two identical or similar π-systems. The effect of π-merization upon the spectrum of a pyridinyl radical is illustrated with the cases of three pyridinyl diradicals (n = 3,4,5) in Fig. 1. The spectrum of the diradical with n = 5 is very similar to that of pyridinyl mono-radical. The dramatic effect of magnesium complex formation is shown for the case of the diradical 3 in Fig. 2. (11). Magnesium complexation does not alter the epr properties of the diradicals, in so far as the spin concentrations are concerned. The diradical 3 is almost entirely spin-paired (spin/radical = 2%), the diradical 4 exhibits 20% spin and the diradical 5, 100% spin. A most interesting case is that of the 1,8-biphenylenyl-bis methyl 4-carbomethoxypyridinyl diradical (formula shown). The uv-visible spectrum of the diradical implies little interaction between the two pyridinyl radical moieties, whereas the epr spectrum shows

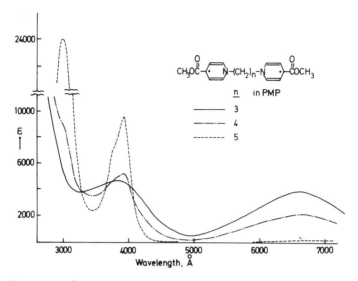

Figure 1. The effect of π-merization on the spectrum of a pyrinyl radical

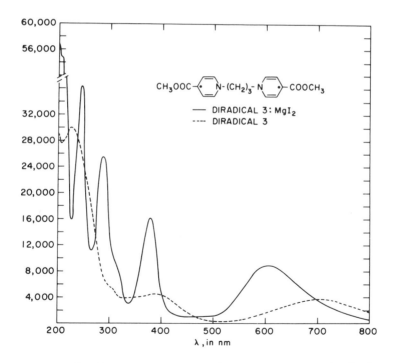

Figure 2. The effect of magnesium complex formation for the diradical 3

exchange narrowing. The magnesium complex has a uv-visible spectrum which shows substantial interaction between the radicals, but the epr spectrum is very similar to that of the diradical in shape and width. Since the distance between the 1 and 8-positions in biphenylene is ca. 3.8A, the pyridinyls must bend towards one another for effective interaction. Apparently, magnesium ion can promote an interaction which does not occur easily in the diradical (<u>12</u>).

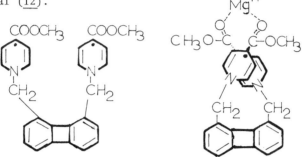

Pyridinyl Radical-Bis-Pyridinium Ion Complexes

In the course of an attempted titration of 1,1'-trimethylene-bis-(4-carbomethoxypyridinium) ion (++) with 1-ethyl-4-carbomethoxypyridinyl radical (·), we discovered that these species formed a strong complex, quite stable in acetonitrile at room temperature. (Eq. 13). Epr spectra exhibited exchange narrowing and possibly two overlapping spectra, but the most significant property of the complex was a near infrared band (λ_{max} 1360 nm (ε_{max} 1500)) which was assigned to an intervalence transition. (<u>13</u>). Addition of another equivalent of pyridinyl radical to the complex results in the formation of a second complex, accompanied by a shift of the near infrared band to λ_{max} 3200 nm. (Eq. 14).

The $+\cdot+$ complex stability is dependent upon the nature of the anion and on temperature, with the activation energy for the disappearance of the complex (followed with the near infrared absorption) varying from 45 kcal/mole for the iodide to 26 kcal/mole for the tetraphenylborate. The epr signal disappears in parallel with the loss of the near infrared band. The mechanism for the loss fo the complex is explained as due to the dissociation of the initial complex to a cation radical ($+\cdot$) which combines with another molecule of the initial complex to yield a diradical complex. (Eq. 15). Further changes in the spectrum ensue, which are presumed to result in the formation of higher complexes, with

(15) $\underset{\cup}{+\cdot+}$ $\longrightarrow$ $\underset{\cup}{+\cdot}$ and $+$

$\underset{\cup}{+\cdot}$ and $\underset{\cup}{+\cdot+}$ $\longrightarrow$ $\underset{\cup\cup}{+\cdot+\cdot+}$

continued heating.

Analogous complexes can be found for tetracyanoquinodimethane (TCNQ or T) including T,T^-, T_2^-, T_2^{--}, T_3^{--}, T_4^{--}. Indicating the pyridinium ring or pyridinyl radical by P, the following complexes have been detected: P,P^+, P_2, P_2^+, P_3^{++}, P_4^{++}, P_5^{+++} and P_6^{+++}. If we bear in mind that considerable variation in the structure of P is possible, the variety of possible complexes is staggering.

Additional Information

We have discussed primarily our own work, especially that which is still in progress. No discussion of pyridinyl radical π-mers would be complete without citing the discovery of the dimerization of pyridinyl radicals by Itoh and Nagakura (14) and the work of Ikegami (15) on different π-mer triplets. The original impetus for the study of pyridinyl radical reaction with halocarbons was given by the discovery by Westheimer, Kurz and Hutton (16) that dihydropyridines were converted into pyridinium compounds via free radical reactions in the presence of tetrachloromethane. An extensive review of pyridinyl radicals has been published elsewhere (1). New ways of measuring the low temperature spectra of pyridinyl radicals are under development (18).

Acknowledgements

No report of our research would be complete without mention of those who have actively contributed to the progress of the work, including William Schwarz, jr., Edward J. Poziomek, Irving Schwager, John Cotter, Leroy Butler, Harold P. Waits, Yusaku Ikegami, Michiya Itoh, Mahboob Mohammad, Joseph Hajdu, J.B. Nagy, A. Lewis, N.F. McFarlane, Avraham Teuerstein and Joshua Hermolin. Senior colleagues from other institutions included A. John Swallow (Paterson Laboratories, Christie Hospital and Holt Radium Institute, Manchester, England) and Hugh D. Burrows (now at the University of Ife, Ile-Ife, Nigeria).

Support for the research was provided over many years by the National Institutes of Health, supplemented by grants from Edgewood Arsenal and the Israel Academy of Sciences.

Abstract

Simple pyridinyl radicals, Py·, like those derived from 1-ethyl-4-carbomethoxy-pyridinium ion through 1e̲ reduction, have proven useful for the investigation of properties of radicals, including complexes, and the study of the mechanisms of their reactions. Discoveries include: (a) distinction between atom-transfer and electron-transfer reactions by means of solvent effect on rate constants. (b) intramolecular radical complexes (π-mers) within pyridinyl diradicals and pyridinyl diradical:metal halide complexes, (c) occurrence of either electron-transfer or dimerization between pyridinyl radicals in water according to structure and (d) complexation between pyridinium ions and pyridinyl radicals. Consideration of actual and potential biological roles for pyridinyl radicals is aided by these discoveries.

Literature Cited

1. Kosower, E.M., Chapter 1 in "Free Radicals in Biology", pp. 1-54, Vol. II, Academic Press, New York,1976(W.A. Pryor,ed.)
2. Kosower, E.M. and Schwager, I., J.Am.Chem.Soc. (1964) 86, 5528-5535.
3. Mohammad, M. and Kosower, E.M., J.Am.Chem.Soc. (1971) 93, 2713-2718.
4. Swallow, A.J., "Radiation Chemistry" 275 pp., Longman, London, 1973.
5. Land, E.J. and Swallow, A.J., Biochim. et Biophys. Acta (1968) 162, 327.
6. Kosower, E.M., Burrows, H.D., Swallow, A.J. and Teuerstein, A., manuscript in preparation.
7. Kosower, E.M., Teuerstein, A. and Swallow, A.J., J.Am.Chem.Soc. (1973) 95, 6127.
8. Neta, P. and Patterson, L.K., J.Phys.Chem. (1974) 78, 2211.
9. Itoh, M. and Kosower, E.M., J.Am.Chem.Soc. (1968) 90, 1843-1849.
10. Kosower, E.M. and Ikegami, Y., J.Am.Chem.Soc. (1967) 89, 461-462.
11. Kosower, E.M. and Hajdu, J., J.Am.Chem.Soc. (1971) 93,2534-2535.
12. Kosower, E.M. and Teuerstein, A., unpublished results.
13. Kosower, E.M. and Teuerstein, A., J.Am.Chem.Soc.(1976) 98, 1586-1587.
14. Itoh, M. and Nagakura, S., J.Am.Chem.Soc. (1967) 89, 3959.
15. Ikegami, Y. and Seto, S., J.Am.Chem.Soc. (1974) 96, 7811-7812.
16. Kurz, J.L., Hutton, R. and Westheimer, F.H., J.Am.Chem.Soc. (1961) 83, 584.
17. Mohammad, M. and Kosower, E.M., J.Am.Chem.Soc. (1971) 93, 2709-2713.
18. Hermolin, J., Levin, M. and Kosower, E.M., unpublished results.

RECEIVED December 23, 1977

INDEX

N

O

P